Ecological Assessment
of
Hazardous Waste Sites

ECOLOGICAL ASSESSMENT OF HAZARDOUS WASTE SITES

James T. Maughan

VNR VAN NOSTRAND REINHOLD
New York

The material appearing in this book as Chapter 8 was written by an employee of the U.S. Government as part of his official duties, and as such may not be copyrighted. All other material appearing in this book is property of the copyright holder.

The views and opinions expressed regarding the two sites described in Chapters 9 and 10 are those of the author and not of the Environmental Protection Agency. The information presented was generated under EPA Contract No. 68-W9-0036 and has been released with permission.

Copyright © 1993 by Van Nostrand Reinhold

Library of Congress Catalog Card Number 92-27826
ISBN 0-442-01091-5

Printed in the United States of America.

Van Nostrand Reinhold
115 Fifth Avenue
New York, New York 10003

Chapman and Hall
2-6 Boundary Row
London, SE1 8HN, England

Thomas Nelson Australia
102 Dodds Street
South Melbourne 3205
Victoria, Australia

Nelson Canada
1120 Birchmount Road
Scarborough, Ontario MIK 5G4, Canada

16 15 14 13 12 11 10 9 8 7 6 5 4 3 2 1

Library of Congress Cataloging-in-Publication Data
Maughan, James T., 1949-
    Ecological assessment of hazardous waste sites / James T. Maughan.
        p.   cm.
    Includes bibliographical references and index.
    ISBN 0-442-01091-5
    1. Ecological risk assessment.   2. Hazardous waste sites-
-Environmental aspects.   I. Title.
QH541.15.R57M38   1992
363.72'872--dc20   92-27826
                                                        CIP

Contents

CHAPTER 4
TECHNICAL APPROACH TO ECOLOGICAL ASSESSMENTS

CHAPTER 5
HUMAN HEALTH RISK ASSESSMENTS
Mary E. Doyle and John C. Young

CHAPTER 6
BIOLOGICAL TRANSFER OF CONTAMINANTS IN
TERRESTRIAL ECOSYSTEMS
Stephen E. Petron

CHAPTER 7
EVALUATION OF CONTAMINANTS IN SEDIMENTS

CHAPTER 8
ECOTOXICOLOGY AND ECOLOGICAL ASSESSMENT
AT HAZARDOUS WASTE SITES
Anthony F. Maciorowski

CHAPTER 9
PINE STREET CANAL ECOLOGICAL ASSESSMENT:
A CASE STUDY
William B. Kappleman

CHAPTER 10
MIDDLE MARSH ECOLOGICAL ASSESSMENT:
A CASE STUDY
Peter M. Boucher

List of Contributing Authors

Peter M. Boucher (Chapter 10) is a Project Biologist and Hazardous Waste Specialist with Metcalf & Eddy, Inc. He has a B.S. in Biology and an M.S. in Environmental Engineering from Northeastern University. He has conducted remedial investigations, ecological risk assessments, environmental impact assessments, wetland evaluations and lake restoration studies. His research has focussed on water resources, particularly in relation to hazardous waste. He has also played a major role in developing sample collection methods and data quality control.

Mary E. Doyle (Chapter 5) is an Environmental Health Scientist for Metcalf & Eddy, Inc., an environmental consulting firm. At Metcalf & Eddy, she is responsible for conducting public health risk assessments and for evaluating the public health implications of a variety of past or proposed human activities. She has an M.S. in Environmental Health Sciences from the Harvard University School of Public Health, and a B.S. in Environmental Health from the University of Massachusetts School of Public Health.

William B. Kappleman (Chapter 9) is a Senior Biologist with Metcalf & Eddy, Inc. He holds a B.S. degree from the State University of New York (Syracuse) and an M.S. degree from the University of Vermont, both in Wildlife Biology. His work involves conducting environmental impact analyses, ecological assessments, statistical analyses, and more recently, ecological risk assessments. His research interests include the investigation of bird-habitat relationships, and modeling of wildlife habitat use. He is also responsible for data management and maintaining wildlife toxicity information.

Dr. Anthony F. Maciorowski (Chapter 8) is the Chief of the Fish Culture and Ecology Laboratory, at the U.S. Fish and Wildlife Service's National Fisheries Research Center—Leetown. He received a B.S. in Biology from East Michigan, and a Ph.D. in Zoology from Virginia Polytechnic Institute and State University. His interests are in evaluating the effects of toxic substances on the environment and management of aquatic resources. He is also involved in research into culture methods for the restoration of natural populations. Dr. Maciorowski has served on committees of several professional organizations, including the Society of Environmental Toxicology and Chemistry and the Water Environment Federation.

Dr. Stephen E. Petron (Chapter 6) is a Project Biologist for Metcalf & Eddy, Inc. A certified wildlife biologist, he specializes in ecological physiology, terrestrial ecology, and environmental regulations. He received a Ph.D. in Zoology from Washington State University, an M.S. in Natural and Environmental Resources from the University of New Hampshire, and a B.S. in Wildlife Management from the University of Minnesota. He has conducted research in mammalian thermoregulation, bird behavior, and wetland restoration.

John C. Young (Chapter 5) writes health risk assessments as an Environmental Health Scientist at Metcalf & Eddy, Inc. He has also worked on environmental health issues at the Massachusetts Department of Public Health, and has participated in a number of environmental investigations. He earned an M.S. in Public Health at the University of Massachusetts at Amherst, and a B.A. in biology at Williams College.

Preface

. Throughout the short history of Superfund, the ecology of hazardous waste sites and the protection of ecological resources during site clean-up has placed a distant second to human health considerations. However, as the sites posing the most immediate public health risks are addressed, and ecologists become aware of biological/hazardous waste cleanup interactions, the ecological assessment of hazardous waste sites is gaining considerable attention. This attention is manifested in the promulgation of legislation and regulations. There is even a field of ecological research developing to supply the base knowledge for assessing ecological concerns associated with site cleanup.

In the early years of Superfund, when ecological concerns were addressed at all, there was often insufficient input from ecologists, and thus a lack of appreciation for the complex nature of biological systems. For example, a system can frequently be resilient and adaptive to low concentrations or isolated areas of contamination, yet extremely vulnerable to even minor physical disturbance. If these characteristics are ignored, or not understood, the result can be (and often was) a cleanup plan that destroyed an ecosystem minimally at risk for its own protection. The classic case of such an action was the massive excavation of contaminated wetland soils, and replacement with different types of material at an altered depth, which resulting in a "clean" site with little or no wetland value for such functions as flood storage, pollutant assimilation, or wildlife/aquatic habitat.

This book addresses these two areas of ecological concern at hazardous waste sites: attention to critical ecological resources and regulations during site investigations, and integration of ecology as a science into the remediation of site contamination. The purpose and objective of an ecological assessment are presented first, followed by a discussion of the legislative perspective, and requirements for the consideration of ecological issues as part of site investigation and cleanup. A thorough discussion of the overall technical approach for assessments is also presented, followed by a summary of human health assessment methodology for comparison to the ecological investigation approach. The discussion of the approach and methods concludes with specific descriptions of the philosophy and methods for addressing three of the most frequently encountered ecological issues at hazardous waste sites (transfer of contaminants in terrestrial ecosystems, sediment quality, and toxicity testing).

The field of ecological assessment at hazardous waste sites is young. As such, much of the guidance was developed from a limited historical perspective and experience base. This book reflects on the early guidance, now that it has been applied at some sites, and assesses the need for refinement in the approach and additional research. This approach of critically assessing the ecological assessment process not only contributes to a deeper understanding of the topic, but provides important food for thought as the next generation of regulation and guidance is developed and future research is implemented.

The need for integrating ecological investigations, concerns, and input into the hazardous waste cleanup process at every stage is a major focus of the book. The most scientifically thorough and accurate ecological investigation will not achieve ecological protection or regulatory compliance if it does not address the overall site issues and is not coordinated with all aspects of site cleanup, including human health concerns, engineering feasibility, and economic considerations. This book describes the importance of coordination and suggests methods for achieving the needed interaction in designing the investigation, developing remediation alternatives, and selecting the cleanup plan.

My background of almost 20 years as an environmental consultant is reflected in the perspective of the book. Problem solving and reaching a workable consensus are major themes throughout every chapter. I have learned that regulators, scientists, potentially responsible parties, and even engineers and lawyers are reasonable people when presented with logical arguments, supported by strong technical investigations, and afforded the opportunity for input and consensus building at early and frequent stages of the process. By following this approach, a workable solution can be identified and the objectives of all parties addressed.

My background is also apparent in two other areas of the book: historical perspective and the liberal use of examples. I frequently describe specific regulations and ecological methods in the context of their development during the 20-year history of the current environmental movement. I also cite hypothetical or exaggerated real examples, to illustrate points and identify potential pitfalls for most of the concepts presented in the approach and method chapters of the book. I have also included two case study chapters, which illustrate how the concepts presented elsewhere in the book can be applied. More importantly, perhaps, the case studies demonstrate that, to successfully complete an ecological assessment, one must understand the concepts involved, and then frequently make conscious decisions that they must be changed and adapted to specific sites and problems.

The book is written for environmental professionals and for those who have such aspirations. For those who have knowledge and experience in ecology, the book describes how the knowledge can be applied to hazardous waste site investigations. Engineers, nonecological scientists, and managers already familiar with many aspects of hazardous waste can gain an overview of critical ecological concepts and how they can be integrated into site investigations and cleanup. The book serves as an important reference for environmental consultants, regulators, academics, and environmentalists. It is also appropriate as a text or reference in an upper-division undergraduate, graduate, or professional course in ecological assessment or general hazardous waste site investigation.

This book benefits from the input and support of many people. Constant interaction among all of the contributing authors resulted in the broad perspective of the book, and thus its applicability to a wide audience. The interaction with the EPA Region I Superfund Environmental Assessment Team or SEAT (particularly Susan Svirosky, the current chairperson), both formally on individual projects and informally on ecological/hazardous waste concepts, was critical to virtually every aspect of the book. I appreciate the feedback and generous contribution of time from Ross Gilleland and Jane Downing on the case study chapters, and their perspective, along with the SEATs, contributed much to my overall appreciation of hazardous waste site cleanup. Finally, without the support and tolerance of my family, Emily, Edwina, and Matthew, I would never have had the strength to complete this book.

1

Introduction

In the 1970s, the concepts of ecological awareness and protection expressed during the first Earth Day became integrated into the national perspective. The National Environmental Policy Act, requiring consideration of ecological and sociological issues before the federal government took any significant action, was passed. With this, a truly national policy of environmental awareness was established. The Clean Water and Clean Air Acts were also passed to protect these national ecological resources from contamination and other abuses. Additional federal, state, and local legislation soon followed to protect other ecological resources. Throughout the 1970s, the regulatory, enforcement, and monitoring structure to implement environmental policy and protect ecological resources was established and matured. Significant public funds were provided to address some of the major problems, such as municipal wastewater management, and limits were placed on the discharges and emissions of industrial pollution sources. Through trial and error, litigation, and regulatory guidance, procedures to comply with the specific regulations and policies were adopted, eventually accepted, and are gradually being implemented.

During the 1980s, we became aware of the extent of hazardous waste in the environment, and the potentially serious threats posed by past hazardous waste management and disposal practices. The perceived magnitude of the problem, and the potential seriousness of the threat associated with the sites, placed hazardous waste problems on the environmental center stage. The legislative and regulatory communities responded by passage and implementation of landmark acts such as the Toxic Substance Control Act (TSCA), the Resource Conservation and Recovery Act (RCRA), and ultimately, the Comprehensive Environmental Response, Compensation, and Liability Act (CERCLA), or Super Fund. The most serious hazardous waste concerns were the

1

threats to public health and environmental resources directly linked to human health, including water and food supplies. Consequently, public health concerns drove the initial hazardous waste cleanup mandate and remediation approach. By the late 1980s, many of the hazardous waste problems and sites that posed the most immediate threat to human health had been addressed, although not necessarily eliminated, and hazardous waste issues began to be viewed from a broader perspective.

The 1990s are emerging as the decade when the ecological awareness of the 1970s and the hazardous waste concerns of the 1980s are merging. From two decades of pollution ecology research and regulation, and a decade of hazardous waste site investigation, it has become apparent that ecological resources must be considered in order to achieve the mandate of complete environmental protection as part of hazardous waste management and cleanup. The original hazardous waste legislation and regulations are being reinterpreted and refined, to more specifically include ecological resources as part of the environment to be protected from hazardous waste releases. Ecological resources are seen not only as individual attributes warranting consideration and protection, but also as an indication of the overall health of the environment and an integral and necessary part of the whole.

PURPOSE AND OBJECTIVES

At this vital and early stage in formulating methods, approaches, and policy for the ecological assessment of hazardous waste sites, it is important to critically analyze and review the current status of the hazardous waste assessment process. The purpose of this book is to provide such a review and analysis. The objective here is to present the review as an instructional tool for performing ecological assessments at hazardous waste sites. The analysis of current practices is also intended to stimulate refinement and improvement of current assessment practices, and to provide thought for future policy development and implementation.

Between 1988–1991, the first guidance documents and symposia for ecological assessment at hazardous waste sites were developed. Much of the information in these documents and discussions is not only summarized in this book, but is evaluated from the perspective of early experiences in implementing the initial, largely theoretical, thoughts and advice presented in the early guidance. By sharing these case studies and other general experiences in evaluating ecological concerns at hazardous waste sites, this book is intended to enrich future individual ecological assessments and the next generation of assessment methods, policy, and guidance.

The focus of the book is on solving ecological problems associated with hazardous waste sites and their cleanup. The intent is to provide information and guidance to address potential issues and, ultimately, if not solve them, at least move the process significantly closer to a solution. Ecological concerns of both an administrative and technical nature can create problems requiring solutions at hazardous waste sites, and both are addressed in this book. The process of incorporating the ecological assessment into the overall site investigation, cleanup, and regulatory perspective are the primary administrative issues covered. Technical issues presented include both the comprehensive approach to problem solving (i.e., ecological assessment methodology) and individual scientific methods available to address specific contaminant-ecological interactions.

CONTENTS

The Superfund law (Comprehensive Environmental Response, Compensation, and Liability Act, or CERCLA, and the Superfund Amendments and Reauthorization Act of 1986, or SARA) and the U.S. EPA's implementation as the National Contingency Plan (NCP) are the driving forces behind hazardous waste site cleanup. The regulations associated with Superfund and the NCP form the mandate and structure for ecological assessments associated with virtually all hazardous waste-site issues. Consequently, the focus of the administrative discussions in this book is on Superfund. Other regulations are included to a lesser extent, but much of the Superfund and NCP process, at least for ecological considerations, has been incorporated officially or by default into other hazardous waste programs. For example, remediation of hazardous waste at federal installations, such as those governed by the Defense Environmental Restoration Program at U.S. Army bases, closely follow the National Contingency Plan regulations. Therefore, the regulatory and agency guidance concepts discussed can generally be applied to most hazardous waste issues.

Environmental and hazardous waste regulations, legislation, and guidance are constantly changing and being reinterpreted by practice and litigation. The discussion of the regulatory framework in this book is a summary of the general status, interpretation, and implementation in current use. No general reference book, including this one, can accurately, completely, and precisely define all of the regulations applicable to every conceivable ecological setting and hazardous waste site. Consequently, to insure regulatory compliance, the actual guidance, regulations, and legislation applicable to each specific site, contaminant, and situation should be consulted at the time of each hazardous waste site investigation and remediation.

The government agency guidance, resource, and reference documents dealing with ecological assessments and hazardous waste are even more dynamic and ephemeral than the enabling regulations and legislation. Often, when the investigations following and referencing the documents are distributed publicly, the subject reference documents are out of print and not readily available. This "grey literature" is covered extensively in this book. The information and concepts presented are summarized, and their effectiveness and usefulness evaluated based on experience using the documents. Detailed discussions and methods of or acquiring the documents are included for the publications that have proven to be especially useful in conducting ecological assessments.

A comprehensive understanding of ecology is necessary to prepare a true ecological assessment. Experience in dealing with the individual ecological resources is important in recognizing the sensitivities and potential of the resources at a hazardous waste site. Perhaps more important is an appreciation and working knowledge of the interrelationships and interdependencies of various resources at a site. This appreciation extends beyond the requirements of specific organisms, species, populations, or even communities. Processes such as the flow of energy, nutrients, or the timing of life stages with the abundance of food sources are more subtle site characteristics that must be integrated into an ecological assessment. This book can not convey this comprehensive understanding. A strong educational and experiential background in the appropriate areas of ecology and other biological sciences is a prerequisite for the understanding. Consequently, the nonecologist can not prepare a comprehensive ecological assessment at a hazardous waste site alone, solely from the information presented in this book.

Just as the ecological assessment at hazardous waste sites is a new and developing area of investigation, so are the terms used in the process. There is not a universally accepted definition and usage of such terms as ecological assessment, environmental assessment, and ecological risk assessment. The following definitions have been adopted and used consistently throughout this book. These definitions may not be consistent with usage in all ecological assessment guidance and site-specific investigations. Hopefully, however, by clearly defining and using them here, the confusion will be minimized.

Environmental Assessment—An evaluation of all aspects of the human environment, including sociological considerations, traffic, air, water, noise, and natural resources potentially affected by some human activity. The assessment usually includes a description of the current conditions and a projection of what conditions would be following implementation of any one of a number of alternative proposed actions. An environmental assessment is used by a federal agency to determine whether their proposed action will have a potentially significant impact, and thus require a formal Environmental Impact Statement, as required under the National Environmental Policy Act. The term is generally not used for hazardous waste site investigations.

Ecological Assessment—An evaluation of ecological resources. The term could apply to the portion of a comprehensive environmental assessment dealing with ecological resources. It is more often used for the ecological portion of a hazardous waste site investigation. In the hazardous waste context, it includes: an evaluation of existing conditions and risks, particularly those resulting from contamination damage; an assessment of ecological potential; an evaluation of the relationship, both past and future, between contaminants and ecological resources; and prediction and evaluation of conditions following site remediation.

Ecological Risk Assessment—The portion of an ecological assessment addressing past or future risk (usually from contamination) to a source (actual or potential). The ecological risk assessment associated with a hazardous waste site usually includes the distinct steps of hazard identification, toxicity assessment, exposure assessment, and risk characterization.

CHAPTER ORGANIZATION

The second chapter of this book describes the needs and objectives of ecological assessments at hazardous waste sites. The needs are described both from an administrative (including regulatory) standpoint, and as they relate to ecological resources at risk from contamination or other hazardous waste site conditions. The primary and intermediate, or working, objectives for conducting the assessment, and then incorporating it into the entire hazardous waste site investigation and remediation process, are also presented. The emphasis of the chapter is the need to fully understand the needs and objectives of an ecological assessment and to make sure that all investigations, analyses, and evaluations address a need and move the process toward the objective.

The regulatory framework of hazardous waste site investigations in general, and ecological assessments specifically, are presented in the third chapter. Where appropriate, the discussion covers the enabling legislation, the implementing regulations, and in some cases regulatory guidance. The specific requirements for the ecological assessment are summarized and presented, as they apply to conditions typically found at hazardous waste sites. The chapter focuses more on the intent and concepts addressed by the regulations than on the specific enforceable statutes. The chapter also presents approaches and examples for complying with the regulations and, where possible, using them to assist in accomplishing the ecological assessment objectives. The chapter separately addresses the three primary sets of regulations mandating and governing ecological assessments at hazardous waste sites: Superfund (or CERCLA); compliance with other environmental regulations; and natural resource protection and restoration.

Chapter four presents the technical approaches for conducting ecological assessments. There is no specific approach mandated by accepted scientific practice or regulatory dictate, so the chapter discusses a model and variations that can be adapted to specific sites and ecological resources. The individual technical steps or elements are described, and the process of integrating all of the investigations into a comprehensive and useful ecological assessment is discussed. The chapter contains some methods for specific scientific investigations, which can be critical to successfully completing the assessment. However, the focus of the chapter is to provide a general understanding of the assessment elements and how they fit together to address the objectives of the assessment and to solve problems.

The fifth chapter presents a summary of public health risk assessment at hazardous waste sites. Such a chapter might seem out of place in a book on ecological assessment, but it really is not, because much of the approach and methodology, at least for the risk portion of the ecological assessment, evolved from the parallel human health evaluation. An understanding of the public health assessment lends a historic perspective, and can generate new and innovative approaches to ecological evaluations at hazardous waste sites. Much of the hazardous waste site investigation, particularly data collection, is driven by the needs for human health risk assessment, so it is prudent for the two assessments to be coordinated. The first step for the ecologist in this coordination is a general understanding of the public health risk assessment.

Chapters six, seven, and eight address specific technical areas proven to be critical for assessments at hazardous waste sites with ecological resources. The transfer of contaminants through the terrestrial ecosystem is addressed in the sixth chapter. The possible pathways, fate, effects, and approaches to estimating these characteristics are discussed. Chapter 7 is a discussion of sediment quality: how it is measured and applied to hazardous waste site evaluations and cleanup. Bioassay techniques and their relevance to ecological assessments at hazardous waste sites are the subject of Chapter 8.

The final two chapters are case studies of actual ecological assessments at hazardous waste sites. One of the sites (Sullivans Ledge, presented in Chapter 10) was reported (HMRCI 1991) as the first in the nation where ecological concerns drove the site remediation. In each of the examples, the regulations, processes, and specific technical approaches laid out in the other chapters can be seen in action. As is usually the situation in any case dealing with natural systems, the examples do not strictly follow the model. However, more can sometimes be learned by revealing areas where a universal model and approach are not exactly suited to a very specific example.

REFERENCES

HMCRI, 1991. EPA decides on cleanup decision based solely on ecosystem, not health dangers. Hazardous Material Control Research Institute. *FOCUS*, Vol. 7:12. Dec. 1991.

2

Ecological Assessment Needs and Objectives

ASSESSMENT NEEDS

The need for an ecological assessment as a part of hazardous waste site cleanup arises from the potential ecological impacts associated with the site and the remediation activities. The most obvious ecological effects are those resulting from a contaminant release at an uncontrolled hazardous waste site. The released contaminants can have toxic effects on the ecological resources within, adjacent to, or downgradient from the hazardous waste site. When these resources are sensitive to the contaminants, and show irreversible effects, the effects can interrupt normal ecosystem functions and thus magnify the impact. Ecological impacts can also be a byproduct of the hazardous waste site cleanup. The extensive physical disruption often associated with site remediation can severely alter the potential for the site to ever again support any significant ecological resource. The treatment and disposal of the waste from a site can also produce impacts on ecological resources not originally damaged by the uncontrolled release of contaminants.

Within the highly regulated and socially sensitive context of hazardous waste programs, the potential for ecological impact from a hazardous waste site does not, by itself, dictate the preparation of an assessment. There must be an awareness and concern over the impact for an assessment to be needed. The national concern over ecological impacts has been generated by the repeated incidence of risk to sensitive resources. Thus, the characteristics of the ecological resource at risk are a driving force behind the need for ecological assessments at hazardous waste sites. When the ecological resources at issue have received much national attention, such as endangered species or wetlands, there is a strong awareness of the resource. Similarly, attention can be focused on the resource due to high local visibility, such as a refuge or prime hunting and fishing area, or contaminant pathways leading to species used for human consumption.

6

In the late 1980s, the awareness and concern of ecological damage at hazardous waste sites reached the action level. The action level was the refinement of the Superfund legislation to more specifically identify and expand the mandate for the protection of ecological resources. The awareness of ecological implications at hazardous waste sites also focused regulatory attention on resources protected by other legislation, which had previously been ignored or at least downplayed at hazardous waste sites. Thus, the real need for ecological assessments was born when it was both required by law and taken seriously enough by regulators to be enforced. The legal and regulatory requirements for the assessment fall into three categories: the cleanup process mandated by the Comprehensive Environmental Response, Compensation, and Liability Act (CERCLA), or the "Superfund law"; compliance with Applicable or Relevant and Appropriate Regulations (ARARs); and protection of ecological resources under the jurisdiction of the National Resource Trustees. Each of these topics are discussed in Chapter 3, by addressing the historic perspective, requirements, and compliance approach.

The tardiness of the ecological concern, and the resulting legislation and regulatory enforcement (considering that Superfund was established in the early 1980s), was due to a number of factors. First and foremost, the major concerns and issues that initially forced action in the area of hazardous waste clean up were human health-related. Consequently, sites with potential human health effects were the first priority. These sites were frequently in densely populated and developed areas, and thus the ecological resources in such areas were limited, tolerant to stress, or not present at all. Therefore, ecological awareness was not inspired. The cleanup of even these top-priority sites in developed areas was greatly delayed by the EPA's slow start in implementing Superfund, so there was a time lag before the lack of attention paid to ecology during the actual remediation activities was obvious to the general public and nonhazardous waste regulatory agencies and personnel.

Later in the 1980s, it became apparent that remedial actions often ignored ecological resources in determining the need for and extent of a cleanup. Also, the construction activities associated with the cleanup often destroyed ecologically sensitive resources within or near the areas of contamination, even when the resources had not been affected by the release of hazardous materials. As the national hazardous waste cleanup initiative progressed, and sites were addressed in the second half of the decade, the list expanded to include more sites outside of developed areas, and thus supporting obvious ecological resources. These resources included those linked to human health, special-status species, wetlands, and areas of wildlife conservation. As a result of these developments, a constituency for attention to ecological issues at hazardous waste sites was formed. The ecological concerns and constituency forced more attention and explicit direction on ecology considerations and assessment, in the 1988 Superfund Amendments and Reauthorization Act (SARA).

The U. S. EPA also responded to the issue of ecological awareness by surveying the attention focussed on ecological concerns during investigation and remediation of the first set of sites under the original Superfund program (U.S. EPA 1989a, b, c, d). The objective of the surveys was to understand the past problems, project future concerns, and to establish guidance. The survey approach was made to examine a cross section of hazardous waste sites, and to determine what type of ecological concerns existed and how they were addressed. An initial list of about 250 sites was identified, and there was sufficient readily available ecological and related information on 52 of these to evaluate the ecological problems and approaches for solutions. The sample sites were not chosen at random, but were biased towards sites with ecological information, and thus most likely to be those potentially affecting ecological resources. Consequently, projection of the results to the full nationwide list of hazardous waste sites may overstate the degree of the national ecological problem at hazardous waste sites. Projection of ecological impacts from conditions found in sites high on the priority list (i.e., the 52 sites surveyed) might also, in some ways, underestimate the magnitude of ecological problems nationwide, because (as discussed above) the early sites tended to have more human-health risk and fewer ecological resources.

From the survey, the U.S. EPA found that approximately 10 percent of the 52 sites posed severe ecological threats. An additional 80 percent may constitute a moderate threat to ecological resources, with the remaining 10 percent having a minor or indiscernible threat. The most commonly threatened resources were freshwater ecosystems, including wetlands. It is not surprising that these systems were identified to be at risk at 90 percent of the sites surveyed. Surface water runoff, eroded sediments, and groundwater discharges from hazardous waste sites are almost always to a freshwater system. If any of these processes transport contaminants from the site (and they almost always do) they will be introduced into the surface waters or wetlands. These systems represent a natural area for containment and accumulation of sediments. If there are contaminants associated with the eroded particles, then the aquatic systems also accumulate the contaminants. In addition, open-water systems, sediments, and wetlands are areas where organisms are in close communication with the environmental medium, and thus susceptible to intense contaminant exposure. The contaminants, therefore, frequently appear in surface water resources and, once there, they generally pose an ecological risk.

Measured or projected effects to terrestrial resources were less frequently reported (50 percent of the sites surveyed). The type of terrestrial effects involved consisted of damaged or killed vegetation, loss of habitat, contamination or loss of food sources, and toxic effects on birds and mammals. The higher-reported incidence of aquatic ecological concerns is strongly influenced by the ease of detection and the more obvious nature of aquatic resources (e.g., a stream compared to a mink denning site) and of damage (e.g., a fish kill compared to a reduction of egg production in robins).

As summarized in Table 2–1, the pattern of organism tissue contamination found at hazardous waste sites mirrors the frequency of ecological risk. Fish and other aquatic organisms were the most commonly reported contaminated organisms. The frequency

TABLE 2-1. **Relative Occurrence of Organism Contamination at Hazardous Waste Sites with Ecological Resources.**

Organism Type	Percent of Sites	
	Observed Contamination	Predicted Contamination
Fish	50	20
Aquatic Invertebrates	15	40
Vegetation	40	10
Mammals	10	40
Birds	5	45
Reptiles and Amphibians	5	20

Source: Summarized from U.S. EPA 1989d.

TABLE 2-2. Relative Occurrence of Various Contaminants at Hazardous Waste Sites with Ecological Resources.

Contaminant	Percent of Sites
Lead	50
Cadmium	40
Arsenic	40
Chromium	40
Zinc	35
Nickle	30
Copper	25
Manganese	15
Mercury	15
Iron	15
VOC	55
PCB	30
PAH	15
Phthalate Esters	15
Pesticides	10
Phenols	10
Cyanide	10

Source: Summarized from U.S. EPA 1989d.

of measured contamination in terrestrial animals was considerably lower primarily because collection and analysis is much more difficult, and is rarely performed as part of a Superfund investigation. The U.S. EPA survey also documented the types of contaminants and the waste management practices at sites with potential ecological risk (Tables 2–2 and 2–3). Individual contaminant occurrence does not necessarily imply the severity of the ecological effect, or even the frequency with which the compound causes an effect. The high proportion of landfills and lagoons in the list reflects the generally large areal extent and isolated nature of such facilities. Ecological resources are more common in these areas, as compared to drum storage areas or tank farms.

As the U.S. EPA (1989a and d) acknowledges, it is inappropriate to extrapolate

TABLE 2-3. Relative Occurrence of Waste Practices at Hazardous Waste Sites with Ecological Resources.

Waste Management Practice	Percent of Sites
Landfills (Including Dumps)	45
Waste Lagoons	35
Direct Discharges	20
Piles	10
Drum Storage	5
Land Farming	5
Tanks	5

Source: Summarized from U.S. EPA 1989d.

from the 52 sites examined in the survey to the full Superfund list. However, it is tempting, and the EPA yields to the temptation. With many caveats and qualifiers, they speculate that there could be between 220 and 420 sites on the National Priorities List (NPL) with significant existing ecological impacts and/or future risk. There is no reason to suppose that hazardous waste sites not yet on the NPL would not also pose threats to ecological resources. In fact, since ecological factors have not been used to place sites on the NPL, the hazardous waste sites not on the priorities list may even have a higher frequency of ecological risk and impact. Even applying percentages from the 52 NPL sites examined in the survey, the total number of hazardous waste sites nationwide with significant ecological risk would be from 2,700 to over 5,000 (U.S. EPA 1989a). If these numbers are further extrapolated (U.S. EPA 1989a), hazardous waste sites could currently be significantly impacting and posing future risk to:

- 12,000 miles of streams and rivers, due to surface water contamination.
- 8,000 miles of stream bed, resulting from sediment contamination.
- 64,000 acres of potential wildlife habitat, from soil contamination.
- the acute survival of fish in 1,700 miles of streams and rivers.
- 22,000 acres of vegetation threatened with defoliation.
- 22,000 acres of wetlands.

Based on the sample of hazardous waste sites, half of the sites nationwide may be threatening unique, vulnerable, commercial, or recreational resources. All of these estimates could easily be off by an order of magnitude, but they do illustrate that ecological risk is not just an anomaly at a few hazardous waste sites, and that the resources at risk are not trivial. In contrast to the high occurrence of ecological risk or impact at hazardous sites detected in the survey, of the Record of Decisions for 20 sites with potential ecological problems, ecological concerns affected the remediation decision at only 10 (U.S. EPA 1989a).

The U.S. EPA acknowledges that, even after ecological concerns at hazardous waste sites were recognized and legislative directives (although somewhat vague) issued, there has continued to be a reluctance to performing comprehensive and meaningful ecological assessments as a part of site cleanup (U.S. EPA 1989b and c). The reasons given for the hesitancy in incorporating ecological concerns are a lack of guidance on what to do, and a lack of uniform and consistent methods regarding how to do it. Specific concerns were raised over the methods used for quantifying ecological damages, the variability of methods used nationwide, and the resulting variability in results.

In December of 1988, the U.S. EPA directors of the Office of Emergency and Remedial Response and the Office of Waste Programs Enforcement issued a memorandum directing a thorough and consistent ecological assessment be performed at all Superfund sites. The directive applies to both removal and remedial actions. Since 1989, the U.S. EPA, both nationally and individually in the Regions, has produced initial guidance documents (see Chapter 4 for specific references) for incorporating ecological issues into the cleanup process, and the 1988 directive is now being enforced. Through these guidance documents, the issue of what to do is becoming clearer. Through experience, research, and the documentation of the results, accepted and consistent methodologies for conducting ecological assessments at hazardous waste sites are starting to emerge. This book represents a significant effort in the process of developing and refining guidance and methods used in ecological assessments at hazardous waste sites.

The need for ecological evaluation at hazardous waste sites, from a resource damage, regulatory, and public concern perspective, is most efficiently addressed by a comprehensive and distinct ecological assessment as a separate but fully integrated portion of the total process and documentation. A similar consolidation of the ecological concerns, presentation of investigations, and analyses into a distinct endeavor as an integral part of the entire process, has proven to be an effective approach in the environmental programs that preceded Superfund, such as the National Environmental Policy Act, the Clean Water Act, and regulation of the power industry. The ecological assessment provides a forum to examine ecological issues, and also to evaluate competing ecological concerns, such as habitat preservation versus the removal of contaminated material.

ASSESSMENT OBJECTIVES

There is only one ultimate objective of an ecological assessment at a hazardous waste site: to achieve the protection of the environment at the site and surrounding areas potentially affected by releases from or remediation of the site. Every sample that is taken, all data analysis efforts, and every paragraph written must move the process closer to this goal. It is an intimidating task to start the preparation of an assessment with the single lofty objective of achieving environmental protection, so it is appropriate to consider the objective in a number of steps or a series of working objectives. The working objectives can be categorized as either technical or administrative. The two types of objectives are not independent, as meeting the technical objectives provides the information needed to address the administrative purposes of the ecological assessment. The three technical and the two administrative working objectives of an ecological assessment at a hazardous waste site are discussed in the following subsections.

Determination of the Nature and Extent of Ecological Damage

This working objective is the most obvious component of an ecological assessment. It has also historically been the primary, if not sole, focus of ecological investigations at hazardous waste sites. The nature and extent of damage to ecological resources has generally been measured or estimated, and used to evaluate the need for cleanup.

There is a parallel objective in the human health-risk assessment, termed the baseline risk assessment. The human health equivalent objective uses a combination of contaminant concentrations measured on-site, potential human health exposure scenarios, and laboratory toxicity data, to calculate the increased risk to the local human population of some undesirable health condition, such as cancer. The determination of ecological damage can follow a similar approach, substituting a more appropriate ecological response, such as egg production, survival percentage, or primary productivity, for the risk of cancer. The ecologist also has the opportunity of minimizing the assumptions and calculations inherent in the classical human health baseline risk assessment. An ecological investigation of damage can include a measurement of actual population characteristics on-site as an evaluation of damage.

The determination of ecological damage as a working objective addresses, but does not by itself answer, the question of the need for cleanup. If no ecological damage is detected, the determination can conclude that no contaminant remediation is required to protect ecological resources. However, to conclude that significant ecological damage has occurred and that cleanup is needed is not as straightforward. First, the site-specific definition of significant ecological damage must be established. This definition relates to ecological goals and endpoints for the site, as discussed at length in Chapter 4. Even if the presence of significant damage is agreed to, the implications of cleanup must ultimately be considered before a final remediation decision is rendered. Engineering, economic, ecological, and social implications must be incorporated into the final decision by weighing the relative benefits and the drawbacks of various remediation alternatives.

The objective of an ecological damage determination is fully addressed by providing all of the information needed to evaluate the location and extent of remediation needed for various levels of protection for ecological resources. The appropriate level of protection does not have to be specified to accomplish the objective. When this information is provided, not only is damage defined, but there is a wealth of site ecological information critical to the other technical objectives.

Definition of Conditions Compatible With Ecological Endpoints

If an ecological assessment was a scientific experiment, it would be completed when the first working objective described above was achieved. The hypothesis of the experiment would have been, "the hazardous waste on-site has caused no ecological damage," and the site investigations, data analyses, and various calculations would have supported or disproved the hypothesis. However, an ecological assessment is not an experiment, it is an essential component in the process of solving the problem associated with protecting the environment from contamination originating at a hazardous waste site. Consequently, establishing the presence of a problem is not sufficient, and the ecological assessment must address the solution to the problem.

The ecological assessment provides input for solving the existing or potential future site contamination problems by defining and evaluating conditions, particularly contaminant concentrations, that do not inhibit the appropriate ecological uses of the site. If the ecological use of the site, or endpoints and their requirements for the site, are unambiguous, and sufficient toxicological information exists, it may be possible to precisely and accurately define the environmental media (i.e., water, air, soil, and sediment) concentrations compatible with the ecological endpoints. Such a complete understanding of the site's ecology, and a wealth of toxicological, information is the exception rather than the rule. Consequently, the working objective of defining compatible concentrations is often a discussion of alternative media concentrations, with comment on the likelihood of ecological protection, and thus on achieving endpoints for each alternative concentration.

The environmental media concentrations developed and evaluated in the ecological assessment are not necessarily remediation criteria or cleanup levels. They are, however, important input for establishing cleanup levels. The concentrations can be considered "ecological effects levels," with the expected ecological condition associated with a range of media concentrations defined. These alternative concentrations should be presented in the assessment, and the implications of exceeding each discussed.

They can then be considered by decision makers within the context of the overall site investigation and remediation process, in determining the ultimate cleanup levels.

In evaluating contaminant concentrations, the ecologist often has the opportunity and responsibility of treating different portions of the site, and sometimes even the same portion of the site at different times of the year, individually. The ecological value and potential functions associated with different areas or seasons frequently have different toxicological characteristics and contaminant exposure pathways. The concentrations compatible with these different functions and characteristics should be considered specifically. For example, if only a small and distinct portion of a site is suitable for a particular species breeding, and the contaminant concentration compatible with breeding is an order of magnitude lower than that for all other species functions, it may be unnecessary to clean up the entire site to the lower level for the protection and maintenance of the population. Not only may it be unnecessary to remediate the entire site to the lower level, but the disruption resulting from the larger-scale remediation may render the site physically unsuitable for any of the species other habitat requirements. In such a case, if the most protective level was used and applied to the entire site, the species would have been "protected to death" or at least to the point of total exclusion from the site.

Even in the most straightforward situations, establishing remediation criteria, like the determination of effects from past contamination addressed by the first working objective, is an inexact science at best. There are numerous uncertainties, and never enough site or toxicity data, so that it is difficult (if not impossible) to achieve a truly comfortable level of confidence in the decisions needed to adequately clean up the site. However, decisions must be made if the ultimate objective of site cleanup for environmental protection is even to be approached. Therefore, the U. S. EPA (1989e), and most scientists, engineers, and managers involved with hazardous waste cleanup, recognize that often the best that can be achieved is an informed risk management decision. The specification of remediation criteria, as well as other aspects of cleanup, must ultimately involve the balancing of a wide variety of factors, and the judicial exercise of best professional and weight-of-evidence judgments (U. S. EPA 1989e and f).

The ecological component of the cleanup process probably has the most uncertainty, because of the extreme natural variability in virtually all ecological parameters, and the scarcity of applicable ecological data. The U.S. EPA (1989f) recognizes this situation and envisions the ecological assessment as reducing the uncertainty associated with understanding the environmental effects of a site and its remediation, and giving specific boundaries to that uncertainty. Ecological assessments of Superfund sites are not intended to be research projects: "they are not intended to provide absolute proof of damage, nor are they designed to answer long-term research needs" (U.S. EPA 1989f).

Provisions for Ecological Site Protection and Restoration

The final technical working objective of an ecological assessment is not unique to hazardous waste sites, it can be applicable to any project producing a physical alteration of the environment. Just as the determination of ecological damage from contamination (the first technical objective) is the most commonly performed activity as part of an assessment, because it can be the most obvious, ecological site protection and restoration can be the most ignored task, because it can be the least obvious. Meeting this working objective is often considered an after-the-fact endeavor, following all of the scientifically interesting and intellectually challenging aspects of the ecological assessment associated with contaminate fate, pathways, and effects.

Meeting this working objective can be the most critical in achieving the ultimate goal of environmental protection. If the contamination causing environmental damage is removed, and the resulting toxicological properties of the media can support all desirable ecological functions, but the physical habitat has been destroyed with no hope of restoration, what has really been achieved? Similarly, if treatment facilities needed to clean up soil suitable for mink dens are constructed in the wetland that is the primary foraging area for the mink, has the mink population been protected or destroyed? Ecological protection afforded by meeting this working objective can frequently be simpler, less uncertain, more likely to succeed, and significantly cheaper than protection resulting from contaminant control, removal, or treatment.

The ecological assessment should address the working objective of site protection and restoration by understanding the resources, their requirements, and comparing them with intended remedial activities. By understanding the resources, critical areas or characteristics that do not require direct excavation or other location-specific remedial actions can be identified and left undisturbed. For example, treatment and storage facilities can be located away from these areas. Similarly, if stream diversion is needed to remove sediments, it can be timed to avoid critical periods of downstream flow requirements, such as trout spawning. The resources that must be significantly altered during remediation must also be understood, so that their critical characteristics and functions can be considered as part of the site restoration. When ecological resources are destroyed during remediation, the site restoration plan should ultimately consider reconstruction of resources. The selection and design of the reconstruction efforts should be based on the ecological endpoints for the site, so that achievement of the desirable ecological conditions is maximized. In some cases, this might include replacement of the specific type of resources lost and, in other situations a different and more ecologically valuable habitat may be created at equal or even less expense. The restoration decision should be based on an understanding of the site ecology and the requirements for meeting the ecological endpoints. Avoidance of resources not specifically requiring remediation and restoration efforts can be relatively simple, and often add little or nothing to the cost of remediation, compared to the potential ecological benefit achieved.

Demonstration of Compliance With All Regulations

Not only must the selected remediation alternative comply with all relevant environmental laws and regulations (see Chapter 3 for a complete discussion of compliance, including important exceptions), but the remediation process must include documentation that compliance was considered and achieved to the extent possible. The compliance with biological-related statutes should be addressed in the ecological assessment, and this documentation of compliance is an important administrative working objective of the ecological assessment. The documentation begins by describing which regulations are applicable or relevant. The ecological assessment must then compare the aspects and implications of the remedial alternatives to the specific mandates of each appropriate regulation intended to protect or enhance ecological resources.

There are laws, regulations, standards, criteria, and even regulations under consideration that must be assessed when evaluating remediation alternatives. These statutes cover all aspects of activities, from designing the site investigation plan to the actual site restoration following remediation. In between this first and last step, there are also regulations that address: areas to be remediated, cleanup levels, waste treatment methods, and assessment of ecological damage.

This administrative objective of the ecological assessment might appear trivial. However, if it is not addressed, the entire cleanup effort could be significantly delayed. The delay could involve not only preparing the documentation after the fact, but also a reevaluation of the remediation alternatives and possible selection of a different method of site remediation. If the originally selected remediation alternative does not achieve the protection mandated by a regulation, or unnecessarily destroys an ecological resource protected by another regulation, it might have to be abandoned or at least reconsidered. In such a case, much detailed work developing the alternative would have been wasted, and the entire remediation process significantly delayed. Consequently, it is important not only to incorporate this working objective into the process, but also to consider and address the ecological regulations early and often in the preparation of the ecological assessment.

Incorporation of the Ecological Assessment Into the Process

The final working objective of the ecological assessment is purely administrative, and it is absolutely necessary to achieve the ultimate goal of the assessment. All of the ecological investigations, evaluations, predictions, and working objectives will do nothing to advance the ultimate objective of achieving protection of the environment at the site, if they are not integrated into the overall site investigation and remediation process. Not only must the ecological aspects of the cleanup be integrated with the other facets of the effort, they must be incorporated at the appropriate time. Issues raised too early will often be forgotten, and those raised too late are frequently strongly resisted.

The process of integrating ecological considerations begins by developing supportable site ecological goals or endpoints, and then reaching a consensus on the endpoints with other technical and management personnel. Reaching a consensus necessitates understanding the technical, financial, and political constraints of remediation, and also appreciating the remediation goals for other considerations, such as public health or drinking water supplies. The coordination must progress from goal setting to establishing a site investigation approach and methods. If site conditions related to the ecological concerns are not measured as part of the comprehensive site investigation, chances are that the data needed by the ecologist will never be collected.

Establishing the need, area, and levels for cleanup are the aspects of the site remediation process requiring the most intense coordination among all of the Remediation Investigation and Feasibility Study interests, including the ecological concerns. The decisions made during these steps in the process are the ones that will ultimately determine the level of environmental protection. The trades-offs among competing issues, such as costs, various environmental resources, public concerns, and future use of the site are also considered as a part of these critical remediation decisions. The ecological concerns must be a part of this process to be legitimately considered and incorporated. For maximum consideration, the concerns should be as precise as possible, and supported by sound scientific information.

A SUMMARY OF ECOLOGICAL
ASSESSMENT NEEDS AND OBJECTIVES

There is a need for ecological assessments at hazardous waste sites from both an idealistic and practical standpoint. There are abundant and sensitive ecological resources at risk from both hazardous waste releases and cleanup. The general mood of the nation is that these resources are important and worthy of restoration and protection. In fact, most people take it for granted that hazardous waste cleanup includes ecological protection. In order to achieve the desired restoration and protection, the resources must be understood and levels of protection investigated. An ecological assessment is the appropriate and accepted approach to provide the needed understanding and investigation.

The practical need for ecological assessments is driven by the idealistic need. There has been legislation, regulations, and directives mandating the consideration of ecological issues specifically at hazardous waste sites. The compliance with general natural resource protection policy and regulation is also formalized at hazardous waste sites. The environmental regulatory, enforcement, and natural resource bureaucracy has structured the response to the legal mandate for ecological consideration as an "Ecological Assessment," which is an integral part of the hazardous waste site remediation process. Thus, not only is there a need for an assessment of ecological concerns, there is a regulatory necessity for an ecological assessment.

The objective of an ecological assessment is simply protection of ecological resources at hazardous waste sites. However, the intermediate objectives required to address this lofty goal are more complex, and must be properly incorporated into the site cleanup and restoration process to be efficient and effective. Although there are several intermediate or working objectives, which address both technical and administrative aspects of the process, they are not, by any means, independent. Addressing, and ultimately achieving, each of these objectives relies on the same site-specific ecological information and toxicological data base.

The most important and often most difficult tasks for ecologists are understanding and incorporating the assessment needs, and focusing the process on the objectives of the ecological assessment. If the process strays into areas of research not related to the narrow objectives of the assessment, as an integral part of site remediation, not only is time and money wasted, but the ecological assessment process loses credibility. Similarly, if the ecological investigations are not focused on supporting remediation decisions directly related to the need, extent, and level of remediation or restoration, the recognition of ecological considerations by decision makers is diminished. If the ecological assessment loses credibility, and ecological issues are not taken seriously, achieving the ultimate ecological objective of the process is less likely.

REFERENCES

U.S. EPA, 1989a. Summary of Ecological Risks, Assessment Methods, and Risk Management Decision in Superfund and RCRA. EPA/230/03/89/046.

U.S. EPA, 1989b. Ecological Risk Assessment Methods: A Review and Evaluation of Past Practices in the Superfund and RCRA Programs. EPA 230–03–89–044.

U.S. EPA, 1989c. Ecological Risk Management in the Superfund and RCRA Programs. EPA 230–03–89–045.

U.S. EPA, 1989d. The Nature and Extent of Ecological Risk at Superfund Sites and RCRA Facilities. EPA/230/03/89/043.

U.S. EPA, 1989e. Guidance for Conduction of Remedial Investigations and Feasibility Studies Under CERCLA. EPA/540/G–89/004.

U.S. EPA, 1989f. Risk Assessment Guidance for Superfund—Environmental Evaluation Manual. EPA/540/1–89/001A.

3

Regulatory Perspective

INTRODUCTION

National attention was drawn to hazardous waste when incidents like those involving Love Canal in New York, dioxin in Times Beach, Missouri, the Valley of the Drums in Kentucky, and Kepone in Hopewell, Virginia resulted in real public health damage and a projection of significant future health risks. Problems were identified with the individual offensive compounds, the handling of hazardous wastes, and the cleanup of past indiscretions from the treatment and disposal of hazardous wastes. Specific federal legislation was passed to govern each one of these aspects of hazardous materials. The Toxic Substances Control Act (TSCA) was passed to regulate individual substances, and the Resource Conservation and Recovery Act (RCRA) was instituted to control the management of hazardous wastes as they are generated. The cleanup of historic hazardous waste sites is addressed by the Comprehensive Environmental Response, Compensation, and Liability Act (CERCLA). Although each of these landmark acts is critical in the national effort to control hazardous materials, the mandates and potential effects of CERCLA are where ecological issues and hazardous waste most often come face to face. RCRA, TSCA, and other federal and state regulations address some aspects of a site's ecology, and thus are briefly summarized below. However, it is almost always within the context or provisions of CERCLA that ecological damage is most apparent, and hence the emphasis of this chapter is on CERCLA, the Superfund law. Consequently, this book focuses on CERCLA and how to achieve compliance with other regulations as part of the CERCLA process.

THE REGULATION OF COMPOUNDS AND WASTE MANAGEMENT

Specific compounds are regulated under the Toxic Substances Control Act (TSCA), which was passed in 1976 as an early attempt to regulate hazardous materials. The legislation was originally restricted by a limited comprehension of the ubiquitous nature of hazardous materials and the potential for associated far reaching effects. The statute is also constrained in effectiveness by last-minute compromises enacted to resolve conflicts between the Senate and House versions (Findley and Farber 1988). There are three basic initiatives set forth in TSCA. First, the Act calls for the development of a database to scientifically describe the environmental effects of manufactured compounds. The production of the database is the responsibility of the chemical manufacturers. The second basic aspect of TSCA gives the federal government the authority to prevent unreasonable risk to public health and the environment from hazardous compounds. The last basic initiative reflects the limitations of the legislation, as it calls for exercise of the government's authority in a manner not to impede or create unnecessary economic barriers to "technological innovation."

The development of the database on testing methods for environmental effects, as required by Section 4 of the Act, has been a slow process. By 1983, the EPA had identified 62,000 existing compounds and proceeded to prioritize environmental testing based on relative risk. The EPA attempted to implement a voluntary testing program with industry for the priority compounds, but was overruled by the court (NRDC v. United States EPA, 595 F.Supp. 1255, S.D.N.Y. 1984), which concluded that the testing must be mandatory, and the EPA has proceeded on this basis. Section 5 of the Act requires that a premanufacture notice be submitted to the EPA before a new compound is manufactured or imported into the U.S. Test results of the environmental effects are not required unless the company has the results on hand. A 1983 study revealed that, of the 740 premanufacture notices issued, only 53 percent contained any information of toxicity. Only 17 percent included actual test information on cancer, birth defects, or mutations, which are the three primary biological effects identified as special concerns in TSCA (Findley and Farber 1988). After a review of the premanufacture notice and any associated information, the EPA is obligated under Sections 5 and 6 to consider placing restrictions on the manufacture or use of the compound, if it will present an unreasonable risk to health or the environment.

TSCA does not require any form of ecological assessment associated with the administration or control of toxic substances. The closest that the Act comes is to provide a mechanism for reporting any preexisting laboratory toxicity testing of compounds. This could be a useful catalog of information, by compound, in conducting ecological assessments when releases to the environment do occur. However, the incompleteness and lack of consistency in the effects data on ecological resources limit the usefulness of the TSCA database.

The Federal Insecticide, Fungicide and Rodenticide Act (FIFRA) is similar to TSCA, in that it regulates individual compounds. FIFRA has a broader scope in some respects, because it regulates the use of an entire category of compounds, not only from the manufacturing standpoint, but also their use. Regulations promulgated under the Act specify application rates, and also provide protection for sensitive environments in specifying acceptable use of the products. Because the intent of the compound is to adversely affect target species, there is always extensive data on the toxicity of the compounds, and FIFRA assessments tend to be data-rich, whereas TSCA assessments are often data-poor (Ramamoorthy and Baddaloo 1991). Thus, the evaluation of potential ecological effects from FIFRA-regulated compounds is usually more precise than comparable assessments for other potentially hazardous materials.

The handling of wastes is governed by the Resource Conservation and Recovery Act (RCRA), passed in 1976 and amended in 1984 and 1986. Subtitle C of the Act establishes a "cradle-to-grave" system for hazardous wastes. Section 3001 requires the

EPA to establish criteria for identifying hazardous wastes and maintaining a list of compounds considered to be hazardous. Sections 3002 and 3003, respectively, mandate detailed recordkeeping and tight controls on the transportation of waste. Regulations on the construction and operations of landfills qualified to receive hazardous wastes are contained in Section 3004, and Sections 9001–9010 contain similar requirements for underground storage tanks. Because of the historical record of old landfills becoming new Superfund sites, the latest amendments of RCRA prohibit the landfilling of most types of hazardous wastes beyond certain dates. Enforcement and control of regulations are provided for in Section 3005, which calls for the permitting of all RCRA facilities.

The RCRA regulations do not address ecological resources directly. Some of the siting requirements for new facilities do require consideration of the potential displacement or risk to existing resources. However, in theory, compliance with all of the stringent RCRA requirements for the transport, treatment, and disposal of hazardous wastes should prevent releases and subsequent risk or damage to the environment. Thus, ecological assessments are not specifically addressed by RCRA or included in the implementing regulations.

THE AUTHORITY FOR
HAZARDOUS WASTE CLEANUP AND
ECOLOGICAL ASSESSMENT

The comprehensive Environmental Response, Compensation, and Liability Act is the "Superfund law," with the lofty goal of correcting damage from over half a century of uncontrolled hazardous waste management, treatment, and disposal. The acronym for the legislation, CERCLA, was no accident (Cummings 1990). The intent of the Act was to encircle and complete the regulation of all aspects of hazardous waste. The Act covers all types of hazardous materials, environmental media, parties responsible for past actions, and financing of the cleanup. CERCLA was also intended to round out the control of hazardous materials initiated by TSCA and RCRA. Finally, the Act would complete the then-open circle of abandoned waste by bringing them back to the responsible parties. The legislation established an administrative structure, financial resource, and authority for the assignment of liability as the principal weapons to tackle the optimistic goals of CERCLA. The broad authority of the Act has proven to be a necessary part of the regulatory arsenal in addressing the objectives of protecting human health and the environment from hazardous waste site releases.

The comprehensive nature of CERCLA, particularly as amended in 1986 as the Superfund Reauthorization Act (SARA), is the driving force behind ecological assessments at hazardous waste sites. The law contains three provisions that result in an unquestionable mandate that the ecology of a site must be considered as part of the hazardous waste cleanup. The first directive is to protect the environment, as well as simply public health. The original emphasis of environmental protection was to limit human exposure to an acceptable cancer risk or other hazard. The protection of the environment was generally limited to preventing any contaminate from entering an environmental medium that could, at some future date or off-site location, present an unacceptable or unquantifiable risk to human health. Most common among the concerns for environmental protection were ground water contamination, if it represented a potential human drinking water supply. Often, when it was not feasible to achieve ground water remediation, institutional restrictions on future use of ground water as a water supply were considered an acceptable protection of the environment.

Similar approaches were often followed for soils and surface waters. If large areas of soil were contaminated and remediation was technically or economically extremely difficult or infeasible, an evaluation of potential future uses of the site was conducted. If the evaluation revealed that frequent or long-term human exposure was unlikely, due to the expected land use at the site, institutional controls or other, less protective, remediation were chosen as being protective of the environment. In cases where surface waters were contaminated at the site, and remediation was not straightforward, the closest downstream surface water use that could potentially present a public health risk was identified. The dilution between the site and the surface water use was then used to calculate the human health risk. A surface water concentration acceptable to the downstream location supporting a human use was considered to establish the remediation criteria at the contaminated site. This approach ignored ecological use, and protection of the stream segment between the site and the location of public use.

The limited consideration of environmental protection previously used is now giving way to a broader definition. This is partly a reaction to past practices, where important ecological resources or their potential existed at a site, and were ignored during removal or remediation. When put into perspective, this approach is clearly understandable. At many of the early sites where gross contamination of multiple media occurred, there was eminent risk to human health and often the time, resources, or inclination to adequately address ecological concerns at these sites did not exist. Also, ecological effects and protection tend to be long-term, with little immediate benefit, so that when cleanup under Superfund finally became a reality, immediate action and expedience in achieving results was desirable. Now that some of the sites with the most immediate threat to human health have been investigated, and a well-structured cleanup process is in place, some of this attention has turned to protection of ecological resources. The regulatory community and private environmental and conservation groups have now recognized that the entire ecology of an area must be considered in hazardous waste site cleanup.

CERCLA, as amended by SARA, has a second requirement for the consideration of the ecology of a site. The Act mandates compliance with the substantive aspects of all federal and certain state environmental regulations. Thus, where congress or the state legislature has seen fit to protect a particular ecological resource, then the resource must be afforded at least the same level of protection when a Superfund site is remediated. Endangered species or clean water are examples of resources afforded protection under federal legislation, so they must also be considered in developing a remediation strategy for a hazardous waste site. The third primary mandate for ecological consideration at CERCLA sites is the provision in the Act for the recovery of damage caused to natural resources by the releases of hazardous wastes.

There is a common thread among the three CERCLA ecological requirements (i.e., protection of the environment, compliance with environmental regulations, and reparation of damage to natural resources). In order to comply with these mandates, the ecological resources potentially on or adjacent to the site must be identified and understood. The effects on the resources from past actions must be investigated, and the potential for future risk to the resources considered. These ecological aspects of the sites are best documented in an ecological assessment that is an integral part of the site remediation process.

The remainder of this chapter describes the three regulatory aspects of CERCLA that are critical to ecological assessments at hazardous waste sites. First, the overall CERCLA compliance process is summarized, emphasizing integration of the ecological considerations and assessment. The requirements for the evaluation of damage to natural resources, and the critical influence of this requirement on a hazardous waste site's ecological assessment, are discussed next. Compliance with other environmental regulations that are applicable or relevant and appropriate, and how the evaluation of compliance fits into a comprehensive ecological assessment, is the final topic in this chapter.

CERCLA PROCESS

Background

CERCLA was passed in 1980, and signed by President Carter a month before he left office. From many perspectives, it is considered the most significant environmental achievement of his administration (Cummings 1990). The original legislation was drafted when there was little national experience in cleaning up hazardous waste sites, and it represented several compromises related to the funding of the cleanup. Consequently, there were several omissions, and what later proved to be impractical aspects of the 1980 Act. Provisions of the Act were clarified and expanded in the Superfund Amendments and Reauthorization Act of 1986 (SARA). The original Act and its amendments, which are commonly referred to as CERCLA, or Superfund, were granted a four-year extension in 1990 by the congress.

The primary objective of CERCLA is to clean up hazardous wastes. There was little debate on this objective, but the responsibility, liability, and financing of the cleanup has received much discussion and national soul searching. The outcome of the debate is that owners, operators, and waste generators are strictly, jointly, and severally liable for the cleanup of hazardous waste sites and the damage caused by releases. This means that, if a party ever owned or had possession of a waste found in a Superfund site, they are potentially responsible for the cost associated with the cleanup and reparation of damages for the entire site. Congress realized that waiting for the ultimate determination of liability, and the high probability of subsequent litigation, could delay the actual cleanup of a site for a decade or more. To minimize the delay, congress established, through CERCLA, a process, administrative structure, and seed funds to initiate site remediation before liability determination was finalized.

CERCLA is composed of four basic initiatives. The first (Sections 102 and 103) addresses the identification and reporting of jurisdictional substances. The EPA is required to issue these definitions and to designate hazardous substances. Section 103 of CERCLA is directed at operators and owners of hazardous waste facilities, and requires them to notify the EPA of the presence, and more importantly, potential release of any hazardous material at their facilities.

The second initiative of CERCLA is the most important to the ecological aspects of hazardous waste sites. Section 104 authorizes the federal government to remove and remediate the effects of hazardous waste releases. The actions are to be in conformance with a National Contingency Plan (NCP) developed by the EPA as referred to in Section 105 of CERCLA (U.S. EPA 1988a). The development and requirements of the NCP are discussed in detail later.

The final two initiatives of CERCLA are the most important from a policy and implementation perspective. They involve the establishment of a Hazardous Substances Trust Fund (or Superfund) for remedial and removal activities. The fund was established for $1.6 billion initially, and increased by 1986 the amendments to $8.5 billion. The

Act, in Section 107, also makes any party responsible for the release of hazardous substances financially liable for: all costs of removal or remedial action; other necessary response actions; and damage to natural resources. The courts have determined that the Act imposes strict liability, not dependent on fault. This provision of CERCLA provides the guiding philosophy of private-party cleanup, either voluntarily or through enforcement action.

The EPA was required to promulgate regulations to implement the cleanup of hazardous waste sites mandated by CERCLA. The EPA responded by developing the National Contingency Plan, which is codified as 40 Code of Federal Regulations (CFR) 300. This plan, its associated regulations, and subsequent guidance developed by the EPA, describes the process for first listing, then investigating and cleaning up hazardous waste sites. Consistent with the overall philosophy of CERCLA, the National Contingency Plan outlines a comprehensive process for dealing with hazardous waste sites. The plan and related regulations cover the nationwide cleanup program, as well as actions and processes at individual sites.

The National Contingency Plan and Ecological Assessments

The National Contingency Plan requires, provides guidance for, and makes possible ecological assessment of hazardous waste sites. The plan requires consideration of ecological aspects by mandating: full consideration of all aspects of the environment; compliance with federal and state regulations designed to protect or enhance ecological attributes; and restoration of natural resources damaged by the release of hazardous wastes. No other legislation or implementing regulations provides such a comprehensive framework for conducting and integrating an assessment of ecological issues at hazardous waste sites. Even the primary national environmental legislation, the National Environmental Policy Act, has been determined redundant to the National Contingency Plan, and thus not necessary, for the consideration of environmental issues as part of CERCLA actions. Without the National Contingency Plan, there would be no regulatory perspective for ecological assessments at hazardous waste sites, and no need for this chapter. In fact, without the Plan in its current form, there would be no regulatory need for a comprehensive ecological assessment and thus no need for this book.

There was little attention to ecological issues during development of the first National Contingency Plan and subsequent compliance. However when the plan was revised in 1990 in response to SARA, there was more consideration given to ecology. The plan guidance that has been developed since SARA has placed increasing emphasis on ecological considerations in assessment of hazardous waste sites. All site investigations now require analysis of at least the presence of ecological resources within the area potentially affected by site releases or remediation.

The National Contingency Plan also recognizes that the consideration of environmental issues cannot be performed in a vacuum, and requires integration of the ecological assessment into the overall site cleanup process. Thus, a general understanding of the process mandated by the Plan is essential in conducting an evaluation of ecological effects and potential as part of hazardous waste site remediation. To be effective, the biological and related investigations and assessments must be coordinated with other evaluations under the National Contingency Plan. The ecological issues must be identified early, so that the data required to assess impacts and potential can be synoptic with other site remediation requirements. Also, the ecological concerns must be voiced during the very earliest stages of remediation planning, so that they can be integrated into the identification of cleanup options.

Removal Actions

When the appropriate federal agency is notified of a release of a hazardous material, they must make a determination of immediance as prescribed in 40 CFR 400–415. The involved agency, generally either the EPA or the Coast Guard, must evaluate the following factors:

• Exposure to humans, animals, or the food chain:
• Contamination of drinking water or sensitive ecosystems:
• Threat of continued release from containers:
• High concentrations in surface soils that may rapidly migrate:
• Impending weather conditions that may hasten releases:
• Threat of fire or explosion:
• Currently available procedures to contain the release:
• Other factors that pose threats to health, welfare, or the environment.

Upon consideration of these factors, the lead agency can determine whether removal action should begin as soon as possible to address the threat to public health, welfare, or the environment. In such circumstances, the removal action proceeds with a minimum of paperwork or approvals.

This emergency type of action or *Response* is limited to the expenditure of $2 million and 12 months, except under extenuating circumstances. When no such immediate threat exists, a full remedial investigation and feasibility study (RI/FS), as discussed later must be conducted, in compliance with 40 CFR 300.420–435. The level of total planning, investigation, paperwork, and approvals is much greater under the remediation condition. Consequently, the opportunity for ecological protection and assessment is also greater. However, even in a response-type situation, ecological issues can not be totally ignored.

The responding agency must consider the protection of ecological resources to the extent that the eminent threat to public health and welfare allows. For example, if there is an active release from a storage tank directly into a wetland that is adjacent to a public water supply, the emergency situation would preclude full compliance with all wetland restrictions. Construction of a berm and removal of the hazardous waste in the wetland would be appropriate without full consideration of all possible wetland protection measures or a complete evaluation of future restoration or mitigation. If, however, there were comparable access routes to the release area, the one with the least direct wetland impacts should be used. The responding agency can easily comply with the intent of ecological protection provisions for response actions by identification of critical ecological resources within the release area. This can be accomplished by contacting ecological staff within the EPA or other natural resource agency, such as the Biological Technical Advisory Group, which (as described in the following section of this Chapter) is established for each region to monitor and assist with Superfund activities. Conducting such a survey can identify additional removal objectives to protect ecological resources from immediate danger. Acknowledging the presence of the resource and confirmation that there are no obvious and equally expedient response alternatives that are less damaging to ecological resources generally constitutes compliance with ecological protection requirements for Removal Actions.

Defining Superfund Sites

The first step in addressing Superfund sites is to determine what constitutes a hazardous waste site requiring remediation. This is done by developing a National Priorities List of sites. The EPA maintains and updates the National Priorities List by a Hazard Ranking System. As of 1990, the EPA had identified over 32,000 hazardous waste sites, but 55 percent of the sites were determined to pose no risk to human health or the environment, and thus no further actions were deemed necessary. Of the remaining sites, 1,236 had a sufficiently high score to qualify for the National Priorities List. Twenty-nine of the priority sites have been deleted because they were considered to be remediated (U.S. EPA 1990). The EPA estimates that sites will continue to be added to the list at a rate of 100 per year (U.S. EPA 1990).

The Hazard Ranking System (HRS) is a screening tool used by the EPA to organize cleaning up the multitude of uncontrolled hazardous waste sites nationwide. After sites have been identified, and basic information characterizing the site entered into the Comprehensive Environmental Response, Compensation, and Liability Information System, the HRS is used to prioritize the effort in addressing the risks caused by the sites. The HRS included in CERCLA does not address the question of feasibility or benefit of cleanup, it is simply a tool to compare the relative risks (almost exclusively to human health) of all sites on a common basis (U.S. EPA 1987).

The information derived from preliminary site assessments, site inspections, and any other available information is used to develop an HRS score. The score addresses risk from contact with waste via ground water, surface water, air, or soil, and is used to determine whether the site warrants placement on the National Priorities List. The original version of CERCLA required two additional HRS scores, direct contact with the waste and the possibility of fire or explosion, which were used in assessing the need for immediate removal action. Scores based on contact, fire, or explosion are no longer required under SARA.

The original HRS placed little significance on the threat to ecological resources in determining the priority of hazardous waste site cleanup. Wetlands and endangered species were the only ecological resource evaluated in the process. Also, only the highest-scoring sensitive ecosystem was included when multiple resources were within the target distance. The ranking system mandated by SARA and codified as 40 CFR Appendix A to Part 300 includes evaluation of a broader spectrum of biology-related resources. Any habitat considered critical by a federal or state resource agency is considered under the revised HRS. Contamination in the aquatic human food chain is also entered into the equation and, although this is designed to protect human health, the significant consideration of ecological resources is an important byproduct. There is also an increased sensitivity to ecological resources in the revised system in the provision for entering the sum of values for all ecosystems, rather than just the single value of the highest scoring resource. The ecological resources are incorporated into the ranking system through the distance from the site to the resource and assigning an appropriate score. Although risk to the ecosystem posed by releases from hazardous waste sites is an increased factor in the revised ranking system, it is virtually impossible that risk to ecological resources alone would qualify a site for the National Priorities List. Risk to multiple resources could, however, increase the score, and thus the cleanup priority of a site.

A site can qualify for the National Priorities List, and thus be eligible for financing through Superfund, in any of three ways. The first is to score high in the HRS. The

EPA originally set the cut-off of an HRS of 28.50 for eligibility on the NPL, because it produced the CERCLA mandate of a minimum of 400 sites on the list. Alternatively, a site can be placed on the list if an individual state determines that the site is its number-one priority. The final mechanism for listing is if a site meets all three of the following conditions: the U.S. Department of Health and Human Services has issued a health advisory for the site; the EPA determines that the site represents a significant threat; and remedial action is determined by the EPA to be more cost-effective than a removal action.

Site Investigations

A hazardous waste site investigation proceeds in stages, with increasing detail collected on environmental and hazardous material release conditions at each sequential stage. The first stage is a removal site evaluation, as described in 40 CFR 410, and is performed to determine the need for an immediate removal action, as described earlier. The removal site evaluation is generally limited to existing information concerning the materials on-site, and details of the potential release. The evaluation can include a site inspection, with the primary purpose of determining overt signs of an actual or potential release. During this first step of site investigation, it is prudent to evaluate the potential for significant ecological resources on-site by such general procedures as viewing U.S. Geological survey topographic maps, or by telephoning a local EPA or Fish and Wildlife natural resource advisor. If this process identifies the potential for ecological resources or damage, and a site inspection is conducted, natural resource personnel should be included in the inspection.

In a nonemergency situation, site investigation begins with a remedial site evaluation (40 CFR 420). The initial phase of the remedial site evaluation, termed the preliminary assessment, is similar to the removal site evaluation, in that it relies largely on existing information. A preliminary assessment can establish that immediate removal is required, that the site warrants no further action, or that a detailed investigation is required. The preliminary assessment should provide the information necessary to evaluate the site under the HRS.

The remedial site evaluation concludes with a remedial site investigation. A site investigation could be conducted to determine the need for emergency action or to rank the site on the National Priorities List. However, earlier phases of the investigation have usually accomplished these objectives. The basic purpose of the investigation is to either eliminate the site from further consideration, or to provide sufficient information to rapidly and efficiently initiate a long-term cleanup process. The first actual environmental sampling is generally conducted during the site investigation. An ecological reconnaissance should be conducted as part of the planning for any environmental sampling. The reconnaissance identifies important resources or locations to sample, as well as preliminarily evaluating potential damage to ecological resources. It is also important at this stage to integrate ecological considerations when developing the initial cleanup strategy.

Site investigation is completed in the remedial investigation (RI) and feasibility study (FS) as described in 40 CFR 300.430, and in EPA guidance publications (U.S. EPA 1988b). A site need not be included on the National Priorities List to be eligible for the RI/FS process. However, the site must be on the list for a Superfund-financed RI/FS, and alternative financing, such as from enforcement, must be used if the site does not qualify for the National Priorities List.

The RI and FS are the foundation of a CERCLA investigation and remediation. The process develops the data to fully characterize the waste, the environment, and the potential for interaction between the two. The RI/FS presents the forum for discussing and evaluating the threat caused by the release and the possibilities for remediation. It is also the forum for different, and sometimes conflicting, objectives and approaches to be aired and compared. Ultimately, the RI/FS develops the basis for designating the recommended plan and issuing the Record of Decision on site remediation. Consequently, ecological considerations must be brought forward in an ecological assessment as part of the RI/FS process, to be incorporated in the cleanup of hazardous waste sites.

The goal of the RI/FS program, as described in the National Contingency Plan, is to select an acceptable remedy to the release of hazardous waste. The remedy must be protective of human health and the environment, must maintain this protection over time, and must minimize untreated waste. In promulgating guidance to implement this objective, the EPA recognizes that the selection of a remedy is an inexact science, and cannot remove all uncertainty. The primary objective of the RI/FS is to provide sufficient information to support the selection of a remedy that appears to be the most appropriate for a site, based on an informed risk management decision (U.S. EPA 1988b).

Even with this recognized limitation, the need to reach a decision is particularly important for the ecological assessment. Ecology is an inexact science even in unperturbed, pristine conditions. When the uncertainties of contamination extent, the effects of contamination, and the potential for exposure are added to natural biological variability, the documentation of site effects can be approximations at best. However, the presence of uncertainty does not relieve the ecologist of the responsibility for being a full participant in the RI/FS process. The ecologist cannot, however, disregard quantitative analysis and rely only on speculation in the hazardous waste site ecological assessment. The ecological data needs must be identified and the collection of the appropriate information considered for the RI. Ultimately, if assumptions, speculation, or subjective comparisons are used, the RI must document the fact that more definitive and quantitative ecological input to the decision was not practicable.

The RI/FS process consists of the following basic steps:

- Scoping and work plan development.
- Site characterization:
  Contaminant location and concentration.
  Environmental description.
  Existing risk determination.
- Determination of remediation criteria.
- Development and screening of alternatives.
- Detailed analysis of alternatives.

The process is usually divided into separate RI and FS documents, with the FS beginning with the determination of remediation criteria. However, the process should be site-specific and flexible, to incorporate a wide variety of conditions. Upon completion of the RI, there should be an understanding of what wastes are on-site, the general locations of the material, and the volumetric extent of contamination of the various media. The risk to human health from leaving the site in the existing condition is also documented in the RI. A parallel assessment of risk to ecological resources from existing site conditions is also a necessary component of the RI. When the RI is completed, there should be a comprehensive understanding of the extent and implication of ecological problems.

In certain incidents, the RI should also address the treatability of the waste present on-site, and the implications of alternative remediation criteria. The alternative criteria are usually expressed as different concentrations of individual contaminants, or combinations of contaminants in different media. The alternative concentrations address different levels of protection or the protection of different resources. For example, one concentration might be acceptable for industrial land use, while a lower concentration would have to be achieved to allow for residential use of the site. Similarly, different ecological remediation criteria could be evaluated for the protection of different species presence on site, and for the breeding and successful rearing of young.

The FS identifies, screens, and evaluates the remedial technologies and alternatives for addressing the hazardous waste risks identified in the RI. The alternatives are evaluated based on nine criteria prescribed by CERCLA:

- Overall protection of human health and the environment:
- Compliance with appropriate regulations:
- Long-term effectiveness:
- Reduction of toxicity, mobility, or volume through treatment:
- Short-term effectiveness:
- Implementability:
- Cost:
- State acceptance:
- Community acceptance.

These criteria can be categorized as threshold, balancing, or modifying. Overall protection of human health and the environment, and compliance with regulations, are considered threshold criteria, and must be achieved by any acceptable alternative. Long-term effectiveness, the reduction of toxicity, short-term effectiveness, implementability, and cost are considered balancing criteria. The acceptance of the remediation alternative by the state and community are considered in modifying the selected alternative.

Scoping and the development of a work plan are the first steps in the RI/FS process. It is critical that there be ecological input in this stage so that synoptic biological, geological, chemical, and hydrological data can be collected. This information is needed to identify damage to ecological resources, conduct an ecological risk assessment, and evaluate remediation criteria to protect vulnerable resources. The input from previous site investigations should be carefully reviewed and, if ecological issues exist, an additional site reconnaissance is usually appropriate before the work plan is finalized.

The objective of the RI is to assess the existing risk to human health and the environment. To accomplish this, a full understanding of the contaminants and environmental resources is necessary. This includes field measurements of the appropriate media and, potentially, a laboratory testing of toxicity. Depending on the biological characteristics of the site, and the toxicity profile of the contaminants detailed, quantitative measurements of biological populations may also be necessary. The objective of the FS is basically to identify a construction project that removes, controls, or contains the hazardous waste. As with any construction project, the existing environmental resources can be directly destroyed or indirectly impacted. To provide maximum protection to the existing resources, while still reducing the risk of hazardous waste, the characteristics of the resources must be documented. Chapters 4, 6, 7, and 8 of this book discuss the types of scientific investigations that should be incorporated in the RI/FS process to insure consideration of the ecological resources. The need and extent of each of these investigations must be based on the site characteristics, the types and magnitude of wastes present, and the investigations planned for other RI/FS objectives.

ECOLOGICAL REGULATION COMPLIANCE AS PART OF SITE CLEANUP

Hazardous waste cleanup involves multiple activities (such as removal, treatment, transportation, and disposal), compounds, and environmental resources. There is a complex set of environmental laws, regulations, and guidances developed prior to or independent of the national initiative for hazardous waste cleanup, which address each one of these aspects individually. The regulatory process for each separate aspect can be extremely lengthy, complex, and interactive. The complexity of the environmental regulation compliance process can increase exponentially when multiple activities, compounds, and resources are involved. The specifications of one regulation can even be in direct opposition to another, or different requirements can be circular in specifying prerequisites for granting permits or conformance. The original Comprehensive Environmental Response, Compensation, and Liability (CERCLA) legislation was basically silent on the relationship of environmental regulations and hazardous waste cleanup, and thus there was the potential for regulatory gridlock. However, in promulgating regulations under CERCLA, the EPA recognized the potential for a lengthy, complicated, and convoluted regulatory compliance process to sidetrack and delay the actual cleanup of hazardous waste sites, perhaps indefinitely. The cleanup was already recognized as a herculean task, because of: serious threats to human health and the environment; significant time pressures because the program was behind schedule; difficulty from a technical standpoint, because of only limited nationwide experience; expense; and extremely complicated, because of all the parties involved. Therefore, the EPA developed the concept of functional equivalence without formal compliance for regulatory considerations at hazardous waste sites, and included this concept in the first National Contingency Plan.

The functional equivalence, as initially codified by the EPA, was legislated by inclusion in the Superfund Amendments and Reauthorization Act (SARA) as a new CERCLA Section 121(e). This new section states that federal, state, or local permits are not required for the portion of any removal of remedial action conducted entirely on a CERCLA site. The SARA legislation addresses nonfederal regulations by including any state statute imposing a more restrictive substantive requirement then the equivalent federal requirement. Even though the permits or other regulatory sign-offs are not required, the intended environmental protection of the various regulations is mandated as part of the RI/FS. Thus, when conducting the cleanup, a process that is the functional equivalent of formal compliance with other regulations must be followed.

The process of evaluating functional equivalence has been formalized as part of the RI/FS process required under CERCLA. The process is termed Compliance with Applicable or Relevant and Appropriate Regulations (ARARs). Simply stated, this concept requires adherence to the substantive aspects of environmental permits and regulations, but eliminates the need for actually obtaining a permit or otherwise complying with purely administrative aspects of a regulation. There has been, and will continue to be, debate on the fine points of what constitutes substantive versus administrative compliance. However, in most cases, the intent is intuitively obvious: if direct environmental protection is afforded, the requirement is substantive; if no identifiable environmental benefit is achieved, the requirement is likely administrative. A clear example of a substantive requirement would be the placing hay bales and silt curtains around a wetland before excavating soil in area adjacent to a wetland. Mandating notification of the local conservation commission or state natural resource agency before beginning the excavation would be a purely administrative requirement. Administrative review, reporting, and record keeping are examples of other strictly non-substantive requirements (U.S. EPA 1988c). The monitoring of an environmental resource after or during remediation does not clearly fall on either the substantive or administrative side. However, the current feeling is that monitoring is a substantive requirement, because by identifying an unanticipated problem, it could result in measurable environmental protection.

Environmental regulations generally revolve around protection of a particular resource. Compliance with the substantive requirements is achieved by demonstrating that the specified level of protection for the resource is afforded by the proposed CERCLA action. For example, if a Superfund site required a groundwater treatment plant, the plant could discharge to a river adjacent to the site without obtaining National Pollution Discharge (NPDES) permit, since the EPA has interpreted the bounds of a site to be any area of elevated contaminant concentration, or the area immediately adjacent to the site. However, to demonstrate substantive compliance, the RI/FS (and subsequently relevant documentation) would have to show that the level and type of treatment meet appropriate NPDES guidelines, and that all appropriate water quality standards and criteria were met. In this example, the resource designed for protection would be the receiving water and all of the beneficial uses of the water, including unaffected aquatic habitat. By demonstrating that the appropriate water quality would be achieved, the resource would be protected and the substantive requirement of the NPDES ARAR would be met. The administrative requirements, such as sign-off of the specific treatment technology, issuance of an actual permit, or holding public hearings on the discharge, could be ignored. If there was uncertainty regarding the reliability of the treatment, the character of the waste, or the receiving water conditions, chemical or toxicological monitoring of the effluent or post-discharge evaluations of receiving water characteristics could be a substantive requirement included in the site remediation plan.

Applicable requirements are defined in the National Contingency Plan (U.S. EPA 1988a) as:

> Those cleanup standards, standards of control, and other substantive environmental protection requirements, criteria, or limitations promulgated under Federal or State law that specifically address a hazardous substance, pollution, contaminant, remedial action, location, or other circumstance at a CERCLA site.

The classic example of applicable requirements are the RCRA landfill design specifications, when a CERCLA action includes landfilling of hazardous waste. From an ecological perspective, the classic ARAR is Section 404 of the Clean Water Act, when wetlands are present on the site.

Relevant and appropriate requirements are not as clearly defined in the National Contingency Plan, and an evaluation is necessary to make the determination. The evaluation is a two-step process, requiring first a determination of relevance, and then applicability. Generally, a regulation is relevant if, although not originally promulgated to address a hazardous waste issue, it concerns a problem or situation similar to one at a CERCLA site. Appropriateness is concluded by comparing the regulation to the characteristics of the site, waste, or remedial action. If full compliance with the regulation would call for any modification of any anticipated action, then the regulation would be appropriate. There is generally considerable discretion in evaluating relevance and appropriateness, and only part of a requirement could be considered in the category. If a regulation is considered to be relevant and appropriate, then it must be complied with to the same extent as an applicable regulation.

The Endangered Species Act could represent a relevant and appropriate requirement. If an endangered species has not been reported on-site, the Act might not be applicable, but if a species habitat could be present on-site, it could be relevant. If relevance is determined, but no action in the habitat area is anticipated, then the Act might not be appropriate.

When an ARAR is identified, the RI/FS or supporting documents must demonstrate compliance with the substantive requirements. This is generally preceded by a determination of what the substantive requirements are, and then (on a point-by-point basis) a demonstration of how each aspect of the remediation achieves compliance. If a remediation alternative does not meet each requirement, it is rejected.

There is a process in CERCLA [§121(d) (4)] that allows for a waiver of an ARAR under a specific set of circumstances. If a measure to remove or isolate a waste causing imminent danger is considered interim, to be followed by more protective permanent remediation, not all ARARs must be achieved. For example, construction of a barrier to minimize the active flow of a pesticide into a water body does not have to achieve water quality standards in the surface water, as long as more permanent remediation will be considered in the RI/FS.

A substantive requirement can also be waived if compliance causes greater risk to human health and the environment than an alternative that does not achieve full ARAR compliance. In considering a waiver under such conditions, a comparison of the environmental benefits and damage must be made. The magnitude, duration, and reversibility of the effects resulting from ARAR noncompliance should be viewed in light of risk to human health and the environment resulting from compliance. For example, a remediation alternative that called for no action in a slightly contaminated wetland with no active exposure pathway, could be in conformance with a wetland protection ARAR. However, a significant storm could erode contamination deep within the sediment, and flush the contamination to a downstream public water supply. The wetland protection ARAR might be waived, because the alternative of excavation or capping might cause less risk to public health. In this example, the risk of a future release would have to be compared to the damage to the wetland caused by the remedial alternatives.

Technical feasibility can be considered in determining the need for ARAR compliance. If removal or isolation of a waste that is in violation of an ARAR is not technically practicable, it is possible to waive the specific requirement. Practicability is predominately judged from an engineering perspective, but the reliability and cost of the measure can also be considered. If an inordinate cost is involved, and there is no history of the remediation being effective, specific ARARs can be waived. ARARs that address specific technologies can also be waived, if alternatives can achieve an equivalent standard of performance.

The final two conditions for waivers are policy oriented. If there are state ARARs dictating an action at a site, and there is evidence that the ARARs are not consistently applied (i.e., only at hazardous waste sites), compliance with the offending regulation can be waived. If the federal reserve of money for hazardous waste cleanup established under CERCLA (i.e., the Superfund) is the source of funding for remediation, the relative benefit of complying with ARARs must be considered. When compliance is very expensive, and only minimal environmental or human health protection is afforded, the generic degree of protection achievable at other CERCLA sites for the same amount of money must be appraised. If there is a gross imbalance in the cost versus benefit ratio, the ARAR can be waived.

In conducting an ecological assessment for a Superfund site, the ARARs dealing with ecological issues must be identified early in the planning process. This is done by first identifying the ecological resources potentially on the site or affected by the release. These resources are then compared with the nature and extent of the contamination, and possible actions associated with remediation alternatives. This process identifies the potential for effects, and thus regulations designed to prevent or control the effect become potential ARARs. In Exhibits 1–9 and 1–10 of the *CERCLA Compliance with Other Laws Manual* (U.S. EPA 1988c), the EPA has published an eight-page list of possible ARARs. Other publications, such as that by Marburg and Parker (1991), also provide extensive lists of potential ARARs and guidance as to their use at specific sites. After the potentially affected resources on-site have been identified, a current update of this list, as well as state and federal regulators and resource agency personnel familiar with the specific site, should be consulted to identify ARARs. However, there are a handful of regulations that cover virtually all ecological issues at hazardous waste sites, and these are discussed individually below.

ARARs fall into three categories: chemical-specific; location-specific; and action-specific. The chemical-specific ARARs relate to ecological issues as they promote protection from toxic effects in the environment. Location-specific regulations are most often those ARARs associated with the ecological aspects of a hazardous waste site. When a protected ecological resource, such as an endangered species or wetlands, is present on a site, the associated ARARs become a major factor in the RI/FS. Action-specific ARARs generally deal with the treatment or disposal aspects of a remediation alternative, such as air stripping, landfill leachate collection, or gas collection. These ARARs are not directly related to the ecological aspects of remediation, and thus are not covered in this chapter. There can be aspects of action-specific ARARs dealing with the protection of ecological resources, but the presence of the resource, and thus a location-specific ARAR, would require consideration of the issue.

The premier environmental legislation, the National Environmental Policy Act (NEPA), and thus a potentially significant ARAR, has been excluded from Superfund site remediation. The Council on Environmental Quality (CEQ) has waived the NEPA requirement for the EPA because the RI/FS process established under CERCLA is the equivalent of full NEPA compliance. However, for federal hazardous waste sites not subject to a full RI/FS process, the applicability of separate NEPA requirements is currently unclear. As a result, there is consideration by the CEQ and the president to more fully integrate hazardous waste cleanup with the NEPA. Many hazardous waste site investigations at federal facilities are performed under the EPA process, simply to avoid separate compliance with NEPA.

The ARARs most often involved in ecological issues at Superfund sites differ from the regulatory concerns of other remediation activities and issues. The RCRA probably contains the most important ARARs for the chemical and engineering aspects of hazardous waste clean up. However, there is very little in the RCRA not covered by other ARARs related to ecological resources or protection. The primary ecological ARARs govern chemical, and sometimes physical, conditions in critical habitats or individual ecological resources. The Clean Water Act contains ARARs related both to habitat protection (e.g., water quality) and specific resources (e.g., wetlands) and is discussed in detail in the following section, followed by discussion of other regulations addressing ecological issues.

CLEAN WATER ACT AND RELATED ARARs

Introduction

Provisions within the Clean Water Act govern both chemical-specific and location-specific aspects of hazardous waste sites. The chemical-specific jurisdiction addresses contaminants in both receiving waters and discharges. Compliance with ARARs related to the surface water ambient concentrations in receiving waters is the more relevant chemical-specific ARAR to ecological assessment at hazardous waste sites. The existing conditions at a site must be compared to the established ambient water quality standards and to criteria to evaluate the potential for ecological damage. A comparison to receiving water criteria is also critical in evaluating ecological remediation alternatives. Finally, when a remediation action involves wastewater treatment, the ecological evaluation of the effluent discharge must incorporate a comparison to the appropriate surface water standards and criteria.

The primary focus of the location-specific Clean Water Act provisions are wetlands or waters of the United States, as defined by section 404 of the Act. Wetlands have received significant national attention because of their ecological importance, and the continuing loss of acreage to support a variety of development projects. As a result, wetlands have been the subject of extensive federal, state, and local protection regulations, including the consideration of a national "no net loss" policy.

The goal of wetland protection is frequently at odds with the remediation of hazardous waste sites. The sites themselves are often within or adjacent to wetlands. Historically, many of the nation's industrial operations were adjacent to waterways because of the need for process water, effluent receiving water, hydropower, and water transportation. Wetlands are commonly associated with the waterways, and often represent the major undeveloped areas in the industrial complexes. When hazardous waste was disposed of on-site within the complexes, the obvious location was the undeveloped area, therefore the wetlands. Also, in urbanized or industrialized regions, wetlands frequently represent the remaining open space away from public scrutiny. Dumpers of hazardous wastes in the past have often taken advantage of these isolated undeveloped wetlands for illegal disposal. In addition to discreet disposal areas, wetlands are also natural areas of sediment accumulation from up gradient sources. Thus, if there is erosion from a hazardous waste site, there is a strong possibility that the contaminated sediments will end up in wetlands. Therefore, many Superfund sites must address the question of hazardous waste in wetlands. Because wetlands are relatively fragile, and very complex in structure and function, massive excavation or other measures to remove or isolate the waste significantly alter, if not completely destroy, critical aspects of the wetlands.

Consequently, the objective of limiting exposure to hazardous wastes to protect human health and the environment frequently conflicts with the wetland preservation objective of the ARARs associated with Section 404 of the Clean Water Act. A likely outcome of this conflict is close scrutiny of the wetland protection regulations, as they apply to the site when a variance of the Section 404 ARARs is under consideration. Major factors requiring ecological assessments at hazardous waste sites have arisen from: the conflict between human health protection and wetland preservation; national attention to wetlands; and the involvement of Natural Resource Trustees at a site with significant wetlands. The presence of wetlands at a site draws early attention, and the Section 404 ARARs provide the regulatory justification for extensive ecological investigations and evaluations at a site.

Intent and Structure

The Clean Water Act was one of the pioneer pieces of legislation in the environmental movement of the early 1970s. The law was passed to address the visible concerns of water pollution resulting from such sources as: electrical power plants, particularly the growing nuclear industry; industrial plants, with an emphasis on the chemical and petroleum industries; and publicly owned sanitary wastewater treatment plants. The approach of the Clean Water Act was to set objectives based on receiving water uses of "fishable, swimmable waters" by certain dates. The objectives were tied to "zero discharge of pollutants." These lofty goals were to be met by establishing treatment requirements and associated effluent quality limits on all discharges to the waters of the United States. The Act also addressed criteria against which the goal of fishable, swimmable waters could be measured. These criteria were to take the form of ambient Water Quality Criteria.

When the Clean Water Act was originally passed, there was very little concern or public attention focused on hazardous waste or hazardous waste sites. The primary emphasis of the Act was to eliminate pollution to the nation's waterways by controlling the effects of the discharge of conventional pollutants by regulating such factors as biochemical oxygen demand, pH, temperature, bacteria, and suspended solids. In implementing the Clean Water Act, an approach of specifying the level of treatment evolved, because different types of treatments were relatively well understood and easily regulated and monitored. For example, secondary treatment, as defined by percent removal and effluent concentrations of biochemical oxygen demand, were required for publicly owned sanitary wastewater treatment plants, and off-line cooling was generally required for electrical power plants. Eventually, waivers or variances from these industry-wide treatment requirements were offered, if the discharger could demonstrate that an effluent of lesser quality would not produce an impact on the ecology of the receiving water beyond some narrowly defined spatial and temporal limitations.

Although priority pollutants were acknowledged when the Clean Water Act was drafted, there was only limited data and less recognition of the potential toxicity and importance of these contaminants. Much of the information on these compounds and much of the concern originated from their presence in industrial discharges or municipal systems receiving significant industrial input. One of the first compounds to be subjected to extensive toxicity research and subsequent regulations was chlorine. This compound received significant attention not because it was accidentally released to the environment, but because (due to its disinfection properties) it was a required addition to sanitary discharges for public health concerns. Similarly, chlorine was added to the heated effluent discharges from power plants to control interference from the growth of algae and other aquatic organisms within the pipes and other structures of the cooling system. Over time, the toxicity of other contaminants were investigated and regulated as problems were identified.

The Clean Water Act, as amended in 1987 by the Water Quality Act, consists of five major sections or titles. Title I—Research and Related Programs provides grants for research and training programs related to water pollution control. Grants for the actual development and construction of full-scale public wastewater treatment works are addressed by Title II—Grants for Construction of Treatment Works. Titles III—Standards and Enforcement, and IV—Permits and Licenses, include the aspects of the Clean Water Act most critical and directly related to ecological assessments at hazardous waste sites, and are discussed in detail later. The final section of the Act, Title V—General Provisions, establishes program implementation, emergency powers, judicial review, citizen suits, state authority, and other administration procedures. Certain aspects of the Clean Water Act either require action by the states (such as developing standards), or provide for the delegation to states under certain circumstances (such as the administration of discharge permits).

Clean Water Act ARARs Pertaining to Ecological Assessments

Surface water quality and wetlands are the two ecological resources at hazardous waste sites that are directly addressed by the Clean Water Act. Protection of these resources is through: the establishment of water quality criteria and standards (Sections 303 and 304); effluent control (Sections 301, 302, and 402); and the control of dredged and fill material discharge (Section 404). Each of these aspects of the Clean Water Act, as they relate to Superfund site ecological assessments, are discussed in the following subsubsections.

Surface Water Quality

As the enhancement and protection of surface water quality is the goal of the Clean Water Act, Congress saw fit to establish a means of measuring attainment of the goal. The mechanism they chose was the development of chemical-specific, numerical standards and criteria. As required by Section 304, the EPA must publish criteria based on the latest scientific information on water quality effects, which the states can use to develop the state-specific Water Quality Standards required under Section 303.

The development of water quality criteria has had a long history, and is continuing today. Early in the process, California assembled the available data on aquatic toxicity (McKee and Wolf 1963), which was a major step in establishing acceptable concentrations of contaminants in surface water. The EPA used the information from McKee and Wolf (1963) and others as they met their obligations under Section 304 of the Clean Water Act to establish criteria for water quality on a compound-by-compound basis.

The EPA published the "Red Book" in 1976, which updated and expanded on the reality available information on aquatic toxicity (U.S. EPA 1976). The 1976 publication also defined concentrations that would be acceptable to both the short-term (acute) and long-term (chronic) survival of aquatic communities. The criteria established were often based on only limited data, and thus several assumptions were required. The EPA tended to be very conservative when assumptions were relied upon, erring on the side of overprotection. This was especially true when safety factors were applied to protect sensitive organisms, for which species-specific toxicity data was not available, or to convert acute toxicity data to protection criteria for chronic exposure.

The toxicity research has continued, and the EPA updated the Water Quality Criteria in the 1986 "Gold Book" (U.S. EPA 1986). The Gold Book defines criteria for: fresh water, marine systems, continuous exposure (for the protection of sensitive life processes from chronic effects), maximum exposure concentration (designed to prevent acute toxicity), and public health (both the consumption of water and aquatic organisms). Based on the amount and consistency of available toxicological data, the values are set as actual criteria (i.e., levels protecting the target species), lowest observed effects level, or simply as a narrative. The Gold Book addresses 81 and 64 compounds for freshwater and marine acute effects, respectively. Sixty-six and 43 compounds are addressed for chronic effects in freshwater and marine systems. The list of criteria is updated periodically by adding additional compounds and by revisions based on recent research. The updates are noticed in the Federal Register and are available through the National Technical Information System. The 1991 scheduled release for the new compilation of all available criteria, commonly referred to as the "Silver Book," was missed, but a 1992 publication is likely. The anticipated Silver Book is expected to increase the number of criteria in the Gold Book by about 10 to 15 percent.

The EPA is also required to address sediment quality in a similar fashion to that for water quality. The requirement for sediment criteria is more recent, and is currently in the initial phases of development. As discussed in detail in Chapter 7, sediment toxicity is a much more complex issue than the parallel processes in water, due to the limited predictability and the highly site-specific aspects of sediments and the associated toxicity. As a result, the development and use of sediment quality criteria has become a very controversial issue.

Water and sediment quality criteria can provide essential input for ecological assessments. First, they represent a quick and easy way to evaluate the potential for ecological effects if surface water quality is the only site-specific data available. They are also a convenient and accepted tool for at least screening-level evaluation of remediation criteria. For the ecologist, they have historically been important in coordinating with engineers and policy makers, because the nonecologists are conditioned to using numerical indicators for evaluations and decisions. At hazardous waste sites, they continue to provide a common language for the variety of scientists, public health specialists, engineers, and bureaucrats involved.

Criteria do have several important limitations. They are chemical-specific, and thus interactions with other compounds in the effluent or receiving water are not addressed. Also, there are many compounds and media where criteria are not available. The criteria are often overly protective for a specific site, because they are developed for species or life stages that are not appropriate for the site. The safety factors and worse-case conditions used in developing some criteria also add to the conservative, overly protective nature of the water quality criteria.

In addition to protecting surface water quality by establishing standards, the states are required under the antidegradation policy (Clean Water Act §131.12) to prevent deterioration of high-quality water. Waters classified as having high quality exhibit chemical and biological characteristics superior to the minimum requirements specified in the standards. In such waters, demonstration of compliance with the appropriate water quality standards for the receiving water is not sufficient compliance with the Clean Water Act ARARs. To qualify under the antidegradation policy, the discharge must be shown not to significantly alter the existing quality outside of a mixing zone. The discharge must not preclude or otherwise adversely affect any beneficial use in place at any time since the promulgation of the antidegradation provision. The beneficial use could be a water supply, shellfish harvesting, or a breeding habitat for a species sensitive to contaminant concentrations or having specific physical condition requirements. The antidegradation requirement can be significant at hazardous waste sites where wastewater treatment and surface discharge is contemplated. If there was no historic discharge into the proposed receiving water, it is likely that there were not even trace levels of certain contaminants reported, and no administrative restrictions placed on beneficial uses. The discharge of even extremely high-quality effluent could produce measurable concentrations or result in policy mandated restrictions on uses, such as water supply or shellfish harvesting. Thus, the discharge could be out of compliance with antidegradation ARARs, and a variance may be required.

Effluent Control

The control of effluent quality is primarily through regulating treatment technology. These "technology-based discharge limitations" are established for categories and classes of point sources of pollutants based on the application of the best pollutant control technology for conventional pollutants (e.g., biochemical oxygen demand and bacteria). Implementing regulations have been developed to define the best conventional technology for all types of industries and other point sources of pollution. For toxic compounds and other nonconventional pollutants, the application of the best available technology economically achievable is required. The Act does provide for more stringent technology-based controls if additional protection of the receiving water quality's intended uses might be impaired.

Enforcement of the effluent limitations and treatment technology is through the National Pollutant Discharge Elimination System (NPDES). Through this program, the discharger is required to obtain a permit. The permit specifies the level of treatment, and also the allowable concentrations of various compounds. Since the 1987 amendments to the Act, many discharge permits also contain requirements for toxicity testing of the whole effluent, and in some cases monitoring of the ecological characteristics of the receiving waters.

The limitation of effluent quality based on treatment technology is sometimes applicable to the remediation of hazardous waste sites. The selection of the treatment method for the soil, groundwater, or surface water is often based on the best technology, and there is a general consideration of site-specific ecological issues. However, the ecological assessment comes into play if the best conventional or economically achievable technology will not protect important ecological uses of the receiving waters, which is a substantive requirement of the Clean Water Act. This substantive requirement can be assessed by using surface water criteria for the protection of aquatic life. The effluent limitations for the discharge can be established by calculating the concentration of a contaminant in the effluent that would result in a surface water concentration less than that of the appropriate criteria, when the volume of effluent is diluted with the receiving water. The ambient concentration of the receiving water must be taken into consideration, as does the volume of dilution water available. In a river, for example, the expected minimum flow during the period of discharge is a reasonable estimate of the dilution water available. If the discharge will continue for a year or more, summer low flow conditions might be an appropriate estimate of dilution. If the discharge is only a dewatering operation that occurs during high groundwater conditions, then using low flow conditions in the receiving water to estimate available dilution may be overly conservative.

When determining substantive requirements of a surface water discharge, the area where dilution is achieved, or the mixing zone, must be considered. Often, the undilted effluent is above levels that are known to be toxic to aquatic organisms. However, after dilution to 10 to 100 parts receiving water per one part effluent, as is frequently available in receiving waters, the concentrations can be below the chronic toxicity value. The dilution is not, however, instantaneous and there is an area, or mixing zone, with concentrations between that found in the effluent and the receiving water concentration after full dilution. The acceptability and thus substantive requirement of the mixing zone can be assessed by considering the ecological resources in the zone and the sensitivity of the resource to the specific contaminant. If the resource will be exposed to the maximum concentration for only a short time (e.g., during a short-duration low freshwater flow condition, or low tide in a tidal situation), the acute, rather than the chronic, might be the operable criterion.

In some situations, alternatives to established criteria may be appropriate. Criteria for all situations (freshwater, marine, acute, and chronic) have not been established for all contaminants. Also, the criteria may have been established for very sensitive species that do not occur within the receiving waters, or there is a mixture of contaminants, and the aggregate toxicity is an issue. In these cases, alternative criteria determined from the toxicity literature can be considered. If the literature is inadequate to define alternatives, whole effluent toxicity testing may be required. The testing can be conducted during treatability studies to establish effluent limits or, if the waste is highly variable, periodic testing of the effluent and appropriate modifications to the treatment during remediation may be required.

The monitoring requirements of the permit can also be ARARs related to ecological issues. In situations where the establishment of effluent limitations has a high degree of uncertainty because of limited toxicity or receiving water data, monitoring can be required to confirm predictions. The monitoring can include periodic or real-time toxicity testing, to evaluate additive or synergistic effects of compounds in the effluent. Receiving water monitoring can also be required to assess interactions with background concentrations or indigenous biota.

Wetland Protection

The original Clean Water legislation included provisions under Section 404 for the regulation of discharged solids, such as dredged or fill material, to surface waters. The primary intent of the provision was to protect water quality and lessen direct interference with beneficial uses. In the intervening years since its original passage, the extent of regulation and the geographical jurisdiction under Section 404 has evolved and expanded. Jurisdiction now encompasses all waters of the United States, defined to include open water and wetlands. After much debate, the primary federal agencies (i.e., the U.S. Army Corps of Engineers, the U.S. EPA, the U.S.D.A. Soil Conservation Services, and the U.S. Fish and Wildlife Service) involved with wetland management, regulation, and protection have agreed on a common definition of wetlands and published the Federal Wetland Identification Manual (Federal Interagency Committee for Wetland Delineation 1989). The manual designates wetlands based on a multiple-parameter technique, which requires an examination of vegetation, soils, and hydrology. In essence, any area that supports vegetation tolerant of saturated soils, has hydric soils, or is hydraulically connected to a surface water body is classified as a wetland, and afforded protection under Section 404 of the Clean Water Act. There has been much controversy over the 1989 manual because of the extensive area classified as regulation wetlands using the methods developed by the manual. The use of the manual was halted in 1991, but not for all applications, such as for Superfund investigations in EPA Region I, and a revised manual is expected to be released in late 1992.

Section 404 is administered primarily by the U.S. Army Corps of Engineers by a permit process. The Corps has promulgated regulation under 33 CFR 320–330, governing the issuance of permits, after an opportunity for public comment, for the discharge of dredged or fill material into the waters of the U.S. The regulations also cover Sections 9 and 10 of the Rivers and Harbors Act (33 U.S.C.§401 and series), which are related ARARs administered by the Corps. The application for a 404 permit must contain enough information so that the Corps can determine the effects of the proposed action in the eighteen specific areas described in 33 CFR 320.4 (a through r). The fist four of these areas, public interest, effects on wetlands, effects on fish and wildlife, and water quality, are directly related to ecological issues.

The public interest review establishes the philosophy for evaluating and potentially granting 404 permits. The Corps considers the public benefits of the proposed activity and weighs them against potential environmental damage. In making the comparison, values to be considered include aesthetics, historic properties, recreation, safety, food production, and general needs and welfare of the people. To make the decision on issuing a permit, the relative value and impact in each area is considered in a subjective manner, which varies greatly from site to site and Corps district to district.

Within the context of the public interest review, the effects on wetlands, fish and wildlife, and water quality must be considered. The regulations specify that the quality, potential, and value of each of these resources must be considered. If the proposed action would degrade any of these aspects, a permit should not be issued unless the benefits of the discharge activity outweigh the damage to the resource.

In general, the evaluation of alternatives required under CERCLA and the National Contingency Plan partially meet the functional equivalent requirement of 404 wetland ARARs in 33 CFR 320.4. As part of the RI/FS, the alternative must be shown to be in the overall public interest. Protection of the environment must be demonstrated, and protection of fish, wildlife, and water quality also considered. There is no specific requirement under the National Contingency Plan regulations for the evaluation or protection of wetlands.

However, to demonstrate compliance with Section 404 ARARs, the presence and characteristics of wetlands within a hazardous waste site must be evaluated. The effect of each remediation alternative on the wetland resources must then be considered, in comparison to benefits anticipated from the alternative.

In addition to the Corps of Engineers guidance contained in 33 CFR 320.4, discharge of dredged or fill material into waters of the U.S. must comply with conditions specified in Section 404(b) (1) of the Clean Water Act. The EPA has been charged with developing and implementing regulations under Section 404(b) (1), and has done so in 40 CFR 230, by establishing what are commonly called 404(b) (1) guidelines. The Corps of Engineers must apply the 404(b) (1) evaluation in reviewing a 404 application, and no permit can be issued unless all of the requirements are met. The 404(b) (1) guidelines have generally proven to be more restrictive than the Corps other regulations on issuing 404 permits.

The expressed purpose of the 404(b) (1) guidelines is to control discharges of dredged or fill material, to restore and maintain the chemical, physical, and biological integrity of U.S. waters. Taken together, this mandate can be interpreted to mean the protection of the ecology of wetlands and open waters by strictly regulating filling and other activities in wetlands. In promulgating the regulations, the EPA acknowledged that activities in aquatic sites, such as filling of wetlands, are among the most severe environmental impacts addressed by the guidelines. The basic presumption in formulating the 404(b) (1) guidelines was that dredged or fill material should not be discharged to wetlands or other aquatic systems, unless certain specific conditions are met. The applicant for a 404 permit must demonstrate, to the satisfaction of the regulators, that the filling or other discharge will not have an unacceptable adverse impact, either individually or in combination with the impacts of other activities affecting the aquatic system of concern. The regulations and guidance published by the EPA go into detail as to how the demonstration of no impact is to be made.

Paragraph 230.10 of the code implementing Section 404(b) (1) of the Clean Water Act specifics four conditions that must be met before a 404 permit can be issued. If activities at a hazardous waste site involve dredging, fill, or similar activities in waters of the United States, these same conditions must also be met to demonstrate compliance with ARARs. The first condition requires an examination of alternatives. The paragraph is very specific about preventing a discharge to U.S. waters if there is a practicable alternative that has less adverse impact on the aquatic ecosystem and has no other significant adverse environmental consequences. For the purposes of this requirement, practicable is considered anything that can be implemented taking into account cost, available technology, and logistics, all in relation to the overall magnitude and purpose of the project. For example, in a total site remediation estimated at $2.8 million and two years duration, one aspect of an alternative with significant effects on 404 waters might be estimated at $150,000 less and one month shorter than an alternative with significantly less impact on wetlands. The alternative with less impact would generally be considered practicable and compliance with the ARARs associated with Section 404 of the Clean Water Act would dictate its selection.

If the remedial activity is directly related to the 404 waters, such as removing contamination from wetland sediments, the activity is presumed to be water dependent. If remedial activities are not water dependent, and produce significant wetland impacts, there are generally presumed to be upland alternatives with less impact, and the activities involving discharge to U.S. waters are out of compliance with the ARARs. Locating a groundwater treatment plant or soil storage area would be examples of non-water-dependent activities common to hazardous waste sites. Even before actual cleanup activities begin, siting a decontamination area or laboratory trailor is a non-water-dependent use with practicable alternatives. Thus, in these cases, wetland areas must be avoided unless the alternatives have other significant adverse impacts, such as a high risk of spreading dangerous levels of contaminants off-site.

The second condition that must be met under the 404(b) (1) guidelines is compliance with other statues. This includes state water quality standards or any applicable effluent toxicity standards under Section 307 of the Clean Water Act. Compliance with applicable sections of the Endangered Species Act and the Marine Protection, Research, and Sanctuaries Act must als be demonstrated as part of a 404(b) (1) determination.

Section 404(b)(1) guidelines also specify that no discharge of dredged or fill material will be allowed if they cause significant degradation of U.S. waters. The regulations are very broad in the definition of degradation, and include loss of fish or wildlife habitat, and the capacity of a wetland to assimilate nutrients. It is unlikely that the filling of any significant area of wetland could be interpreted as not taking habitat or diminishing the primary productivity of the system. The final condition of the 404(b)(1) guidelines requires that all appropriate and practicable steps be taken that minimize and otherwise mitigate adverse impacts on the aquatic ecosystem.

In order to evaluate compliance with the 404(b)(1) guidelines, the potential effects on the affected wetlands or other water bodies must be documented. Paragraph 230.11 of the guidelines specifies that the following eight predischarge characteristics must be documented and the effects on each evaluated:

Physical substrate
Water circulation, fluctuations, and salinity
Suspended particulates and turbidity
Contaminant concentration
Aquatic ecosystems and organisms
Area of disposal
Cumulative ecosystem attributes
Secondary ecosystem characteristics

The guidelines specify what areas are "Special Aquatic Habitats," and what types of evaluation and testing should be performed for each resource type. The guidelines also specify, in Subpart H (40 CFR 230.70–230.79), activities that should be considered to minimize the effects of the proposed actions on the ecological resources associated with wetlands and open waters. Most of the examples are not applicable to hazardous waste sites, but the intent is clear. All measures that result in an avoidance or lessening of adverse effects on any value of the wetland must be considered. The implementation of the measure may not be feasible, either from a cost or engineering standpoint, or due to interference with the primary objective, but the infeasibility must be demonstrated.

Many ecological assessments for hazardous waste sites are prepared, which do not specifically demonstrate compliance with 404 regulations. General reference to compliance with wetland ARARs is often included in other portions of the RI/FS, but no specific attention is given to the obvious ecological intent of the wetland protection regulations and the individual demonstrations required for ecological considerations. Similarly, the RI/FS contains an evaluation of alternatives that, in a general sense, addresses the consideration of relative effects on wetlands, but the specific requirements of wetland protection regulations are often not addressed.

Before the strict enforcement of Section 404 of the Clean Water Act, two Executive Orders (11988 for flood plains and 11990 for wetlands) were issued to minimize damage due to actions by federal agencies. The orders basically required the agencies to avoid any damage to wetlands and floodplains if there was a practicable alternative. The requirements for wetlands have been expanded on and detailed through promulgation of the 404 regulations by the Corps of Engineers and the EPA.

The approach used, of not specifically addressing wetland ARARs from an ecological prospective, can leave the RI/FS process vulnerable. If a remediation alternative is selected that involves the excavation or fill of a wetland, it is in obvious noncompliance with wetland ARARs. Unless the RI/FS document clearly shows that alternatives were considered, and none were found to be feasible in protecting human health and the environment, the selection of the alternative over no action could be successfully challenged by the potentially responsible party asked to pay for the expensive excavation and treatment alternative. Similarly, if there is not a full evaluation of wetland mitigation included in the RI/FS, compliance with the ARAR mandate of

CERCLA leaves the process open to challenge by an environmental group. The lack of attention to wetlands and the associated fish and wildlife resource value in the hazardous waste site ecological assessment can also inhibit the positive participation of Natural Resource Trustees in the negotiation process with potentially responsible parties.

ECOLOGICAL RESOURCE ARARs

Introduction

Just as the Clean Water Act is centered around a natural resource (i.e., waters of the U.S.), there are other ARARs addressing specific ecological concerns. The regulations for the other resources are generally less broad in scope and do not have the extensive regulatory structure and national significance of the Clean Water Act. However, if a particular resource afforded protection by one of the regulations is present at a Superfund site, the associated ARARs can be critical in site remediation.

The Endangered Species Act

The Endangered Species Act was enacted in 1973 to conserve various species of fish, wildlife, and plants that are threatened with extinction or likely to become threatened within the foreseeable future, throughout all or a significant portion of the species range. The Act is one of the few pieces of federal environmental legislation that actually precludes a federal action if certain impacts are anticipated. Most federal Acts and the associated regulations either require the decision maker to consider the impacts and mitigation (e.g., the National Environmental Policy Act) or mandate specific limits or treatment (the Clean Water Act and Clean Air Act). However, by passage of the Endangered Species Act, congress intended to prevent federal action that would be a threat to the continued existence of a species. Consequently, no provision was established to allow consideration of costs to implement a nonimpacting alternative or of the impacts to other aspects of the human environment, such as the local economy, if the species is in danger of extinction. The Act does provide for exemptions, but the process is lengthy and detailed before an Administrative Law Judge, and has only been successfully carried out in a few cases in the twenty-year history of the Act (U.S. EPA 1989).

The Act and promulgated regulations (Endangered Species Act of 1973, 16 USC§1531 et seq. ;50 CFR 402) establish a formal procedure for evaluating the effects of proposed federal actions on the continued existence of species and their habitat. This procedure, as summarized below, must be strictly followed and documented for federal activities outside of CERCLA sites. As with other ARARs, CERCLA exempts on-site activities from compliance with the administrative requirements of the Endangered Species Act. Any CERCLA activity that occurs off-site must comply with all formal administrative and documentation requirements of the Endangered Species Act. The CERCLA exemption for nonsubstantive requirements applies to all actions taken on-site, even if the potetential threat to a special-status species occurs off-site. Therefore, the exemption to administrative procedures would apply if remedial activity on-site potentially transported contaminated sediment to a downstream location of an endangered species critical habitat. Similarly, if a bird obtained prey on-site, and the associated ingestion of a contaminant could conceivably affect egg shell thickness (and thus hatching success) at a distant location, the administrative aspects of the Endangered Species Act would not apply. In each of these examples, the RI/FS must demonstrate that the remedial action would not threaten the continued existence of the species, but a formal endangered species consultation process would not be necessary, and official approval from the federal agency with juridsiction over the species would not be required.

Through CERCLA's ARARs variance procedure, the strong mandate for species protection and potential prevention of federal action afforded by the Endangered Species Act can be waived. However, endangered species are a critical responsibility of the Natural Resource Trustees, and a viable subject for complex Natural Resource Damage Assessments, and thus (as discussed elsewhere in this book), when endangered species are a potential issue at a site, the RI/FS process should include a detailed consideration of the issue. Therefore, the decision to waive the substantive requirements of the Endangered Species Act should not be taken lightly. When there are endanger species issues, it is important that the RI/FS clearly and completely satisfy the substantive requirements of the Act, to prevent delaying the implementation of the recommended alternative. Also, if a waiver is required for the recommended alternative, the RI/FS should fully evaluate alternatives, including no action, that would have less or no effects on special-status species.

The first step in the formal Endangered Species Evaluation process is to determine the potential occurrence of a threatened or endangered species on the site. In most non-Superfund projects, this is generally accomplished by formally notifying the agencies with endangered species responsibility (Fish and Wildlife Service, of the Department of Interior, and the National Marine Fisheries Service, of the Department of Commerce). A letter is sent specifying the location of the project and some general description of the type of activity that is to take place. The agency responds by listing the special-status species (i.e., threatened or endangered), if any, that could be affected by the process. For CERCLA projects, this can normally be accomplished efficiently by early coordination with the biological technical advisory group, which typically includes representatives from the Fish and Wildlife Service and National Marine Fisheries.

The identification of potentially affected species is usually more detailed and helpful if performed by the biological technical advisory group, than if attempted through written notification to the regional office of the agency. Members of the advisory group are encouraged to visit the site during the initial planning stages of an RI/FS, and thus they can consider the site-specific habitat when assessing the potential affects on listed species. Also, members of the advisory group are familiar with the type and extent of activities that can be involved in site remediation, so they can better assess the potential for threats to endangered species. The more precise identification of the potential occupancy of special-status species allows the RI/FS to focus on a limited set of issues. Thus the process is more efficient, both with respect to time and funds. Also, the possibility of an unidentified endangered species issue arising late in the RI/FS process, when substantial effort has been expended, is reduced.

Although the use of the biological technical advisory group is an excellent resource for identifying and assessing endangered species issues, it should not be the sole method employed in the RI/FS. Documentation of regional occurrence of special-status species must be reviewed as part of the RI/FS. Also, site visits, reconnaissance surveys, and (in some cases) sampling should be considered by the Remedial Project Manager, if there is an indication of endangered species issues at a site.

If input from the biological technical advisory group, a review of the literature, and preliminary site activities fail to identify any potential endangered species issues, the substantive requirements of the Endangered Species Act are met. The RI/FS should document the effort to identify the occurrence of special-status species and state the findings. If potential impacts to covered species or their habitat are identified, the requirements of the next step in the process, the Biological Assessment, must be considered.

The Biological Assessment provides the scientific information required to determine a threat to the species. The site-specific conditions, as they relate to habitat require- ments, is information that must be considered. This can include specific biotic con- ditions, such as vegetative cover type. Abiotic conditions, including the size of the site in relation to the species home range, access to the site for specific life stages, the microclimate, and hydraulic conditions, must also be considered. The type, level of detail, and specific methods required to determine the important biotic and abiotic site conditions should be coordinated with the appropriate agency (the National Marine Fisheries or Fish and Wildlife Service) during the Work Plan Phase of the RI/FS.

As part of the Biological Assessment, the activities associated with the remedial alternatives are compared to the critical endangered species resources present on-site. When an important habitat element will be altered or destroyed by remediation, the potential for recreation and the duration of the alteration must be assessed. Also, the ability of an alternative to remediate the contaminaant release and other impacts to endangered species and their habitat must be evaluated.

The condition of the population of an endangered species in the site area is the final information needed for the Biological Assessment. The collection of such information is almost always beyond the scope of an RI/FS because of the extensive temporal and spatial data base required to make an assessment of the status of a biological population. Consequently, the resource agency with jurisdiction over the species (the National Marine Fisheries or Fish and Wildlife Service), and sometimes state agencies, must be consulted. When sufficient information is available for the population and predic- tions of remediation-related impacts on individuals are possible, an evaluation of the threat to the continued existence of the species can be attempted. Generally, there is not sufficient data or predictive capability to make a totally objective projection, and the evaluation is largely subjective, relying heavily on coordination and concensus with the resource agencies. Thus, the outcome of the Biological Assessment is a forecast of the threat to the continued survival of the population under conditions resulting from each remedial alternative. This information is considered when com- paring and selecting remedial alternatives.

When there is concern over special-status species at a site, all of the information required for a Biological Assessment should be developed during the RI/FS process. If pertinent information is not available during the development and evaluation of remedial alternatives, the process can be seriously hampered. If the selected alternative requires a variance from an ARAR, and there is not sufficient documentation of the impact on special-status species from other alternatives, the variance may be denied. Also, if natural resource damages are claimed from endangered species effects re- maining after remediation, the negotiations with potentially responsible parties can be significantly complicated.

General Ecological Resource Regulations

There are several pieces of federal legislation similar to the Rare and Endangered Species Act, which were enacted to protect ecological resources considered important and potentially threatened by future federal actions. Of these statutes, the Wild and Scenic Rivers Act, Fish and Wildlife Cofordination Act, Wilderness Act, Coastal Zone Management Act, and Migratory Bird Treaty are of primary concern for ecological assessments. As summarized below, each of these pieces of legislation afford different levels of protection for various resources. Procedures for compliance with resource-specific legislation at hazardous waste sites should follow the process outlined above for the Endangered Species Act. The existence of a listed or potentially jurisdictional resource should be investigated by contacting the appropriate agencies. If an eligible resource occurs in the vicinity of the site, the actions included in each remedial alternative must be evaluated to determine their potential effects. If a potential impact is identified for the selected alternative, mitigation should be employed to reduce the effect, and the action should be coordinated with the appropriate agencies.

The Wild and Scenic Rivers Act (16 USC§91271, *et seq*) was enacted to designate and protect water resources that are free flowing and support exceptional scenic or natural value. The Act is administered jointly by the Department of Interior through the National Park Service and the Department of Agriculture, primarily by the Forest Service. Water bodies that have officially been placed in the National Wild and Scenic Rivers System or have been listed on the National Rivers Inventory for evaluation and potential inclusion in the National System are covered by the Act and the implementing regulations (36 CFR 297). Activities upstream, downstream, or on tributaries of the designated water bodies must also consider effects on the listed resource. The Act prohibits federal involvement via grant, loan, license, or other activities in water resource projects that would have adverse effects on any attributes of a designated water body that the Act serves to protect and preserve. For the purposes of the Act, a water resource project has been broadly interpreted as virtually any activity that could effect the quality or quantity of the water resource. Activities considered water resource projects likely to be considered during the RI/FS process at hazardous waste sites include: discharges, water diversions, dredging, filling, and shoreline protection. Impacts are considered to include the effects on the free-flow character, or unreasonable reduction of the recreational, scenic, or fish and wildlife habitat value of the resource. Much of the water quality, biological habitat, and dredging activities originally included in the Act are now addressed in a more detailed fashion by the Clean Water Act. However, the aesthetic and free-flow aspects of a Wild and Scenic River are important ARARs, which must be considered at hazardous waste sites under the procedures of the Act.

The Fish and Wildlife Coordination Act (16 USC §661 *et seq*) is very similar to the Wild and Scenic Rivers Act, in that the intent is to minimize the effects from federal actions on natural water bodies. However, the Coordination Act differs in that it only covers biological resources, and it includes all water bodies, not just those designated as wild and scenic. When a site condition or remedial alternative could affect a water body, then consultation with the U.S. Fish and Wildlife service and/or the National Marine Fisheries Service (if marine resources are involved) is required. Under the Act, the appropriate state wildlife resource agency must also be consulted if effects on water bodies and associated biological resources are anticipated. Since, by definition, the focus of the Act is "coordination," many of the requirements could be considered administrative, rather than substantive, and thus not applicable to Superfund sites. However, the measures called for in the Act are, in most cases, applicable to the Clean Water Act. Also, the resources protected by the Coordination Act are under the jurisdiction of the Natural Resource Trustees, so the agencies are involved through the biological technical advisory group, or otherwise by the coordination required in CERCLA. Consequently, the debate over administrative or substantive requirements for the Fish and Wildlife Coordination Act, for purposes of CERCLA compliance, is a largely academic argument.

The intent of the Wilderness Act (16 §USC 1131 *et seq.* and the implementing regulations at 50 CFR 35.5) is to first designate arease for inclusion in the National Wilderness Preservation System, and then maintain their pristine wilderness traits. Applicability of the Wilderness Act would generally only occur if the Superfund site was within or adjacent to a designated wilderness area. In such cases, the remediation of hazardous waste would generally be within the intent of law, provided the remediation minimized permanent modification of the land or other resources within the jurisdictional area. The Wilderness Act could also be an ARAR if the incineration or transportation of wastes removed from the site could result in the inadvertent introduction of the waste or its byproducts to a designated Wilderness area.

The Coastal Zone Management Act (16 §USC 1451) and associated regulations (15 CFR 930) constitute a Federal program administered by each individual state. The intent of the Act is to develop a plan for the coastal zone, and then insure that all federal activity is consistent with the plan. Much of the emphasis of the Act and associated plans deal with land use and development. Although CERCLA action can be applicable or relevant to such plans, they are not generally related to ecological resources or assessments. The plans can designate specific areas as a wildlife or marine biota habitat. Therefore, the appropriate Coastal Zone Management Plan should be reviewed to identify any such areas potentially affected by site remediation, as part of the ecological investigation for the RI/FS.

The Migratory Bird Treaty (16 USC§ 703 *et seq.*) was originally enacted to protect against the unpermitted hunting of migratory birds and the wanton destruction of their nests and eggs. The ARAR prevents the "taking" of migratory birds, their parts, or eggs, except as permitted by any authorized regulations (such as hunting licenses). The actions covered by the term "taking" have been debated, but there is general consensus that effects on migratory birds caused by the release of contaminants constitutes a "taking." Most song birds, water fowl, and raptors in North America are migratory birds, and are thus covered by the Act. Consequently, the Migratory Bird Treaty is a potentially important ecological ARAR at many hazardous waste sites, because the animals and their habitat are ubiquitous and susceptible to damage from contaminants and remediation activity.

NATURAL RESOURCE RESTORATION

Authorization for Natural Resource Restoration

CERCLA, particularly with the SARA amendments, addresses more than the cleaning up of hazardous waste sites. Section 107(f)(1) of the law requires rectification of damages to natural resources caused by a release of contaminants. The requirement is restricted to residual damages after implementation of an EPA-approved remedial action. The damage can be in the form of injury to, destruction of, or loss of a natural resource.

The terms injury, destruction, and loss were not defined in CERCLA or the implementing regulations, but subsequent case law has resulted in an extremely broad definition of these terms. Damage can occur through the death of a biotic resource, or by forcing an organism from an area normally inhabited prior to the release. The interpretation also applies to the physical alteration or chemical contamination of a habitat such that the biological community can no longer exist in a prerelease condition.

Natural resources requiring restoration under CERCLA are not restricted to those within the site. Any resources that have been injured, destroyed, or lost by the release of contaminants off-site must also be restored. Damage to off-site resources could occur by the release of contaminants through the air, surface water, groundwater, or soil. Injury could also occur from an animal feeding on the site and then experiencing reproductive impairment or the transferring of the contaminants to other organisms through the food chain.

CERCLA and subsequent regulations provide an extremely expansive definition of natural resources. In Section 101(16) of CERCLA, resources are defined as: land; fish and wildlife; biota; air; water; ground water; drinking water supplies; and other resources belonging to, managed by, held in trust by, appertaining to, or otherwise controlled by the government. Resources or property that are purely privately owned are specifically excluded from the CERCLA definition of natural resources subject to restoration due to release from hazardous waste sites. However, the resource need not be owned in a conventional sense by a government entity. Any type of governmental jurisdiction over a resource affords that resource restoration from injury under CERCLA. Such jurisdiction could be through promulgation of protection standards, as for air and water quality. Classification of an entity as a natural resouce under CERCLA could also result from government management of the resource, which is often the case for water supplies or the management of fish or wildlife for recreation, hunting, fishing, or commercial harvesting. Certain lands and plant communities managed by government entitis for similar purposes also qualify as CERCLA natural resources. The Migratory Bird Treaty greatly expands the list of jurisdictional natural resources, and thus the authority of federal agencies at hazardous waste sites.

The CERCLA definition of qualifying natural resources has been made even broader by the implementing regulations. As discussed later, the National Contingency Plan identifies individual resources for specific Natural Resource Trustees, such as anadromous fish as a National Oceanic and Atmospheric Administration resource. However, the listings in the National Contingency Plan are clearly presented only as examples, and are not intended to be restrictive. The plan also identifies not only the specific resources, but also "their supporting ecosystems" as natural resources that must be restored if injured by a release from a hazardous waste site. A supporting ecosystem could be interpreted to include virtually any biotic or abiotic parameter.

CERCLA designates the President of the United States as Trustee for Natural Resources on behalf of the public. As the Trustee, the President is required to: assess damages to natural resources from releases of hazardous substances; pursue recovery of damages and the cost of assessing the damages; and use the sums from recovery to restore, replace, or acquire resources equivalent to the affected resources. By issuing Executive Order 12580 on January 29, 1987, the president designated the Secretaries of Interior, Commerce, Agriculture, Defense, and Energy as the primary federal Trustees of Natural Resources. The designation of the Trustees and their role are also addressed in Subpart G of the National Contingency Plan at 40 CFR 300.600. The assignment of natural resource responsibility in the presidential order also includes a catchall designation: for resources located in the U.S. but not otherwise described in the section, the head of the federal agency or agencies authorized to manage or protect those resources will serve as the Natural Resource Trustee.

The roles and responsibilities of the various federal departments varies according to their jurisdiction and the susceptibility of their resources to damage from releases at hazardous waste sites. The Departments of Interior and Commerce are generally the most active Natural Resource Trustees. Subelements of the Department of Interior that can act as Trustees, and examples of resources under the jurisdiction of the Department of Interior are listed in Table 3–1. The National Marine Fisheries Service is the subelement of NOAA that is generally the most involved in the restoration of natural resources. National Marine Fisheries has jurisdiction over virtually all marine resources, including anadromous fish and their habitat when they migrate into inland waters. In general, other federal agencies serve as Natural Resource Trustees only when lands owned by a subelement of the agency are directly involved. In cases where a government agency is a Natural Resource Trustee because lands of the agency are contaminated, the agency is often one of the potentially responsible parties. Consequently, the agency's role as a Natural Resource Trustee is complicated, if not compromised.

The most obvious examples of a protected natural resource that can be damaged from releases are fish, wildlife (particularly birds), and wetlands. The contaminants from a hazardous waste site most often migrate downgradient until they reach surface water bodies. The migration can occur by direct runoff of water-solube compounds, through erosion of contaminated soil, or from the discharge of contaminated groundwater to the surface water. Once the contaminant is in the surface water, aquatic organisms are exposed through numerous pathways, including dermal contact, respiratory tissue contact, the food chain, and the ingestion of contaminated water and sediment. In certain cases, such as contamination by coal tar or fly ash, the release can alter the physical habitat by sealing the crevasses in the substrate that animals require as cover or that plants require for attachment. Thus, a release can significantly affect fish and the ecosystem required to support the fish at the site, and at downgradient locations some distance from the site, and for an extended duration. Birds are often vulnerable because many are piscivorous (fish eating) or otherwise dependent on aquatic resources. Wetlands are often involved because, as are other aquatic systems, they can be downgradient of the release and represent areas of deposition of sediment, and

TABLE 3-1.   U.S. Department of Interior (DOI) as a Natural Resource Trustee: Bureaus and Services that Function as Trustees and Examples of Jurisdictional Ecological Resources.

DOI Subelements

Fish and Wildlife Service	Geologic Survey
National Park Service	Mineral Management Service
Bureau of Mines	Bureau of Reclamation
Bureau of Land Management	Bureau of Indian Affairs

Jurisdictional Resources

Migratory Birds	Anadromous Fish
Endangered Species	Marine Mammals
Wildlife Refuges	Fish Hatcheries
National Parks	National Forests

thus potentially, contaminated sediment. The wetlands are the habitats and sources of nutrients for many species afforded protection by the Natural Resource Trustees. The wetlands are also a qualifying natural resource, because they play a critical role in water pollution control, by retaining sediments and controlling nutrient releases.

Although groundwater and drinking water supplies are considered qualifying natural resources, they generally are not a major concern of the Natural Resource Trustees. These resources are of primary public health concern, and are a major emphasis of Remedial Investigation and Feasibility Study effort, and are thus remediated as part of the cleanup, and not a primary candidate for a Natural Resource Damage Assessment. An exception to this generalization is when the groundwater contamination occurs on federal lands, such as a military base.

CERCLA and Subpart G of the National Contingency Plan designated the governors of each state, and authorized agents of indian tribes, as well as the President of the U.S., as Natural Resource Trustees. Most governors have delegated the responsibilities to the appropriate state agenies in a manner similar to that described above for the federal government. The state trustees have the same responsibilities and authorized authority as the federal trustee. This can be important when federal lands are involved or the release is caused by a federal agency. In such causes, the state trustee can force the restoration of a damaged natural resource, or sue for appropriate restoration, replacement, or acquisition of an equivalent resource to the affected resource. Originally, municipalities were designated as Natural Resource Trustees, but the passage of SARA in 1986 did not automatically include municipalities. However, the governor of a state can designate a municipal official as a trustee.

Procedures for Natural Resource Restoration

The designation of protected natural resources and establishment of Natural Resource Trustees, as described above, has little practical significance unless there is an enforceable procedure for the restoration of damage. CERCLA provides for the restoration of injury to natural resources resulting from releases of contaminants. In practice, there are two methods that trustees may employ to rectify injury resulting from a release and related activities at hazardous waste sites. The most direct method, although not necessarily the most efficient or simplest, is through conducting and implementing a Natural Resource Damage Assessment. The damage assessment is based on the damage remaining after execution of the remediation developed during the RI/FS process. A quantification of the damage is determined and assessed to the responsible parties. The second method for the Natural Resource Trustees to fulfill their responsibility is to actively participate in the RI/FS process and subsequent negotiation with the responsible parties. In this role, the trustees identify elements of a remedial action that would insure that the selected remedy address the injury to the natural resources under their stewardship. If the complete remedial action and the agreement negotiated with the responsible parties addresses the injury caused, there is no residual damage, and it is not necessary to restore, replace, or acquire the equivalent to the affected resources. Each of these methods, and the relationship between the two, is discussed below. The methods for conducting Resource Damage Assessments is a broad and well-covered topic, beyond the scope of this book. However, presented herein is an overview of the process, and a listing of references that address various aspects of Damage Assessments in detail.

Restoration by Coordination With RI/FS

Coordination with the RI/FS is the clearly preferred procedure for the restoration of damage to natural resources, and must be persued before attempting to recover damages through the Damage Assessment Process. SARA recognizes that a remedial action selected during the RI/FS process may include significant restoration, rehabilitation, or replacement of damaged natural resources. Thus, to prevent expenditures to assess damage that may not persist, SARA generally bars any action toward recovery of natural resource damage at a site with an active RI/FS process before the designation of remedial actions. Similarly, Congress reports that Damage Assessments at hazardous waste sites should, to the extent possible, be integrated with the remedial action (House of Representatives Report No. 253, October 31, 1985). The EPA and NOAA are currently negotiating a Memorandum of Understanding, which formalizes and details the coordination process through all stages of the RI/FS.

In 104(b)(2), CERCLA addresses the need for the coordination of all aspects of the process with the Natural Resource Trustees by calling for active participation in the assessment, investigation, and planning of site remediation. The trustees' primary goal should always be to achieve restoration or replacement of injured resources, or the acquisition of the equivalent to such resources, either as part of the remedial action, or as an adjunct to the implementation of such action. The process of Natural Resource Trustee coordination during the RI/FS begins with EPA notification to the trustees of the potential presence of or other vulnerability of natural resources on or adjacent to a site. Once notified, the trustees are expected to be active participants in site investigations and other RI/FS activities. In practice, this often results in the trustees making an initial site visit to assess the potential presence of natural resources under their jurisdiction in the affected area. This site visit not only allows them to identify the extent of jurisdiction, but also to observe overt signs of potential ecological or natural resource damage. The information gained during an initial field visit can also be invaluable during the planning of the investigation.

The next major activity of the trustees is participation in the work plan for the RI/FS, which includes the objectives and methods of the site investigation. It is critical that the site investigation, as outlined in the work plan, inventory the protected natural resources on the site and document any possible damage. During development of the work plan, trustees should identify what restoration or replacement measures may be needed, so that this can be factored into the data collection and the types of remedial alternatives to be considered in the FS. The work plan should also provide basic information needed to calculate the value of the damage, in the event that it becomes necessary to do so. The trustees maintain an active role during the RI/FS to insure that the data collected is adequate for their needs from a quality, as well as quantity, standpoint.

The selection of the remedial action is probably the most important step in the Natural Resource Trustees' role in RI/FS coordination. The objective of the coordination is to select a remediation that achieves the maximum restoration of natural resource damage. The trustees should understand what, if any, residual damage to protected natural resources will remain after the implementation of each remedial alternatives. There are four areas of potential effects that must be considered: historic or ongoing damage to biological populations caused by toxicity; the destruction of habitat caused by the contaminant release; the destruction of habitat resulting from the physical alteration associated with remediation; and potential toxic effects from contaminants left on-site after remediation. When the trustees provide input to the selection of a remediation alternative, they should identify which alternatives would rectify natural resource damage to the degree that negates, or minimizes, any further investigation or restoration of such resources for Natural Resource Damage claims.

The final step in RI/FS and Natural Resource Trustee coordination is during the negotiations with the potentially responsible parties. During the negotiations, the trustees can identify the extent of remediation required to rectify natural resource damage. If the potentially responsible parties agree to such a level of remediation, then (in accordance with CERCLA 122(j) (2)) the trustees can agree in writing that there will be no residual damages. The consent decree between the EPA and the potentially responsible parties can then contain a "covenant not to sue" for natural resource damages. This is an important tool for the federal government in negotiating a settlement with the potentially responsible parties. Avoidance of a full Natural Resource Damage Assessment, which, as described later, can identify and claim a significant value of damages, and the lengthy and expensive litigation generally required to reach a settlement, can be attractive to both parties. Reaching a covenant not to sue can also eliminate the sometimes long delays in cleanup actions due to conducting the Damage Assessment, and protracted negotiations. The potential for a covenant not to sue also provides incentive to the EPA to select a remedial alternative with minimal residual natural resource damage.

In summary, numerous advantages accrue to all parties by the early and close coordination between the RI/FS efforts and the Natural Resource Trustees. First, coordination is in compliance with the intent of the legislators and regulators, and thus can prevent any procedural deficiency claims during remediation negotiations or litigation. Coordination also maximizes cost efficiency, because the data needed for the RI/FS and a Damage Assessment are generally the same. The intent of CERCLA (i.e., the cleanup of hazardous waste sites) can be achieved more expeditiously through coordination, because there is less chance for delay in remediation due to a lengthy Damage Assessment and any resulting negotiations or litigation. Finally, because of the covenant not to sue carrot, the EPA has a much bigger stick during negotiations with the potentially responsible parties.

A biological technical advisory group is frequently the forum for coordination with the Natural Resource Trustees. This group consists of representatives of the trustee agencies, as well as resource specialists from within EPA groups, such as the Environmental Services Division, the EPA laboratories, or the Wetland or Water Sections. Most EPA Regions have such a group, but they are under different names, such as SEAT (Superfund Ecological Advisory Team) in Region I or BAG (Biological Advisory Group) in Region III.

The biological technical advisory groups can provide an excellent forum for coordination and efficiency. The Remediation Project Manager knows where to go for notification, and does not have to determine which specific protected resources may be involved. The group also establishes the logistics and mechanisms for appropriate participation in site visits, input to the work plan, and the review of relevant documents. The group is an excellent structure for trustees with different resources to protect, and sometimes for competing objectives to reach a consensus on investigation and site remediation. For example, one trustee might want to insure that there is no remaining contamination on-site that could affect aquatic communities. Another trustee might want to minimize disturbance to a wetland habitat because it is important to a special-status species. These objectives could be in conflict because the removal or isolation of contaminants that might reach the aquatic environment would involve work in wetlands. Regular meetings and discussions of a biological advisory group provide an opportunity for consensus building and compromise, such that a remedial alternative amenable to a covenant not to sue can be defined.

Restoration by Natural Resource Damage Assessment

The role of the RI/FS process, as defined by CERCLA, is to implement an action that protects the environment from further harm. When Congress established this role and the sequence of the RI/FS, they acknowledged that, although coordination of resource restoration during the process as described earlier is an objective, site remediation may not eliminate past harm from contaminant releases. Consequently, additional measures to restore or replace injured ecological resources may be required. The first step in reclaiming the damage is to determine the extent and value of the damage ramaining after remediation. In Section 301(c), CERCLA requires the president to establish regulations for the evaluation of natural resource damages. By issuing Executive Order No. 12316, in August 1981, the president delegated this authority to the Department of the Interior, which issued regulations for conducting damage assessments codified as 43 CFR 11. Anticipating that releases could be either simple or complex, congress required the president to promulgate regulations to address two distinct types of Damage Assessments [CERCLA §301 (c)]. Type A assessments (43 CFR 11.40–41) are designed to be a standard simplified procedure requiring minimal field investigations. In contrast, Type B regulations (43 CFR 11.60–84) are for site-specific assessments, and generally involve unique situations.

The complete description of these regulations, procedures, and associated methods is beyond the scope of this book. However, since an RI/FS and the associated ecological assessment strive to eliminate the need for a damage assessment, a general understanding of Resource Damage Assessments is important in conducting a full and comprehensive ecological assessment of a hazardous waste site and is summarized herein from various references. Documents used to summarize the Damage Assessment process and available references for conducting damage assessments are cataloged in Table 3–2.

The Department of Interior regulations require an initial preassessment phase. This phase is initiated by notification of the trustees of a release that may potentially injure natural resources. If the release is an emergency, the trustee can take limited action to protect the resource. The trustee then screens the information to determine if the release justifies the completion of a damage assessment. This process consists of reviewing existing data to: identify the substances released; estimate the areas and pathway affected; and identify the natural resources potentially at risk (43 CFR 11.24 and .25). The trustee continues with the damage assessment under the following conditions: if the discharge is covered by CERCLA or the Clean Water Act; if natural resources are or likely will be adversely affected; if the characteristics of the released substance are sufficient to potentially cause injury to a protected natural resource; if there is adequate data to pursue an assessment or it is reasonably possible to obtain such data; and if emergency response actions will not sufficiently remedy the injury.

If there is justification to continue with the Damage Assessment, an assessment plan is required. The objective of the plan is to insure that the assessment is conducted in a well thought out, systematic, and efficient manner. A part of addressing these objectives is maximum coordination with the RI/FS process for the site. This includes

TABLE 3-2.   References Describing the Natural Resource Damage Assessment Process and Methods for Conducting the Assessment.

U.S. Department (1987) of the Interior Type B Technical Information Documents[1]—
U.S. Department of Interior, Washington, D.C. June 1987.

Application of Air Models to Natural Resource Injury Assessment	PB88-100128
Approaches to the Assessment of Injury to Soil Arising from Discharges of Hazardous Substances and Oil	PB88-100144
Injury to Fish and Wildlife Species	PB88-100169
Guidance on the Use of Habitat Evaluation Procedures and Habitat Suitability Index Models for CERCLA Applications	PB88-100151
Techniques to Measure Damages to Natural Resources	PB88-100136

Various Damage Assessment Methods	References (see Reference Section for Full Citation)
Title	Author and Date
Measuring Damages to Coastal and Marine National Resources: Concepts and Data Relevant for CERCLA Type A Damage Assessments	Economic Analysis, Inc., 1987
Rehabilitating Damaged Ecosystems	John Cairns, Jr., 1988
Wetland Creation and Restoration, the Status of the Science	J. A. Kusler and M. E. Kentula, 1990
Ecological Engineering: An Introduction to Ecotechnology	W. J. Mitsch and S. E. Jorgensen, 1989

[1]Obtainable from the National Technical Information Service, telephone 703-487-4650.

participation of the trustees in the work plan development of the RI/FS, to insure the availability of the data necessary to document the effects to the resource. It also means that the pathways or other links from the release to the injured resource be investigated using methods that could ultimately be used in identifying responsibility for the damage.

The potentially responsible parties must also be afforded a role in the Assessment Plan phase. Their legal right is established in 43 CFR 11.32, and provides for thirty days of comment on the plan before the work is initiated. It is to the advantage of all parties to have the potentially responsible party as an active participant in the planning of the damage assessment. They often have data available that can be used, in at least a qualitative fashion, to identify compounds and media of concern and potential sampling locations. There is often the opportunity for the potentially responsible party to conduct some of the investigation, which can be a benefit to the potentially responsible party, because they can be assured that the data will be collected in a timely fashion, and methods and procedures that could limit their responsibility at the site would be identified and strictly followed.

The obvious advantage to the Natural Resource Trustee is that the potentially responsible parties will pay for a substantial portion of the planning and implementation. In such an arrangement, the trustee or other federal government agency must closely scrutinize the proposed plan, implementation procedures (including quality assurance and quality control), sampling, analysis, validation, and data interpretation. However, it is sometimes easier to find gaps, flaws, or unclear objectives in plans developed by others, particularly if they are familiar with the site, then it is to develop an original plan. If close supervision of a potentially responsible party's sampling and analysis program is not deemed to be sufficient oversight, the trustee or other federal agency can implement a parallel program of reduced magnitude. If the data from the two programs are comparable, the data base is drastically increased, with minimal expenditure and effort by the trustee.

The Assessment Plan must identify the basic type of Natural Resource Damage Assessment that must be employed. A type A Assessment would be anticipated if the release was not complex and is in an area that is typical of the region. However, the Department of Interior has only been able to establish Type A methods for releases in marine environments, and has only developed the Natural Resource Damage Assessment Model for Coastal and Marine Environments (Economic Analysis, Inc. 1987). The approach for a Type A Assessment is a computer model that draws on several data bases and algorithms. The data bases include: an encyclopedia of chemical characteristics for potentially released contaminants; an inventory of marine resources within designated geographic regions; an estimate of the quantity and value of the marine resources; and input factors, by region, used to calculate the physical fate of the release. The time, location, quantity and quality of release, and the weather conditions at the time of the release, are input to the model. The algorithms in the model then calculate the monetary value of damage to the marine resources caused by the contaminant release. The calculated value is then used in the Damage Assessment and, ultimately, to obtain damages from the potentially responsible parties.

A Type B Assessment is conducted when site-specific information is required. A Type B study is always necessary if the release occurs outside of the marine environment, because no standard methodology has been established for terrestrial or fresh water systems. An injury determination is first required to link the injury to the release. The rules promulgated by the Department of Interior provide definitions of injury for each of the major categories of resources: surface water, groundwater, air, geologic resources, or biological resources. The definitions generally refer to quantifiable characteristics, such as exceedance of drinking water standards, for a resource serving as a drinking water source before the release. Once the presence of a resource injury, loss, or damage has been confirmed, the pathway from the release to the resource is established either by fate modelling or the detection of a concentration gradient from the release to the injured resource. Determination of the value of the injured resource is the final step in a Type B Assessment.

In practice, a specific study, or Type B Assessment, generally begins with a review of existing information available from ongoing regional baseline monitoring studies conducted by federal or state agencies. Other sources of data on site-specific conditions are often available from Environmental Impact Statements, Assessments, or other government-required environmental investigations for proposed facilities or activities in the immediate area. Facilities with active wastewater discharges or air emissions are other sources of environmental data, as they are often required by permits to maintain an extensive monitoring program. The existing site-specific data is generally supplemented by a survey to determine what environmental resources could have been present before the event in the potentially affected area. The information resulting from the survey is also used to assess the condition of affected resources, in comparison to similar resources in an unaffected area. This information, along with the chemical and physical characteristics of the release, is analyzed to estimate the damage caused by the release. As in the Type A Assessment, the estimate of damage is then used to obtain funds to restore or replace the lost or injured environmental resources.

In a marine situation, the trustee is encouraged to use the Type A procedure, unless limitations on the simplified method preclude its use. Such limitations could include: a compound or combination of compounds not adequately covered by the data base; the absence of appropriate geographical information; or an inappropriate treatment of critical natural resources, because of local distinctive factors not covered by the large-scale general Type A procedure. Potentially responsible parties can require implementation of a Type B Assessment if they assume financial responsibility for conducting the Assessment and all associated investigations. If the potentially responsible party feels there are unique circumstances that make the affected resource less valuable than the average regional value of the resource, it could be to their advantage to conduct a site-specific evaluation. Similarly, if specific spatial, temporal, or contaminate characteristics of the release might have resulted in less-than-expected damage to the resource, a Type B Assessment could predict a lower damage cost than the generic Type A Assessment.

When the Damage Assessment is complete, the trustee must prepare a Restoration Methodology Plan. This plan must detail how the money recovered for the injury to the resource resulting from the release can best be used to restore or replace the resource. If restoration or replacement is not deemed feasible, the plan must outline the procedure to acquire additional resources. The Restoration Methodology Plan must also document that all money recovered by the Natural Resource Trustees will be spent on a combination of: the costs of assessing the damage; the rehabilitation of the damaged resource; the restoration of the resource to prerelease conditions; or the acquisition of replacement resources.

Methods for Injury Recovery

Superfund legislation authorizes federal and state Natural Resource Trustees to recover damages to protected natural resources [CERCLA §107(a)(4)(C)]. As discussed earlier, the preferred method of recovery is to incorporate restoration, replacement, or acquisition of the equivalent to the affected resources into the selected remediation alternative. If this is not achieved, the recovery of injury can be offered to the potentially responsible parties as an adjunct to remediation during negotiations, in return for a covenant not to sue. If this fails, then the trustees are forced into a full Natural Resource Damage Assessment, and then monetary recovery through negotiation or litigation.

When injury to the protected natural resource remains an issue following the RI/FS process, the trustee must establish that damage to a protected natural resource has occurred from a release. The damage can be established empirically, such as by the reduced hatching rate of fish eggs in the affected area at the time of the release, in comparison to similar location or at a different time. The damage can also be demonstrated by comparison to a previously determined criteria or standard. For example, if a release exceeded Water Quality Criteria (U.S. EPA 1986), damage has occurred even if there is no evidence of aquatic toxicity or other effects. The trustee must also demonstrate that a potentially responsible party is liable for the damage, by describing the link between the release and the resource injury. However, the responsible party need only be shown to have contributed to the damage. Thus, in this respect, the CERCLA standard of proof for the cause of damage is less stringent than common law.

When the occurrence of damage to a protected resource, and liability, have been established, and the injury to a protected resource must be redressed by monetary recovery, the value of the lost or damaged resource must be determined. The value of the injury can be calculated using one of two basic approaches, or some combination of the methods [CERCLA §301(c)(2)]. One approach is to estimate the commercial or recreational value of the resource that was damaged or destroyed. For example, if a stand of trees was destroyed by landfill leachate, the lumber value could easily be calculated. However, to be complete, the estimation of lost value would also have to take into account such factors as the loss of wildlife habitat, the increased erosion resulting from the vegetation-free soils, and the effects on the aquatic habitat caused by sedimentation from the increased erosion.

A second approach is to develop a construction estimate for the replacement cost of the damaged resource. This approach has attraction, because it is straightforward and the construction methods used are generally standard practice with a well-documented cost history. The degree of success, location of the replaced resource, and timing are drawbacks for this approach. There is only a limited amount of experience in reconstructing damaged natural resources, and the success rate has been less than complete (Cairns 1988, Kusler and Kentula 1990). Consequently, it can be extremely difficult to agree on the extent of reconstruction necessary to achieve full replacement of the damaged resource.

The location of the replacement is also an uncertainty in the estimate of the replacement cost. Reconstruction in the original location may not be possible, because a waste treatment plant may be required on the site, or an impervious cap may so alter the soil and hydrology that appropriate revegetation is not possible. If an off/site location is required, it is never straightforward as to the dependence of the resource value on surrounding resources. For example, a component of the value for an acre in a large undeveloped wetland system derives from the aggregate value of the entire system. A similar acre in an urban area does not have the same value. It has some decreased value, because the total value of some functions in a large wetland generally exceeds the sum of the individual acres. For the isolated acre example, other functions have an increased value, because of location-specific uniqueness. It is very difficult, if not impossible, to determine the relative worth of these two different attributes when assessing replacement costs.

There is generally a significant lag in the time between the initial injury until repair or replacement is achieved. This requires that the value of the lost resource be determined for this lag period, and the uncertainties associated with estimating the value cannot be avoided. Also, the issue of when and if full replacement value is achieved must be addressed as part of the Damage Assessment. This generally involves long-term monitoring and an agreement as to what criteria are used to determine success.

There are limitations imposed on the recovery of damage to protected natural resources, and some of the limitations can have a bearing on ecological assessments. Claims of damage can only be made by state or federal trustees acting on behalf of the public, or by trustees acting on behalf of involved indian tribes. Consequently, damage to private lands or resources are not subject to natural resource damage, and must be redressed through civil or other action. The monies or other resources derived from the Damage Assessment and resulting settlement must be used to restore, replace, or acquire the equivalent to the affected resources [CERCLA §107(f)(1)]. Thus, the monies recovered may not be diverted to the general fund of the responsible agencies, but must be used for the replacement or restoration of the resource. CERCLA places a cap of $50 million on the recovery of natural resource damage, unless willful negligence or a violation of a federal safety standard can be demonstrated.

There are also time restrictions and certain exemptions on the recovery of natural resource damages. Damages before 1980 were excluded from monetary recovery. However, the courts have ruled that, even if the release occurred before 1980, if damage continued to occur after that date the appropriate Natural Resource Trustee could pursue an assessment and recovery of damages. The trustee is also limited to three years after the completion of remedial action to file a claim of resource damage. If an impact was identified in a Federal Environmental Impact Statement or other National Environmental Policy Act document, and the responsible federal agency acknowledged the impact and proceeded with the action, then any natural resource damage related to such an impact is not subject to recovery under CERCLA. However, the damage or loss must be demonstrated to have occurred after the authorizing permit was granted or other federal action occurred. Similarly, natural resource damages resulting from a federally permitted release (e.g., a wastewater discharge in compliance with a National Pollution Discharge Permit) or application of a federally registered pesticide are exempt from liability under CERCLA.

There are also limitations on the source of funds used to restore or replace natural resources damaged by releases from hazardous waste sites. The fund established by CERCLA (i.e., Superfund) can not be used to assess or restore the damage. Under the original CERCLA, the fund represented a potential avenue for recovery of some costs associated with natural resource damage, but SARA (revised section 517 of CERCLA) has been construed to prohibit all use of the fund for the restoration of natural resources. Therefore, if a potentially responsible party can not be identified or forced to pay for remediation and natural resource damage, the injury will likely not be corrected or otherwise addressed. Also, if a federal agency is the responsible party, another agency acting as the Natural Resource Trustee can not sue for damages. However, if the injured natural resources are within the jurisdiction of a state or indian tribe, the state or other qualifying trustee can recover damages from the responsible federal agency.

Recovery of damages through Natural Resource Damage suits has not been extensively applied as part a of hazardous waste site cleanup. For the same reasons that ecological assessments were not previously major components of early RI/FS processes (i.e., the early focus was on public health), the damages were not purused. Now that the focus has expanded to ecological resources, the trustees are actively involved, and there is some relevant research and case history information to conduct Damage Assessments and estimate monetary compensation. Consequently, the number of sites where natural resource damage assessment and compensation are being considered has expanded since 1989. When these are completed, probably in 1992 or 1993, a precedent will be established, and there could be a drastic increase in the collection of natural resource damages from CERCLA sites.

REFERENCES

Cairns, John, Jr., 1988. *Rehabilitating Damaged Ecosystems*. Boca Raton, FL:CRC Press.
Cummings, P. T., 1990. Completing the Circle. The Environmental Forum. Nov/Dec pp. 11–18.
Economic Analysis, Inc., 1987. Measuring damages to coastal and marine national resources: Concepts and data relevant for CERCLA Type A Damage Assessments PB87–142485. Springfield, VA: National Technical Information service.

Federal Interagency Committee for Wetland Delineation, 1989. Federal Manual for Identifying and Delineating Jurisdictional Wetlands. U.S. Army Corps of Engineers, U.S. Environmental Protection Agency, U.S. Fish and Wildlife Service, and U.S.D.A. Soil Conservation Service, Washington, D.C. Cooperative technical publication.

Findley, R. W., and N. A. Farber, 1988. *Environmental Law*. St. Paul, MN: West Publishing Company.

Kusler, J. A., and M. E. Kentula, 1990. *Wetland Creation and Restoration, the Status of the Science*. Washington, D.C.: Island Press. Also published as U.S. EPA Publication 600/3–89/038.

Marburg Associates and W. P. Parker, 1991. *Site Auditing: Environmental Assessment of Property*. Vancouver, B.C., Canada: Specialty Technical Publishers, Inc.

McKee, J. E., and H. W. Wolf, 1963. Water Quality Criteria. California State Water Resources Board, May 1963. PB82–188244.

Mitsch, W. J., and S. E. Jorgensen, 1989. *Ecological Engineering: An Introduction to Ecotechnology*. New York: John Wiley and Sons.

Ramamoorthy, S., and E. Badaloo, 1991. *Evaluation of Environmental Data for Regulatory and Impact Assessment*. Amsterdam and New York: Elsevier. 465 pp.

U.S. EPA, 1990. Superfund: Focusing on the Nation at Large. EPA/540/8–90/009.

U.S. EPA, 1989. Risk Assessment Guidance for Superfund—Environmental Evaluation Manual. EPA/540/1–69/001A. OSWER Directive 9285.7–01.

U.S. EPA. 1988a. National Oil and Hazardous Substances Pollution Contingency Plan. 40 CFR Part 300. 53 Federal Register 51395.

U.S. EPA, 1988b. Guidance for Conducting Remedial Investigations and Feasibility Studies Under CERCLA. EPA/540/G–89/004.

U.S. EPA, 1988c. CERCLA Compliance with Other Laws Manual: Interim Final. EPA/540/G–89/006.

U.S. EPA, 1987. Identifying Superfund Sites. WH.FS–87–005R.

U.S. EPA, 1986. Quality Criteria for Water. Office of Water Regulation and Standards. EPA/440/5–86/001.

U.S. EPA, 1976. Quality Criteria for Water. U.S. Environmental Protection Agency, Washington, D.C. 20460.

4

Technical Approach to Ecological Assessments

INTRODUCTION

An ecological assessment is a strange mixture of: responses to regulatory mandates; scientific investigations; and inputs to decisions. Each of these needs must be addressed when the assessment is prepared. To be useful, however, the document must be logically organized, and the objectives and conclusions apparent. If the objectives are clear at the onset of the investigation, and the information is collected and analyzed in support of the objectives, the logic of the process will usually be reflected in the ecological assessment. The result will then be a document that addresses the multiple objectives in a readable and usable fashion. This chapter describes the components or elements of the ecological assessment and explains how they can be integrated into the overall RI/FS process for maximum effectiveness.

To meet the mission assigned to the ecological assessment as part of the RI/FS process (U.S. EPA 1988a), it must address the three basic technical working objectives discussed in Chapter 2:

Identify the nature and extent of the ecological damage at the site;
Establish ecological goals and endpoints, and then evaluate contaminant levels compatible with the endpoints; and
Define ecological resources to be avoided or restored as part of the remediation process.

The approach used in addressing each of these objectives, and thus the overall approach to the ecological assessment, must reflect the ecological potential and plausible damage at the site. The simpler the resources and the less extensive the potential for damage, the simpler the ecological assessment should be. The ecological potential is dictated by the resources that exist or could exist at the site and in the surrounding area. Most commonly, the resources are apparent and consist of wetlands, other important fish or wildlife habitats, or the presence of important or special-status species. However, there can be less obvious, but equally important, ecological resources associated with a hazardous waste site. There can be seemingly insignificant or nonapparent functional resources at a site that are critical to biological communities downgradient from or adjacent to the site. For example, a site can provide an important function by preventing erosion and sediment deposition in a downstream trout breeding area. Similarly, a site can represent an edge or transitional habitat, which adds significantly to the value of other habitats, in the area. The site can also represent significant potential ecological value not immediately recognized because of past contamination or physical alteration of the site.

The plausible damage to be considered in developing the assessment approach encompasses the nature and extent of the contamination, and the range of remediation alternatives under consideration. If the contamination is limited in concentration and extent, far-reaching and detailed ecological assessments are often not necessary even if important resources occur in the area. Even if the level of contamination is great, but migration is controlled and removal is easily accomplished, the assessment can be limited. For example, if the contamination is confined to a nonvegetated area with no other ecological significance, and the remediation is off-site treatment with removal by way of existing roadways, the assessment could be limited to restoration following removal. At the other extreme, there could be only a minimal extent and degree of contamination, but remediation necessitated new road construction to reach the site, which altered drainage patterns and required stream diversion. In this case, the ecological assessment approach must include at least a broad consideration of off-site as well as on-site resources and an analysis of the ecological functions and interdependences of the resource.

Logic dictates, and current EPA guidance recognizes, the need for flexibility in ecological assessment methods and an approach based on the ecological potential and plausible damage at the site. The more complex and valuable the ecosystem, and the greater the potential for ecological damage, the more elaborate the ecological assessment should be. The flexibility needed to determine the appropriate level of complexity is achieved through a tiering or phasing of the ecological assessment, which is a philosophy ubiquitous in the Superfund program. The objective of the early phases is simply to determine the need for an ecological assessment based on resources present and the potential for damage. If a need is identified, the results of the early phases are critical in establishing the methods and approaches of the ecological assessment. Subsequent phases are at increasing levels of detail, and designed to address one or more of the overall ecological assessment objectives or, alternatively, to identify the need for more detailed study.

The technical approach to ecological assessments is a combination of phased investigations and the addressing of multiple objectives. An important component of this approach can be visualized as a matrix, with the various phases on one axis and the overall objectives on the other axis. Each matrix cell consists of the activities required to either complete the objective of that cell or to move to another cell requiring a more detailed investigation. There is a third dimension to this theoretical ecological assessment matrix: the actual investigations or elements of the assessment. Each individual investigation can provide information for more than one objective. For example, an evaluation of the on-site aquatic community could reveal ecological damage when compared to a reference site, and at the same time identify pathways for contaminant transfer and thus be essential in establishing cleanup criteria. The same investigation could also provide the information needed to reconstruct the aquatic habitat, if remediation called for extensive dredging. Each investigation could be conducted at increasing levels of detail until all of the information needed to complete the matrix cell for each applicable objective can be filled.

The ecological assessment does not conclude with accomplishing individual objectives, or filling in the cells in the above analogy. Often, comparisons must be made between competing ecological objectives and trade-offs considered. Also, the ecological considerations must be integrated into the complete RI/FS process, to achieve the ultimate objective of a site remediation that achieves protection of the environment. Integration of the ecological concerns into the RI/FS and the recommended remedial action is perhaps the most complex task and final test of success for an ecological assessment.

This chapter is structured around the creation of matrices comprised of ecological assessment objectives, individual investigations, and phases. The organization of the chapter reflects the needs, and thus the potential organization, of an ecological assessment for a specific hazardous waste site. The backbone of the assessment is the set of individual elements of the investigations and evaluations, and these are described first. The order of presentation roughly reflects the chronological order generally followed in conducting an ecological assessment. However, many of the elements can be iterative and dependent on other elements, so an ecological assessment is not a strictly linear process neatly flowing from one element to the next. The presentation of individual elements is followed by discussions of phasing and the integrating of the results in an ecological risk assessment.

ELEMENTS OF AN ECOLOGICAL ASSESSMENT

A basic fundamental of ecology, expressed colloquially as "everything is related to and dependent on everything else," also applies to an ecological assessment. Thus, an assessment with multiple objectives, purposes, phases of investigation, areas of investigation (e.g., contaminant concentration, biological characteristics, and the physical environment) can not be structured solely on any one of these areas. Such an approach would be ripe for unending duplication, and susceptible to a substantial number of missing pieces. If, for example, an independent investigation is performed to determine the ecological resources present on-site, the information could be useless or even misleading to decision makers. The conclusion of no resources present could indicate that no consideration of ecological resources is needed in the RI/FS process. Alternatively, the lack of resources could reflect severe ecological damage due to contaminant releases. Without the companion investigations and consideration of general site conditions, contaminants present, and potential ecological pathways, it is impossible to determine the correct interpretation of the "no resources present" conclusion. Thus, to avoid ignoring a potentially serious situation or unnecessarily spending time and money on nonissues, the coordination and integration of all investigations are needed to make the information on ecological resources useful and timely for the remediation process.

For practical purposes, the ecological assessment must be structured in a rational, logical fashion centered on individual activities or elements. Although distinct, each of these elements generally has multiple purposes and there are often overlapping data requirements. The elements can address specific ecological assessment or RI/FS objectives, feed into the design of subsequent investigations, or evaluate regulatory applicability or compliance. Before designing or implementing an ecological assessment, or any of the individual elements, it is crucial to understand the basic components of each element and how the elements fit into the overall picture.

The specifics of the elements of an ecological assessment are highly dependent on the specific site conditions and the nature and extent of contamination. However, there are certain general elements that should be considered for all ecological assessments. In many cases, if no ecological resources are at risk, many of the elements are not necessary, and the assessment can be accomplished simply by documenting the absence of risk. The specific content and organization of the elements of an ecological assessment can vary not only by site, but also by the style of the preparer and the overall approach and style of the site RI/FS. The amount of information already available when the assessment is initiated also effects the structure and level of detail for the elements of the assessment. An organization commonly used for the critical ecological assessment elements is presented below. The discussion includes the general contents and approach for each element, and how the element contributes to the whole process. The sequence of presentation is roughly intended to parallel the chronology of a typical ecological investigation. However, an ecological assessment, as with other aspects of a hazardous waste site investigation, is frequently a phased and iterative process. Therefore, the elements of an assessment discussed below are sometimes repeated at increasing levels of intensity and detail. Table 4–1 also presents a summary of the elements and their relation to the ecological assessment and RI/FS.

Preliminary Site Description

A preliminary description of a hazardous waste site is the applied equivalent of the observational first step in the scientific method. Just as the scientific method calls for an observation of the objects and events in the appropriate physical setting before generalizations can be made or procedures designed to test the theories (Keaton 1972), a hazardous waste site must initially be described. The underlying objective of this first step is a general understanding of the potential relationships among various ecological resources and contaminants. To avoid premature conclusions, or validating a theory before it has been objectively formulated, the observations should be general and focused on describing conditions rather than effects.

In general, the site descriptions should address structural attributes, rather than functions or effects. For example, at a specific site, a product of this element could be the identification of a stream present adjacent to the site. The preliminary description could also go so far as to distinguish the stream as a suitable habitat for cold water, warm water, or no fisheries. It would generally be inappropriate, as part of the site description, to portray the stream as supporting a breeding trout population. Similarly, the presence of contaminants in ground and surface water could be an appropriate observation as part of this element of the ecological assessment. However, to conclude that ground water is the source of contaminants to the surface water would be unnecessary and premature during the preliminary site description.

The site description should also focus on qualitative rather than quantitative information. Presence versus absence is an important conclusion from this first step of the ecological assessment. In contrast, statements of population density or rates of fecundity, nutrient transfer, or primary production would not contribute to the objectives of the initial site description. Documentation of the dominant vegetation types, as opposed to a calculation of percent cover by type, would be the more useful terrestrial

TABLE 4–1.   The Objectives and Remedial Investigation and Feasibility Study Uses of Ecological Assessment Elements.

Ecological Assessment Elements	Objectives or Element	Use in the RI/FS
Preliminary Site Description	Qualitative description of site ecology. Identification of site contaminants. Identification of contaminant pathways.	General site description. Input to define site investigation methods.
Ecological Endpoints	Establish desired future ecological characteristics of site. Define conditions at site unaffected by contaminants.	Determine nature and extent of ecological damage. Establish cleanup criteria.
Ecological Assessment Work Plan	Define nature and methods of ecological investigations and analyses required.	Incorporation in RI/FS work plan.
Ecological Field Investigations	Determine the nature and extent of ecological damage. Describe ecological resources and potential of the site. Quantify contaminant pathways. Establish relationship between contaminant and effects.	Establish area for remediation. Evaluate cleanup criteria. Design site remediation.
Bioassay	Establish ecological dose response relationship. Quantify contaminant pathways. Evaluate the cause of effects.	Evaluate cleanup criteria. Establish the cause of and responsibility for damage. Evaluate alternative treatment methods.
Hazard Identification	Determine ecological resources at risk. Determine contaminants posing a risk.	Provide input to cleanup extent and criteria.
Exposure Assessment	Identify and quantify contaminant–ecological resource interface.	Provide input to cleanup and extent and criteria.
Toxicity Assessment	Identify the relationship between contaminant dose/exposure and ecological resource.	Provide input to cleanup extent and criteria.
Risk and Ecological Effects Levels	Establish toxicity values compatible with endpoints.	Provide input to cleanup extent and criteria.
Remediation Alternatives and Site Restoration	Characterize effects of remediation alternatives on ecological endpoints. Evaluate ecological objectives of site restoration.	Provide input to selection and design of remediation alternatives.
Compliance with ARARs	Compare specifics of ARARs with conditions resulting from remediation alternatives.	Establish regulatory compliance of RI/FS.

ecological site characteristic at this stage of the assessment. Confining the observations to qualitative and structural site aspects minimizes the potential for a premature diagnosis of problems or solutions and a dismissal of nonproblems. It also fosters broader and more objective future assessments of potential relationships between resources and contaminants.

Although the initial site investigation for an ecological assessment should concentrate on biological resources, there are additional considerations that are often crucial. Nonbiological site characteristics such as hydrology, soil type, and ground water regime are important in developing a general impression of the site's ecology. This information is generally available from other previous or ongoing components of the hazardous waste site investigation, but should be confirmed as part of the RI/FS and ecological assessment coordination. The surrounding land-use and regional ecological characterization can also be essential to determine the ecological potential for the site. Off-site but nearby critical ecological resources, such as special-species breeding habitats or wildlife refuges should also be noted as part of this element of the assessment.

The site description for aspects of the RI/FS parallel to the ecological assessment tend to concentrate on the type of contaminants potentially present, their general concentrations, and the media where they might occur. Although this information is important to the ecological assessment, the ecologist should not dwell on contaminant information. The public health assessor will generally assess the toxicity concerns of the contaminants, and this is frequently directly usable by the ecologist, at least in this early stage of the investigation. Similarly, the geologist and hydrogeologist will evaluate the information initially needed to assess media and contamination pathways of ecological concern. The ecologist should develop a general understanding of these issues from other participants, but concentrate on the ecological aspects of the site. No one else will identify, recognize, understand, or perhaps even care about these resources.

The specific objectives of a preliminary site description are to determine what needs to be investigated and how it should be done. When this element is complete, it should be apparent what ecological resources warrant investigation. The specifics of a quantitative investigation could be assigned from the information developed for the site description, or the need for a subsequent qualitative assessment could be identified as an intermediate step to define the specific scope of a complex investigation. It may also be possible to conclude that some resources are not of concern and need not be considered in the RI/FS process. It could also be determined that there are no ecological resources, ARARs, or pathways associated with a particular hazardous waste site, and thus no ecological investigation at all is warranted. It is common to reach the conclusion that a resource associated with the site exists, and some form of semiquantitative investigation is required to assess the effects of a contaminant release. It is also frequently determined that there are potential exposure pathways linking the contaminant to a resource, and the subsequent investigations should collect the physical, chemical, and biological data necessary to quantify the pathway and assess the risk. The investigations identified from the findings do not always provide the final answers or solutions. Sometimes, they only elucidate the need for more detailed investigations. Occasionally, the investigations identified from the initial site description fail to confirm or dismiss a problem or define the pathway producing an ecological effect. Such a situation is analogous to a negative finding during hypothesis testing in the scientific method and a new theory is called for.

An ecological site description usually proceeds in three sequential steps. The first is a review of data previously collected for the contaminant investigation of the site. The next step is a site reconnaissance, followed by contacting local academic and agency personnel. There is always some site-specific contaminant information developed as a part of listing the site on the National Priorities List. This could be as minimal as a site history, or citizen reports documenting chemical use and disposal practices on the site. In other cases, it could be well- documented concentrations of contaminants in a water supply well, which could identify the chemicals of concern, the concentrations present, and any fate and transport mechanisms. In rare cases, such as high tissue concentrations in sport or commercial species, there could be existing information on ecological effects or resources present. In many incidents, the ecological assessment is conducted as an add-on to an already initiated RI/FS. Obviously, in these cases the existing information is not only sufficient for a large portion of the site description, but additional data may not be needed to complete the ecological assessment.

A reconnaissance survey is required to document the presence and general nature of the ecological resources on-site, such as wetlands, other critical habitats, the potential for special-status species, and resources particularly sensitive to contamination effects. The survey is typically completed in one day, and includes a site walkover and a drive-through of the surrounding area, to gain an appreciation of the regional ecological setting and identify potential significant downgradient resources. Sometimes, the existing information is adequate to identify the general types of ecological resources (e.g., aquatic, wildlife, or wetlands) that are likely to be of concern, prior to the reconnaissance survey. In such cases, ecologists with the appropriate technical expertise should participate in the reconnaissance. Otherwise, an experienced biologist with a sufficiently broad background to recognize the types of resources, general conditions, and potential of the resources on the site is needed for the reconnaissance. The participation of such experienced ecologists in the reconnaissance frequently has additional benefits, because (unfortunately) senior-level staff may not take other opportunities to visit the site, yet they are often responsible for reviewing all products and making critical decisions. In ecology, probably more than any other science, a general appreciation of the character of the site, achievable only from being on the site, is crucial to all phases of the study, from design through implementation to evaluation. Thus, the presence of the senior ecologist, who is ultimately responsible for the assessment, on the reconnaissance goes far towards developing an appreciation for the site by the person who will make critical decisions.

Benefits from the reconnaissance can be maximized by including other technical, regulatory, and management personnel. On-site identification and discussion of potential issues during the reconnaissance among the biologists, chemists, geologists, and engineers will help insure an early start to an integrated and coordinated approach. Also, it increases the awareness of possible investigation constraints and areas of concern. Inclusion of regulatory and Natural Resource Trustees in the reconnaissance not only provides an additional avenue for incorporating their concerns at the first stage of the investigation, by establishing some common reference points, it also aids in subsequent discussions and product review. With all of the parties on-site, there is an excellent opportunity for the candid expression of ideas and feedback, which can make development, review, and approval of subsequent elements in the ecological assessment much more expedient, efficient, and productive.

The most important product from the reconnaissance is an assessment of the actual or potential presence of ecological resources on the site. Any characterization of the resources, such as condition or requirements, that can be made from the reconnaissance will also be useful in subsequent elements of the ecological assessment. It is helpful to develop a checklist of information to be collected during the reconnaissance, prior to the actual visit. Table 4–2 presents a generic checklist that can be used as a starting point. If there is significant existing information on the site prior to the reconnaissance,

TABLE 4-2.   Example Checklist of Ecological Information Obtainable from an Initial Site Visit and Field Reconnaissance.

ECOLOGICAL RESOURCE CHECKLIST

Characteristic	Forested	Upland Vegetation Grassy	Shrub	Other
% of Site/Acres				
Continuous (Y/N)				
# of Stands				
Dominant Species				
Overstory				
Understory				
Other Species				
% Cover				
Stressed Vegetation				
Present/Absent				
Indication				
Area				
Successional Stage				

		Aquatic Resources				
Wetlands		Lakes/Estuary		Streams/Estuary		
% of Site/Acres	_____	Area	_____	Length	_____
Classifications	_____	Tributary Area	_____	Width Range	_____
Number of Areas	_____	Depth Range	_____	Depth Range	_____
Species Present	_____	Substrate Type	_____	Substrate Type	_____
Soil Type	_____	Surrounding Vegetation	_____	Velocity	_____
Standing Water	_____	Water Color	_____	Shoreline Vegetation	_____

the list can be substantially modified for anticipated site-specific resources. An extensive photographic record of the reconnaissance, allows subsequent input from individuals not on the visit. In addition, it sometimes provides a second chance to look at site characteristics not initially noted. Pictures of the site taken during the reconnaissance can also extend the seasonal coverage of the overall investigation, by providing at least a visual characterization during a month when no other sampling was accomplished. The use of fieldglasses on the visit can provide additional site detail and, on large sites, save a lot of walking. Field identification guides are usually included in the reconnaissance equipment, but are most effectively used on the trip back, rather than using valuable field time for species identification in a survey intended to generally characterize the site. If aquatic resources are known to exist on-site, a kick net designed for aquatic invertebrate sampling is useful to generally characterize the substrate and habitat, even if no animals are collected.

Another possible source of information for the preliminary site description would be discussions with agency or academic personnel familiar with the ecology of the region or site. This third step in the preliminary site assessment can occur prior to, during, or after the site reconnaissance. If enough existing information is available to generally characterize the site, the appropriate agency and other personnel should be contacted before the reconnaissance visit, to anticipate specific-site features. If no ecological site information is available, it is generally prudent to contact locally knowledgeable people after the reconnaissance. In this way, the general habitat type can be described, and the local ecologists can identify the expected species. If the interviews are conducted after the reconnaissance, the site photographs can be a useful tool. It is sometimes appropriate and productive if local biologists participate in the reconnaissance. If they are likely to be involved in a subsequent review, their early direct involvement can be an added benefit. Ecologists or other scientists with experience in the vicinity of the site can contribute to the initial site description, both from personal knowledge and by identifying information sources in the "grey" literature dealing with site-specific conditions or regional ecological conditions, which may be applicable to the site. These sources can include graduate research theses, agency monitoring reports, or environmental impact reports from nearby projects.

The completion of this element in the ecological assessment does not constitute the completion of site description as a scientific endeavor. Most subsequent elements in the assessment fill in the overview description of the site's ecological functions and structure generated during the preliminary site description element. As more information is collected and evaluated, the understanding of the site is enlarged and frequently revised. However, the initial description is required to formulate generalizations and to structure the investigation plan.

Ecological Endpoints

Endpoints identify desired achievements and provide a means for measuring attainment following all hazardous- waste related testing, evaluation, and cleanup. Establishing "Endpoints" has become popular jargon, in both hazardous waste ecological assessment (Suter 1990) and ecological risk assessment (Warren-Hicks and Parkhurst 1991), to describe a process somewhat analogous to identifying remediation goals and objectives as part of an RI/FS. However, an ecological endpoint typically has a broader scope than simply setting cleanup concentrations or establishing an acceptable level of risk for some generic undesirable event. When these expressions of desired results are formulated, the intent is generally to address function, as well as structure, and the site's potential. Therefore, the endpoints must be tied together in some analytical way to the desired product or ecological use of the site. Useful ecological endpoints must also be realistic, both with regard to achievement and an assessment of success, and they must consider future site constraints. These differences from conventional remediation goals are illustrated by the three separate but totally interdependent faces of a complete ecological endpoint. These faces (site use, assessment endpoint, and measurement endpoint, discussed later) must be developed in sequence, with each building on the goals established for the preceding endpoint.

It is crucial that the endpoints are considered before the third, or Work Plan, element of the ecological assessment is completed. There should be refinement and development of additional detail for each of the endpoints throughout the investigation. If an endpoint is found not to be achievable, this should be acknowledged and provide input to decision making. Alternative endpoints, or the relative benefit of partial achievement, can then be evaluated, but such an objective process is preempted if the endpoints are constantly bent to fit each newly identified constraint.

Site Use

The most basic and initial endpoint is the future use of the site. This is generally not part of the ecological assessment, but a much broader topic to be determined at or above the RI/FS management level. However, the decision should be coordinated with those individuals conducting the ecological assessment. The ecologist's role is to insure that the ecological potential of site is not ignored. Designation of a future site use that precluded ultimate protection of existing, or development of potential, ecological resources, without prior knowledge and an appreciation of the consequences, would be a serious oversight. It could also be costly and embarrassing to designate the site for some totally inappropriate ecological function, given the regional characteristics or surrounding land use. Common future site use endpoints are residential, commercial, or industrial land use. Open space designations, such as conservation land or recreational areas, are also common future uses of remediated hazardous waste sites, and are the future uses usually most closely linked to ecological resources.

There are other aspects of the future site use endpoint with little immediately apparent impact to the ecologist, yet they should be considered in formulating the ecological endpoints. For example, future groundwater use for a water supply could be prohibited at a site with historically high use. This situation could result in changes in ground water levels or surface water discharges. This, in turn, could change the site potential for a wetland habitat, by allowing a larger area or longer seasonal duration of saturated soil. The increased discharge to the surface water could similarly create the potential for a superior aquatic habitat. The restriction of human access or future development could also, in some instances, improve the wildlife habitat value of the site. These factors must be understood by the ecologist, because their input is significant to establishing ecological endpoints for site remediation.

It is not unusual that political considerations, status of ownership, general land use trend uncertainties, or cleanup uncertainties prevent establishing firm site use endpoints. In such cases, the ecologist must make assumptions for the assessment and measurement endpoints. The assumptions must be coordinated with other participants in the RI/FS process, to insure that the future site use applied in the ecological assessment is possible and perhaps even likely. If there is no guidance as to future site use, a common default endpoint for ecological purposes is an estimate of the site use prior to any contaminant release. Often, a consensus on the assumed future use of the site for the ecological assessment can not be developed. In such cases, multiple uses must be evaluated by developing separate ecological endpoints for each alternative site use. This approach should be used as a last resort, because it can drastically increase the effort required. It is also contrary to an evaluation against fixed objectives, and thus it can be conducive to substituting endpoints for achievable goals.

Assessment Endpoints

In a very general sense, assessment endpoints relate to the future ecological use of the site. They are somewhat analogous to the human health goals of not causing a specified increased cancer risk if the site is used for residential development. The assessment endpoints represent the RI/FS team ecologists' statement as to what is considered achievable, important, and compatible with the future use of the site. They also identify the ecological structural and functional site aspects that are sensitive to or dependent on site contamination or remediation activities. The ultimate goal in establishing the endpoints is not only to set the desired ecological character of the site, but also to identify the structural and functional requirements critical to achieving the designated ecological site use. This ultimate goal is frequently elusive and, as discussed later, surrogates or assumptions for ecological requirements must often be used. Assessment endpoints consist of a subject or ecological entity and an action or effect. For example, rainbow trout might be the subject, and the action portion of the endpoint could be breeding.

Establishing the assessment endpoints is the first major ecological input to the RI/FS process. By stating these endpoints, the ecologist sets the standard for evaluating the area and the extent of potential remediation. Thus, the endpoints agreed upon ultimately establish the basis for evaluating the ecological need for remediation, how much of the site must be considered for remediation, and the intensity of the cleanup effort. The assessment endpoints also provide the guidance and direction for the site investigation and evaluation. Without preestablished endpoints, the ecological assessment can become an undirected data collection program. In such cases, much of the expensive and detailed information is never used in the evaluation, comparison of remediation alternatives, or consideration of mitigation measures, and thus the ecological efforts are viewed as a waste of time and money. With early and thoughtful attention to the endpoints, the ecological investigation is more directed, efficient, productive, and useful. It also makes the ecological assessment more credible and objective in the sometimes skeptical eyes of the other technical and management personnel working on the RI/FS. The endpoints also provide the scale against which the ecological success of the remediation will be measured. Thus, establishing the assessment endpoints should not be taken lightly, and full consensus with other technical, management, and regulatory RI/FS team members should be developed before the endpoints are finalized.

Assessment endpoints come into play during four stages of the ecological assessment. First, they form the basis and direction for designing the data collection program during the work plan development. If a data requirement can not be related to an assessment endpoint, it can not be justified in the work plan, either because it is not needed or because there is insufficient consideration of how the data will be used. The second use of assessment endpoints is for a determination of contaminant effects. The presence, severity, and significance of ecological effects can only be objectively established by relating measured existing conditions to the assessment endpoints. The results of this comparison establish the need for ecologically based remediation. Determining the effects levels for input to remediation criteria for ecological protection is the third use of the assessment endpoints. The ecological effects levels are calculated so that the ecological value of a site can be related to specific contaminant concentrations, and informed decisions on remediation criteria can be made. The final application of the endpoints is the evaluation of the mitigation needs from damage anticipated from either past contaminant releases or remediation activities.

Assessment endpoints must strike a balance between broad, sometimes abstract, goals and a realistic tool to be used in the RI/FS process. One end of the spectrum is illustrated by abstract endpoints such as: the site should 1) support a well-balanced self-sustaining assemblage of populations typical of unaffected similar ecosystems in the region, and 2) the community should achieve all energy transfer and other ecological functions ascribed to similar systems in the region. The other end of the endpoint spectrum could be: the egg production of a particular species of zooplankton must be a set number. In the first case, the endpoint is admirable and meaningful to nonecologists, but impossible to use in designing an investigation, determining contaminant effects, establishing remediation criteria, or evaluating site restoration measures. The other extreme is a useful ecological tool and very practicable, but unless it is related to a more universal objective it will not generally impact the overall RI/FS process because it cannot readily be incorporated into a decision or action. Useful endpoints fall between these two extremes and incorporate appropriate aspects of both. Measurement endpoints, as discussed later, are useful to balance the two extremes.

Establishing multiple assessment endpoints and selecting representative sensitive ecological functions are also methods of incorporating both abstract goals and practical objectives into the endpoints. These approaches require a prior and relatively complete understanding of the existing and desired ecological community for the site. From this understanding, the critical components of the community can be identified, and separate assessment endpoints established for the protection and maintenance of each community segment. For example, a site may have distinct terrestrial and aquatic components, so endpoints for each would be established. If each segment is protected, the successful relationship and interaction between them could then be assumed. This approach would be much more practical than attempting to measure and evaluate the interactions between the two segments of the community. The process of segmenting the community to lower levels of organization could continue until an endpoint that is practical and still related to a broad objective is identified. Once suitable segments are determined, a characteristic of the segment, which indicates the health of the entire segment, can be used as an assessment endpoint. The number of endpoints to be considered, and thus the complexity of and effort needed for the ecological assessment, can be simplified by designating endpoints that are both sensitive to effects and critical to the function of the community segment.

The U.S. EPA (1989a) list of criteria for establishing assessment endpoints includes biological relevance, measurability or predictability, susceptibility to contaminants, and their relation to a decision. The biological relevance is the importance of the endpoint to the structure or function of the ecological community as a whole. To be a useful endpoint, its achievement must be demonstrated to relate and contribute to higher organizational levels or functions of the ecosystem. The measurability of the endpoint is a practical consideration of the endpoint in the spectrum of abstract goals to realistic tools, discussed earlier. The selection of endpoints should not be unduly constrained by the measurability criterion because, as discussed later, measurement endpoints can be used to interpret abstract assessment endpoints. The sensitivity or susceptibility criterion reflects the relationship of the assessment parameter to the anticipated types of effects. For example, if heavy metals are the primary contaminant of concern, the toxicity of metals to the organism or function being considered as an endpoint should be considered. Similarly, if modification of the ground water regime is a potential remedial action, some characteristic of wetland plants dependant on the ground water level might be a useful endpoint.

The assessment endpoint must be an integral part of the entire RI/FS process, and thus related to decisions. This criterion has both a practical and a philosophical component. On the practical side, the quantifiable relationship of a measured existing condition or predicted future condition to an assessment endpoint should be analytical, objective input to the decision regarding the nature and extent of remediation. The philosophical aspect of the criterion relates to the perceived relevance of the endpoint to the regulatory framework, public reaction, and overall site cleanup objectives. If an endpoint is not perceived as relevant and worthwhile in a forum broader than the ecological assessment, it is unlikely that it will be used in the decisionmaking process. The U.S. EPA (1989a) considers this aspect of selecting endpoints as its "social relevance," but the impact of such a criterion is only its influence on decisions.

There are additional criteria that should be considered in establishing assessment endpoints. One is compliance with ARARs. Such criteria have predetermined and mandatory "social relevance" and input to decision making, since CERCLA mandates their consideration. They are also generally presumed to have biological significance and accepted measurement techniques. Site-specific and regional conditions should also be considered in selecting assessment endpoints. For example, interaction of the site with nearby wildlife preserves or other conservation area, or the regional uniqueness of any potential habitat on-site, could be important criteria for endpoint selection.

Typical assessment endpoints can reflect any level in the ecological hierarchy, from populations to communities to ecosystems (Suter 1990). This range correlates to the spectrum between abstract goals and realistic tools, discussed earlier. Assessment endpoints directed at individuals or lower levels of the ecological organization generally represent the realistic tool end of the spectrum. By themselves, they are frequently inappropriate, because the occupancy or survival of an individual is not significant to the ecosystem structure or function. If there is population integrity, an affected individual organism will be replaced, with no ecological repercussions. However, as discussed later, individuals cannot be ignored, as they are frequently used to measure the attainment of assessment endpoints. Community-level endpoints generally reflect abstract goals that are highly "socially relevant." To be useful, appropriate and multiple measurement endpoints must be established for each community-level endpoint.

In a theoretical sense, the higher the endpoint in the ecological organization, the more useful the assessment endpoint is, because it is a more complete indicator of the overall system. Achievement of an endpoint targeted at a high organizational level also indicates an absence of significant effects at all lower levels. Higher-level endpoints also avoid the often unanswerable question of "is this specific aspect really important to the site's ecological integrity?"

The objective of selecting a broader and higher-level assessment endpoint must be balanced by the measurability/predictability selection criterion. If damage to a higher-level ecosystem parameter can not only be conclusively demonstrated, but also be related to a specific stress (such as a contaminant concentration), it is difficult to establish the need for remediation. Such a demonstration is usually much easier on a population or lower level, by showing absence, decreased density, abnormalities or significant reduction in physiological rate function. In contrast, the conclusive expression of ecosystem effects, such as reduced energy transfer efficiency among multiple trophic levels, is much harder to determine. Similarly, it is much simpler and straightforward to formulate remediation criteria on the protection of lower-level organizational characteristics.

Population-level assessment endpoints are generally most appropriate if there are species potentially on-site that are highly visible, of special status (e.g., rare or endangered), or recreationally/commercially important. For such species, there are usually established evaluation techniques and often existing information on the condition of the regional, if not site-specific, populations. Also, the life history, behavior, and environmental requirements of species in this category are frequently well-documented in the literature. Population endpoints can also serve as indicators of overall system effects. However, such a substitution requires documentation that the protection of the designated population has broader implications.

If an important population is the subject or ecological entity of the assessment endpoint, there still must be an action or effect. The action or effect component of a population-level assessment endpoint can vary according to the level of protection warranted, and the assessment capability. Probably, the most protective action is a successfully breeding, self-sustaining population within the confines of the site. A narrower effects characteristic might be sufficient shelter or food, for a transient, migratory species.

Assessment endpoints targeted at the community or ecosystem level are generally most appropriate for sites supporting a distinct, relatively well-understood habitat. Beaver ponds, saltmarshes, or trout streams are typical examples. The methods for evaluating requirements and characteristics of such systems are generally available, and values for unaffected areas documented, so meaningful measurement endpoints can usually be easily established. Also, particularly for aquatic systems, it is often feasible to compare community or ecosystem characteristics with appropriate references site, to detect damage from the release of contaminants.

Measurement Endpoints

Measurement endpoints link the existing or predicted conditions on the site to the goals expressed by the assessment endpoints. Achievement of the assessment endpoints is determined through measurement endpoints, thus there must be at least one measurement associated with each assessment endpoint. Often, the assessment and measurement endpoints are so closely linked they are identical, for all practical purposes. For example, density of trout larvae relative to a reference site could measure the assessment endpoint of unaffected trout breeding. In other cases, due to measurement difficulties or abstract assessment endpoints, the link between the two endpoints is less direct, and can involve significant assumptions. Probably the most common example of an indirect link is the use of water quality criteria as an indication of a balanced indigenous aquatic community. An evaluation of contaminant concentration in food sources as a measurement of an unaffected predator population assessment endpoint is another example of assuming that a somewhat abstract condition exists, based on the measurement of only one requirement for the assessment endpoint condition.

There are criteria that should be considered when selecting measurement endpoints. The first and foremost criterion is that of a defensible relationship to an assessment endpoint. The second mandatory consideration for selection is the ability to be measured. If the characteristic defined by the measurement endpoint can not be determined and expressed in terms of an effects level (i.e., in relation to water quality criteria or another ecological benchmark) or in comparison to an unaffected condition (e.g., the population density in the region or at a reference site), it is not truly a measurement endpoint. Some flexibility can be built into the measurement endpoint by establishing a degree of qualitative evaluation and professional judgment. For example, if the assessment endpoint addresses the pelagic fish segment of the community, and the measurement endpoint relates to a reference site, because of the transient nature of pelagic animals and variability in sampling, the assemblages at the reference and hazardous waste sites would never be identical. Consequently, a degree of professional judgment, considering such factors as the relative success of field collection efforts and similar requirements and sensitivities of different species collected at the two sites, must be integrated into the interpretation of the results. Other measurement endpoint selection criteria include: the availability of existing data; the relationship to known contaminants and pathways; the degree of natural variability; and the temporal and spatial scale of the parameter (EPA 1989a).

Measurement endpoints can address various levels of ecological organization, reflecting the level of the associated assessment endpoint. It is not, however, unusual for the measurement endpoint to be at a lower level, and thus serve as a specific indicator of a general condition. Consequently, measurement endpoints based on individual organisms, or even lower-level biochemical, physiological, or behavioral characteristics, can serve as appropriate endpoints. In practice, these lower-level endpoints are most common because they are readily quantified and related to reference site or literature measurements.

For the lowest levels of ecological organization, biomarkers have gained recent popularity in measuring ecological effects from contamination (McCarthy and Shugart 1990). Such measurement endpoints include the presence or increased levels of stress-related enzymes or other proteins, gene or DNA alternation, or histopathologic conditions. Often, these endpoints can be measured with a relative high degree of precision and, once the methods are standardized, they can be compared to a large data base. Biomarkers are also amenable to comparisons with reference sites, because the high precision of the measurements and relatively low natural variability are well-suited to statistical comparisons. The largest drawback to biomarkers has been relating the results to assessment endpoints or other more general site characteristics. Once some of these relationships have been documented and quantified, biomarkers could be very important standard measurement endpoints at hazardous waste sites.

Measurement endpoints for individual organisms fall into two categories. The first is the condition of the organism on-site, which can be measured by the health of the individual (often as indicated by suborganism parameters) or the concentration of a contaminant in the organism's tissues. The life history condition and behavior of the individual on-site are also endpoints measured at the individual level, but can be important indicators of population or even community endpoints. For example, lactating mammals can indicate successful breeding, or frequent foraging can signify normal trophic relationships. The response of an individual to contaminant exposure is the second type of organism-level measurement endpoint. The most common of these measurements are toxicity tests and, as discussed in detail in Chapter 8, the endpoints can include death, fecundity, growth, or even behavioral modifications.

At least for animals, at the individual level, presence or absence is generally a measurement endpoint that must be carefully considered. The mere sighting of an animal or its sign is not necessarily an indication of anything other than a transient occupancy. However, repeated observations or specific signs related to feeding or breeding can indicate population characteristics.

Population characteristics suitable as measurement endpoints are generally restricted to conditions of the populations as measured in the field. The presence of a population on-site is the most basic measurement endpoint. The presence of a population differs from the presence of an individual. For a site to qualify as supporting a population, there must be some indication that an important life function (e.g., breeding or rearing) is occurring on-site at an intensity and success level comparable with the reference site or with regionally expected levels. More sophisticated population measurement endpoints involve detailed field measurements, such as sampling and evaluation of density or biomass. Growth or reproductive rate are endpoints that can be measured for some populations. These more intense measurements are generally limited to annual plants, or to indicators of animal reproduction such as number of offspring or eggs per female.

Measurement endpoints at the community level were commonly used for aquatic and terrestrial plant systems long before hazardous waste sites were a concern. Plant species composition, density, and biomass per individual are endpoints historically used to measure the successional stage of terrestrial communities. This same approach can easily be used to measure the condition of a plant community at a hazardous waste site. The limitations of this approach are natural variability and establishing quantifiable endpoints related to some effect or protection level. Species diversity, domination by tolerant organisms, and density are measurement endpoints commonly used in aquatic systems to evaluate community-level effects (Hynes 1970). The natural variability for these parameters is often small relative to characteristics of populations, and there is a large body of literature for typical or unaffected communities. Community-level measurement endpoints related directly to terrestrial animals are limited, and generally restricted to observations of trophic relationships determined by feeding behavior.

Ecosystem-level measurement endpoints are generally limited to process or functional parameters. Productivity, energy transfer, and nutrient dynamics are examples of such parameters. The measurement of these functional ecosystem characteristics is usually difficult, lengthy, and expensive. Also, the temporal and spatial natural variability is almost always quite large. Endpoints at the community level, if properly measured and evaluated, are the ultimate for an ecologist, but it is difficult to tell an engineer or manager that they must spend $10,000,000 for site cleanup because the production of forbes is down 30 percent.

Measurement endpoints are not restricted to characteristics related to contamination. There are potential physical and biological alterations of ecological resources that must be addressed as a part of the ecological assessment. There must also be endpoints useful in evaluating site alteration due to remediation activities, such as soil removal and the construction of treatment, storage, or disposal facilities. Typically, measurement endpoints associated with these types of effects are evaluations of the unaffected area of certain habitats, the soil types, or the surface water hydrology.

Ecological Endpoint Summary

Endpoints provide a direction to ecological investigations and evaluations. In the 1970s, megavolumes of ecological data were collected on baseline conditions associated with projects such as regional planning and the development of nuclear power plants and hydroelectric projects. It was never really clear how this data was to be used, other than the vague question of whether things changed after the project was implemented. The overall objectives of hazardous waste site cleanup are well defined (i.e., protection of human health and the environment) and, if present, the ecological effects are relatively easily detected and cause-assigned. Consequently, ecological assessments at hazardous waste sites are amenable to a process streamlined and directed by the early development of ecological objectives or endpoints.

The ecological endpoints represent a continuum from site possibilities (future site use) to ecological desirability (assessment endpoints) to an evaluation of success (measurement endpoints). The endpoints approach is also directly applicable to identifying past effects and establishing future goals. Since these are the essence of the Superfund, endpoints are well-suited to ecological assessments at hazardous waste sites.

Finally, endpoints represent an effective mechanism for communicating ecological information. They are useful in translating a large body of scientific information to site-specific conditions or objectives. Because they can be related to generally understood concepts, and they are measurable, ecological endpoints are also logical tools for inputting ecological considerations into the decisionmaking process.

Ecological Assessment Work Plan

There are two basic components of the work plan: a summary of what is currently known about the site, and a description of the needs to finalize the remediation strategy. It is thus analogous to the introduction and methods section of a scientific report. The work plan is also a planning document that serves to clearly assign responsibility, schedule, and cost.

The ecological investigations and analyses presented in the work plan must be structured around the ecological endpoints. Thus, the relation of the proposed activities to the measurement and assessment endpoints must be clearly described, and the expected use of the resulting data discussed. The ecologist must use the work plan to present a case for the proposed investigations, by tying them to critical decisions or to specific evaluations.

The site summary and description of future work in the work plan includes all aspects of the RI/FS, from community relations (e.g., Superfund jargon for human public participation, not an ecological paradigm) to ecological assessments to engineering considerations. It therefore establishes an integrated approach to site investigation and remediation, with broad and coordinated objectives. In the first portion of the work plan, the site is described in terms of the physical, geological, hydrological, land use, and contaminant characteristics. The ecologist must add a general description of the biological attributes of the site to this overall description. The preliminary site description, as described above, is the major ecological input to this portion of the work plan.

The project team should consider the condition and characteristics of the site from the aggregate of individual descriptions, and develop a first attempt at a conceptual model to describe, at least qualitatively, the relationships among contaminants, environmental media, and potential receptors. The objective of the site model at this early stage is a qualitative evaluation of processes and relationships on the site. The processes considered should focus on the nature and fate of contaminants, but also include site characteristics that could affect contaminant exposure or migration pathways. Typical topics addressed in this qualitative site model include the direction of ground and surface water movement, human exposure points and pathways, and general contouring of contaminant concentrations (hypothesized or gleaned from any existing information). The ecologist should add a discussion of the relationships of potential contaminants to the biological resources. A general picture of the trophic relationships and habitat—animal associations and requirements should be included in the discussions.

The site summary portion of the work plan should conclude by establishing general investigation objectives. These objectives should address the types of remediation decisions that must ultimately be made. The major ARARs that need to be considered should also be identified in this portion of the work plan. The ecologist's role in this process is to incorporate the ecological endpoints into the overall site strategy. The incorporation of the ecological endpoints may well involve negotiation and compromise to insure that there is consensus on the endpoints, and that they are realistic given other site objectives and constraints. At this stage of work plan development, there should be a preliminary definition of general remediation considerations. These objectives could well necessitate a reevaluation of the ecological endpoints. For example, if groundwater remediation is not to be considered, ecological endpoints associated with the alteration of groundwater flow or level may not be necessary. More appropriate endpoints associated with groundwater may be contaminant concentrations, with the objective of determining the need for remediation. Similarly, treatability studies could be identified as an objective, and additional ecological endpoints considered to evaluate the toxicity of the products and byproducts of the treatment techniques being considered, so that the reuse of the treated material can be evaluated. For example, if sediments or soils are to be treated and returned to a water body or wetland, their toxicity to indigenous organisms would be an appropriate topic for investigation.

Once the conceptual site model is formulated and the major objectives agreed to, the scope of the RI/FS must be developed. The scope is developed by a combination of the technical considerations of the site, cleanup objectives, and coordination with management and regulatory personnel. The scope first presents the overall approach to the investigation, relating the general investigations to the established objectives. The bulk of the work plan is a detailed description of the investigations that are to be conducted. The description includes the sampling and analysis methods, sample locations, and (where appropriate) the data evaluation and statistical analyses to be used. This is done for all areas of investigation, including water, soil, air, and biological media. Much of this information is included in a sampling and analysis plan, which also details the quality assurance plan and discusses the quality of the data required for each type of investigation.

The initial work plan should provide sufficient detail for the first tasks of the investigation so that someone not involved in the work plan development could start the investigation. Later tasks may be dependent on initial findings, and thus may not be described in detail. Tasks associated with the feasibility portion of the RI/FS often are not described in detail in the initial work plan, because they are highly dependent on the findings or the RI. For example, the initial work plan might contain a task to describe measures to recreate wetland habitats following soil remediation. It would be futile to provide specifics for this task until there was some indication of the need for the extent and methods of soil excavation and remediation proposed.

Consequently, development of the work plan is commonly an iterative task. The ecological portions of the plan should basically be laid out initially, with a detailed description of the first investigations. Possible subsequent steps should be identified, and the approach envisioned to determine the need for, and extent of, additional investigations should be discussed. The information needed to make these decisions is a combination of the ecological findings and the results of other investigations. There are usually periodic amendments to the work plan, identifying the need for additional investigations or providing the details of the investigations that were not included in the original work plan. To the degree possible, the first iteration of the plan should identify the fact that such amendments are anticipated, and generally describing when they will be issued, what they will contain, and the critical information needed to finalize the amendments. This approach is consistent with the general philosophy of a phased investigation, discussed later.

Ecological Assessment Field Investigations

The objectives of the ecological field investigations are to provide input to the measurement endpoints. The endpoints are structured to address either the assessment of existing contaminant effects, the establishment of ecological remediation criteria, or the development of mitigation for remediation impacts. Thus, the field studies should address these issues. However, the connection between the endpoints and these overall objectives is sometimes overlooked when the investigators are focused on the individual scientific studies. It is therefore a good practice to periodically step back and ask the question "how will the data from this study help determine effects, establish criteria, or design mitigation?" Although the field studies are discussed as an early element of the ecological assessment, at least some work on other elements, particularly defining chemicals and pathways of concern, must be considered before the design of the field studies is finalized.

The ecological field investigations are not conducted in a vacuum. Sampling of the various environmental media is probably the single largest effort in a typical RI investigation. The total program often includes evaluation of surface and ground water hydrology, meteorological conditions, soil and other geological characteristics, as well as the nature and extent of the contamination of air, water, sediment, and soil. The sampling program is elaborate in that it usually entails setting up field trailers, on-site decontamination facilities, mobile laboratories, extensive QA/QC efforts, personnel and equipment mobilization, and an extensive logistical effort to support the program. The ecological sampling effort must be well-synchronized with the total program, to take advantage of economies of scale and the availability of equipment and logistical support. Most importantly, the total effort must be coordinated to produce synoptic data, so that variations can be related to location or some other parameter of interest, rather than to changes with timing or other artifacts of the sampling program.

It is essential that chemical and biological data collection be closely coordinated in time and space. The evaluation of data collection from various media in a coordinated fashion is an important tool in defining contaminant pathways. Such data is also essential in any attempt to determine transfer rates, or to establish a quantitative relationship between environmental media and biological characteristics. The relationship between contaminants and biological effects is frequently difficult to firmly establish within the scope and duration of a typical RI/FS. In cases where the assessment concludes that site remediation is required for ecological resources, the relationship between contamination and effects is often the weak point in the argument, and thus subject to challenge. Without synoptic biological, chemical, and transfer-related (such as hydrologic conditions or soil characteristics) data to verify the relationship, it is very difficult to refute the challenge.

The design of ecological field studies is dependent on three factors. The first factor is the characteristics, particularly sensitivity, of the ecological resources on or adjacent to the site. If there are no existing or potential resources of concern, the ecological assessment generally consists of a simple documentation of this fact. At the other extreme, if there are potential endangered species on the site, the dependence of the individuals on the site resources and their sensitivity to contaminants must be documented, and sometimes quantified. The second factor governing the design of field studies is the nature and extent of contamination. If the site contamination is minimal or restricted, such as within a building or bedrock, ecological field studies beyond reconnaissance are generally inappropriate. The final factor to be considered in establishing the need for field investigations is the anticipated extent of remediation. If extensive earth moving, construction, or alteration of the hydrology is likely, investigations may be necessary to determine the effects of the activities on the surrounding ecological resources. Similarly, if the creation of habitats or other mitigation of resources is anticipated, the critical requirements of the resources must be documented. If, on the other hand, the remediation anticipated is limited to groundwater pumping, treatment, and on-site discharge, and there are no groundwater-sensitive resources on site, the ecological field studies would most likely be minimal. The phased investigation approach, discussed below, is often necessary for an understanding of these three factors sufficient to determine need and extent of the ecological investigations.

There are generally three basic purposes, and thus potentially three types of field studies comprising an ecological assessment. The first and most obvious is an investigation of the extent of effects from contaminant releases. The second type of investigation is dependent on the findings of the effects evaluation. This type of investigation relates the observed effects to the contaminant by evaluating contaminant pathways, the biological fate, and the correlation of contaminant concentration and biological effect. The third basic type of field investigation is the determination of the presence and/or condition of an ecological resource on-site for evaluation of ARARs, Natural Resource Trustee jurisdiction, or potential future mitigation. Most of the information needed in these three areas is the same, or can be obtained from the same ecological investigation by measurement of all of the appropriate parameters. It is a mistake to conduct the actual investigations independently for the three purpose, but the study design must recognize that there are sometimes subtly different data requirements for the three purposes.

Contaminant Effects

The investigation of contaminant effects generally follows classic applied ecological techniques. The most common basic approach is to evaluate the structure of some segment of the community in terms of species composition, density, and derivatives of these functions, such as diversity or percent cover. There is a wealth of field methods documented for these investigations, and some examples are examined in the discussion of phased investigations later in this chapter. These community characteristics measured on-site can then be compared to equivalent parameters reported in the literature or to measurements taken from a predesignated reference site as part of the field investigations. Characteristics of individuals, such as abnormalities, reduced size, or a change in egg production can also be used to indicate effects from contaminant release. Where significant differences from "typical" values or other anomalies are detected, the relationship between contaminants and the observed conditions could be a topic for more detailed specific investigations.

Cause of Effects

The second basic type of field investigation is initiated only if some contaminant effect is identified. This type of investigation involves evaluating transport mechanisms and contaminant concentrations in various media, to establish cause and effect or to document pathways. The general methods used to investigate transport mechanisms include: field measurements of physical processes or characteristics, such as ground water flow, surface water hydrology, or soil erodability, and contouring of contaminant concentrations. The results of the field investigations can then be used to construct predictive models. Transport mechanisms are also generally an issue for human health assessment and other aspects of the RI/FS. Consequently, the ecologist's role is to identify the answers needed for the assessment, and then coordinate with those conducting the site modeling or other transport evaluations. For example, the source and characteristics of ground water discharged to an apparently affected wetland might be a critical input to establishing the cause of the effect. The ecologist would then identify to the hydrogeologist the information needed, and then the ground water measurement and model output could be designed to provide the needed information.

Simultaneous measurement of the contaminant concentration in the biological resource demonstrating an effect and the water, soil, sediment, food, or other environmental media can also be used to evaluate cause and effect, as well as contaminant pathways. A correlation between contaminant concentration in the media and degree of biological effect is strong evidence for contaminant effect. The most common effect used in this type of analysis is tissue concentration in the organism of concern. However, other measurement endpoints, such as reproductive success, biomass, presence/absence, or community indices can often be more useful because they are more directly related to assessment endpoints. Contaminants are the most common and obvious parameter to use in establishing pathways or linking cause and effects, because they generally represent the largest data base and precision of measurement for site contaminants. Other parameters can be used to establish and sometime quantify pathways if there is a large historic data base, or if they are easily measured. For example, if field observations reveal a strong correlation between site occupancy of piscivorous birds and the rearing of juvenile fish, a potential pathway can be established without measuring the levels of contaminants in the fish and birds.

Simultaneous measurements are an optimum method of quantifying contaminant transfer to biota or biological concentration factors. The ratio of contaminants in the organism of concern is compared to the concentration in the media of concern, such as the soil, water, or food. This evaluation can identify the media most closely linked to the concentration in the biota, and thus identify the critical areas for remediation. The same data can be used to quantify the transfer from the media to the organism, and is thus used to evaluate remediation criteria. This approach is discussed in more detail in the later presentation on remediation criteria.

Resources On-Site

If the type of field investigations discussed above have been conducted, the final type of investigation (i.e., the presence and/or condition of an ecological resource on-site) should be largely completed. Additional investigations required might be directed at sensitive resources that are not affected by the contamination, but could be impacted by remediation activities. The most common example is the investigation of wetlands. Since these are protected and sensitive ecological resources, even if they are not currently effected by contamination, they should be understood to avoid destruction during other field studies. The wetland characteristics is also necessary input when considering remediation technologies and alternatives. If wetlands are potentially at risk, the wetland investigation should, at a minimum, be a detailed delineation and specification of the wetland type present. If destruction and subsequent creation or restoration of the wetland during remediation is anticipated, much more information on the structure and function of the wetland will be required. An investigation of resources on-site is also necessary to identify the location-specific ARARs that must be considered in the RI/FS.

The level of wetland investigation should be tied to the degree of threat to the resource. A general delineation from existing maps, photographs, and field reconnaissance is more than adequate initially. If sampling or other RI activities are anticipated near the wetland areas, field delineation may be required so that the wetland can be avoided where possible. If remediation is not required in the wetlands, field delineation may not be necessary. However, if the contamination in a wetland area is close to the remediation criterion, the exact location and functional value of the wetlands in the area may be important in deciding the need for and benefit of remediation.

Bioassay

In the broadest context, a bioassay is any measurement with a biological characteristic as the endpoint. Drug testing was one of the early large-scale uses of bioassays, where organisms were given drugs at various rates to determine the drug's effectiveness and optimum dose. In these cases, some adverse reaction, such as death or behavioral modification, was the endpoint used to define the upper limit of an acceptable dose. More recently, bioassay endpoints have been used in a broader context, such as species interactions, including predator-prey relationships (Cairns and Pratt 1987), or ecosystem-level parameters, like community structure and system productivity, used to evaluate ecosystem enrichment (Oviatt et al. 1987). Almost all ecological investigation at hazardous waste sites fall within this broad definition and range of bioassay. However, common convention has restricted the definition of bioassay at hazardous waste sites to two general areas: toxicity testing and tissue analysis. Toxicity testing is similar to a bioassay for drugs, where an organism is exposed to a contaminant (or mixture of contaminants) of concern, and the reaction of the organism is used to evaluate the effect of the contaminant. The most common use of tissue analysis is an evaluation of contaminant uptake, by measuring the concentration of contaminant in an organism collected on-site or following toxicity testing. Many biologists only use these two separate terms (toxicity testing and tissue analysis), and do not invoke the classical bioassay term at all. However, the two are considered together here under the collective term, because they are closely related and both generally employ the classical organism response to a chemical compound as the measurement endpoint.

Toxicity testing and tissue analyses alone do not establish hazardous waste site remediation goals or even ecological endpoints. They simply define the relationship of a contaminant and a sample population or, in some cases, a single individual. The information from these bioassays is critical input for remediation goals and assessment endpoints, but they must be used in combination with the more subjective preestablished overall objectives for the site. Similarly, these tests do not by themselves define the need for remediation. They must be compared with the agreed-upon site objectives. For example, if on-site sediments are shown to preclude normal hatching rates for trout eggs, yet trout breeding is not an assessment endpoint, due to an unsuitable habitat, then the bioassay results do not necessarily dictate the need for remediation. Similarly, if the field mice on-site have tissue concentrations of contaminants significantly higher than literature values or reference populations, unless these concentrations are shown to preclude an assessment endpoint, remediation may not be warranted.

Bioassay tests were the first, and generally still are the principal ecological component of hazardous waste site investigations. Their predominance arises from their extensive historic use in pollution control and human health. Toxicity testing was the mechanism used to establish water quality criteria (U.S. EPA 1986) and is currently proposed for sediment quality criteria (see Chapter 7). Therefore, when chemists, engineers, and toxicologist set about evaluating biology at hazardous waste sites, they employed the same mechanisms. The approach was, and continues to be, very effective in establishing cause and effect relationships and evaluating remediation criteria. It represents the simplest, most direct approach, and is thus always appropriate, until proven otherwise on a case by case basis. Similarly, tissue analyses of on-site organisms used for human consumption were a primary tool used for public health assessments. It was possible to use the same data for ecological evaluation, so the same methods were employed when nonhuman health ecological concerns were raised. However, in the current view of ecological assessments, which attempts to address an ecosystem approach, tissue analysis and toxicity testing must be considered as available and important tools that can supplement other types of investigations to address broad objectives. Bioassays can no longer be considered as stand-alone complete investigation techniques.

The use of bioassay at hazardous waste sites is an extensive topic and is treated as a separate chapter in this book (Chapter 8). That chapter deals in detail with the methods and use of data, but a conceptual discussion of the topic is appropriate in this chapter, to view bioassay in relation to field investigations and other elements of ecological assessments.

There are three general approaches to using bioassays as a part of ecological assessments at hazardous waste sites. The first is the measurement of on-site concentrations in various media, and a comparison to published ecological benchmarks. This approach does not necessarily involve actually conducting the bioassay, but rather makes use of data from previously conducted bioassays. The use of such benchmarks is most commonly used in aquatic systems, where measured surface water is compared to Water Quality Criteria (U.S. EPA 1986) or other toxicity results. On-site concentrations can also be measured directly for biological media and compared to tissue levels reported in the literature associated with various effects. The results can then be compared to endpoints, or used in ecological models to determine significance.

The second approach to bioassays is in situ measurement of toxicity. This approach subjects appropriate organisms to on-site media and measures the organism's response, in comparison to a similar exposure at a reference site. A typical example of this approach is caged aquatic animals up- and downstream of a leachate seep from a landfill. This approach is particularly useful in determining the presence of an effect, and relating the effect to a source. It is less useful in establishing ecological effects levels or remediation criteria, unless combined with other information or investigations. The final approach is to take on-site media, or simulated concentrations, into the laboratory and conduct conventional bioassays under controlled conditions. This approach is applicable to toxicity testing, and if extended tests are conducted, the approach can also be used to evaluate tissue concentrations from various exposure concentrations and durations. By controlling the contaminant concentration, ecological effects concentrations can be determined directly using this method.

Bioassay continues to be the most important single tool for ecological assessments of contaminants at almost all hazardous waste sites. Bioassay, in the form of in situ tests, laboratory investigations, or reliance on published bioassay results, is essential in effects identification and establishing remediation criteria. The topic is so important, and so well-established in hazardous waste site evaluation methodology, that it warrants its own chapter in this book. However, a common mistake is to use bioassay too early in the assessment process, alone, or as a substitute for assessment endpoints. It is nonproductive to conduct toxicity testing using on-site media or laboratory simulations of on-site concentrations, until there is some indication of potential effect. It is also inappropriate to directly apply bioassay results (either site-specific or from the literature) directly, without consideration of the resources and pathways potentially on-site.

Hazard Identification

The objectives of a hazard identification are simply to determine the conditions at the site that pose a hazard or risk to ecological resources, and to identify the resources that may be at risk. A preliminary identification of potential hazards to ecological resources is inherent in the preliminary site description. This screening of hazards is used to conceptualize the field studies and bioassays. This initial screening is reiterated as a more detailed element of the ecological assessment. The hazard identification as part of the assessment becomes more specific as ecological and other site information becomes available. This identification is a milestone or check point, which follows data collection in the ecological assessment process. Based on ecological and chemical data for the site, the contaminants that pose a potential hazard, and the critical resources possibly at risk, are identified as part of this element of the ecological assessment. This identification takes into account the nature and extent of contamination, the potential transport of the compounds, and the susceptibility of the resources present. Hazard identification is also an opportunity to revisit the ecological endpoints. It may be that some endpoints have been defined too broadly, and measurement is impracticable. In such cases, intermediate or surrogate assessment endpoints may be warranted. In other instances, it may be apparent at this stage that certain preestablished assessment endpoints are currently achieved and not at future risk. In these cases, no additional effort, such as an exposure assessments, is required to evaluate the endpoints. Resources can be at risk from remediation activities and site restoration, in addition to form contaminants, and these potential ecological impacts can not be ignored when identifying hazards.

There are two possible outcomes from any attempt at hazard identification. One is an outline or plan of action to fully evaluate the nature and extent of the hazard. The other possible product from the identification is the recognition that more data is required to determine the presence of a potential hazard. In the latter case, the needed information is collected, and eventually the nature and extent of the hazard is evaluated to the extent necessary to meet the objectives of the ecological assessment.

The aspects of the initial hazard identification are the identification of contaminants of concern, and resources potentially at risk. These two aspects are not totally independent, as an organism's sensitivity to the contaminants helps to define the compounds or concern, and the type of compounds present determine which resources are at risk. The resources included in the hazard identification usually fall out of the ecological field studies. The organisms or habitats affected by past contaminant releases are almost always included in the hazard list. The resources addressed by the assessment endpoints must also be closely considered and consciously dismissed or included as resources at risk. The contaminant transport and fate mechanisms, which are actually more relevant to the exposure assessment, as discussed later, are aspects of the site that must be considered in establishing which ecological resources may be at risk. Although the complete ecological exposure assessment should not be conducted prior to the hazard identification, there is generally sufficient consideration of the means of transport early in the process or other aspects of the RI/FS to provide input to the hazard identification. Through this process, off-site resources could be identified at risk, and thus become critical input to the ecological assessment. Other ecological resources could be included in the hazard identification, based on their sensitivity to classes of contaminants, high visibility, public concern, or general sensitivity to human activity or disturbance.

The hazard identification must also specify the contaminants that are a concern on-site, and thus must be considered in the exposure assessment. An initial comparison should be made for each environmental media, between the highest concentrations measured on site and relevant ecological benchmarks, such as water quality criteria or literature values in terrestrial systems. The investigations done to determine appropriate benchmarks are an initial stage in the toxicity assessment, and thus these two elements of the ecological assessment must occur somewhat in parallel. Another avenue for identifying contaminants of concern is by considering the sensitivity of the critical resources on-site to various compounds, or even type of compounds. Even if concentrations are below the benchmarks, if an endangered species potentially on-site is particularly sensitive to a contaminant, it would be appropriate to address such a compound in the hazard identification. A quick assessment of contaminants in this category can be made from the U.S. Wildlife Service Hazard Synoptic Reviews, discussed later in the section on Toxicity Assessment.

The entire contaminant identification process can be cut short by tiering off the similar tasks done for human health assessment. An early step in the human health assessment is to review virtually every compound detected on-site or historically associated with the relevant operations, and consider the potential human health threat (see Chapter 6 for a full discussion). Since most of the human health toxicity data is based on laboratory animal tests, there is generally a strong correlation between compounds producing a human threat and a concern for animals in the wild. Thus, the contaminants of concern for human health generally represent a short list of on-site contaminants for risks to animals. The compounds rejected for human health consideration, and the reasons given, should also be considered for ecological assessments, because some would have been rejected due to the lack of a pathway, but this might not be true for ecological receptors. Contaminants that might be of concern for plants and microorganism may not be considered in the human health assessment, and thus may have to be considered separately for the ecological hazard identification. The conventional wisdom is that multicelled animals are the most sensitive, and thus by considering animals, plants and microorganism are protected. This wisdom is largely based on the limited data on organisms other than animals, and so may be selfserving. Bodex et al. (1988) and the references cited therein represent a toxicity data base beyond the commonly considered mammals and aquatic invertebrates.

Exposure Assessment

An exposure assessment is the evaluation of contact between a receptor and a contaminant. In the case of an ecological assessment at a hazardous waste site, this involves determining the interface between ecological resources and contaminants of concern identified in the hazard assessment. Stated simply, the approach to ecological exposure assessments is to describe and, where warranted and practicable, quantify the contact or interface between the contaminant and an ecological receptor. This involves a detective-like investigation of tracking the contaminant from the source to the target. The objective of the exposure assessment is to determine how much of a contaminant an organism is exposed to, so that the potential for harm can be assessed in subsequent elements of the ecological assessment. The investigation uses a combination of evidence, in the form of: sampling data, studies reported in the literature, model predictions, and assumptions. The assessment thus involves a number of steps to evaluate the transport and fate of the contaminant from the source to a point of exposure potentially harmful to a predesignated ecological resource of concern.

The assessment of exposure is a technical investigation that should be conducted on a comprehensive basis, incorporating all relevant site data. It is not a subsection

of an ecological assessment directed only toward ecological ends or using only ecological data. It should include information from all potentially contaminated media
and consideration of all possible transport and transformation processes. All of the
available information and concerns should be identified and incorporated early in the
process, to insure the most complete exposure assessment.

The habitat, behavior, and other requirements of an ecological resource of concern
must be considered in structuring the exposure assessment. For example, groundwater
or soil at depth might not be a directly required resource for a reptilian species of
concern, so the initial reaction might be to ignore such exposure pathways. However,
winter behavior of the animal might include burrowing to depth, and thus exposure
to deep soils. Although the animal would not burrow into soils below the groundwater
table, during wetter periods of the year the burrow depth could have been within the
groundwater saturated zone. If the groundwater were contaminated, residue could
remain in the soil when the level receded and thus there could be a reptilian exposure
pathway from groundwater.

The exposure assessment is frequently initiated by a screening process. The screening is used to identify the potential pathways from contaminant to resource. It is based
on a subjective consideration of the type of contaminant and characteristics of the
resource. There is usually a large element of experience and professional judgment
inherent in the initial identification of exposure pathways to consider in detail. It is
not unusual for additional pathways or variations of pathways first described to become
apparent as the exposure assessment proceeds, and thus expansion of the analysis
would be required. As the exposure assessment proceeds, the most critical pathways
become apparent, and it is often possible to identify with confidence those that will
be critical to defining remediation criteria. When the most significant pathways can
be identified, it is often productive and efficient to pursue the evaluation of selected
pathways in detail and ignore the others. This approach serves to conserve and focus
resources available for the ecological assessment.

The evaluation of exposure, both qualitatively for a screening-level analysis, and
quantitatively for more detailed analyses, begins with the consideration of a potential
contaminant release from a source. The primary avenues of release critical to ecological
resources are the same as those described for human health exposure (U.S. EPA 1988b)
and include:

Volatilization—This mechanism of release is primarily applicable to soil and water.
   It can be of ecological concern for animals in close proximity to the soil, particularly
   in confined areas such as burrows.
Overland Flow—If surface water is contaminated, the release of the contaminants to
   other media can occur as surface water migrates. This is an obvious form of release
   to downstream portions of a river. Overland flow can also be a release mechanism
   during high-flow conditions, when a stream overtops the natural banks and contaminates adjacent wetlands or flood plains. Surface runoff through a contaminated
   area is another example of release via overland flow. This mechanism can contaminate surface waters or, due to erosion associated with the runoff, adjacent soils or
   downgradient sediments.
Direct Discharge—A release of contaminants can occur from the direct and often
   intentional discharge of contaminants to various environmental media. This occurs
   illegally, and also in the form of permitted discharges during events when treatment
   is not appropriate for the quantity or quality of waste.

Leachate—The generation of leachate is a common contaminant release mechanism from landfills. Groundwater is generally the media most directly affected by leachate, and it is often critical for human health analysis. Groundwater releases are usually less significant for ecological resources, because exposure only occurs after the groundwater enters a surface water body. Usually, dilution is very great by the time the leachate enters a river, large lake, or estuary. If the ecological resource of concern is an isolated wetland, vernal pool, headwater stream, or small pond with no outlet, a release via leachate and ground water could be of significant ecological concern.

Fugitive Dust—Dust from areas with contaminated soils can transport contaminants of concern to other media. Surface water, adjacent soils, and biological media, particularly leafy plants, are media of concern. Dust as a release mechanism is usually of greatest concern during excavation or other remediation operations.

Combustion—Combustion, either as a part of remediation or resulting from uncontrolled fire at the site, poses release threats similar to fugitive dust.

The Superfund Exposure Assessment Manual (U.S. EPA 1988b), and the references contained therein, presents detailed methods for evaluating and quantifying the release (via these various methods) to different environmental media. The evaluation of releases should not be an independent analysis as part of the ecological assessment. Release evaluations must be a coordinated effort among all disciplines and across all data bases. Extensive on-site data are generally required if sophisticated modeling of the contaminant transport or fate is necessary. For example, the relative groundwater elevations and soil conditions must be measured, to understand the rate and direction of groundwater transport. It is also generally necessary to monitor meteorological conditions, if transport via volatilization or fugitive dust is a concern. Evaluation of surface water transport requires measurement of flow, drainage area, and often watershed response to storm events.

Once the avenues of contaminant release have been considered, and those critical to ecological resources quantified, the release must be linked to the resource. If the release is close to the resource in time and space, the process is generally simple and direct, because the exposure is in a concentration and form similar to that of the release. However, if the contaminant must migrate some distance, or there is a lag between the times of release and exposure, the characteristics of the potential toxic compounds can be altered. There can be physical or chemical transformation, and there is usually an alteration in concentration due to these factors, and to dilution with uncontaminated material. If the form or concentration is altered during transport, the characteristics of the contaminant at the time and point of exposure must be evaluated. The evaluation should include direct measurement, where possible. However, it is generally desirable to also construct some form of model to explain the fate and transport, so that environmental conditions in addition to those present during measurement can be evaluated. For example, as the concentration at the source changes, the model can be used to predict what the concentration at the point of exposure will be. Also, if the transport mechanisms are understood and modelled, the effects of some possible future event, such as a flood, can be estimated. It is also frequently necessary to use the model to analyse different release scenarios, to evaluate alternative remediation criteria.

The most thorough approach for evaluating types and concentrations at the point of exposure is to construct models and then to verify them with field measurements. Any prediction method attempts to estimate the mean or typical value, and verification of the prediction with field measurements is a well-established and very useful method for abiotic media such as groundwater. However, for biota there is tremendous variability in the population concentration or any other biological effect used to verify the model. The variation is typically of one to two orders of magnitude. Thus, it is not surprising that the measurements of concentration in biological media may differ considerably from the predictions.

A number of mathematical models are available to simulate the transport and fate of contaminants. The models are generally media specific, and include various combinations of physical transport, chemical alteration, and concentration prediction. Models for terrestrial systems applicable to exposure assessments at hazardous waste sites are limited, generally only address specific components, and are largely unverified (U.S. EPA 1988b). Consequently, modifications of models developed for other purposes, such as the TEEAM model used for pesticide application (U.S. EPA 1989b), or other methods such as those discussed in Chapter 6 of this book, are generally necessary. Models useful in exposure assessments for aquatic systems are much more advanced, due to the relatively simple transport mechanisms and the greater attention to water pollution, standards, and treatment early in the environmental movement. A narrative summary of the methods available for evaluating the transport and fate of contaminants in aquatic systems is given by Waddell (1989). Ambrose and Barnwell (1989) also describe computer models available for the analysis of the environmental fate of contaminants. The more common models cited by Waddell (1989) include:

TOXI4—This model, as described by Ambrose (1987), incorporates alterations of contaminants by biological and chemical processes, as well as transfer between phases, to produce an output of time-variable sediment and water column concentrations.

EXAMSII—Contaminant loadings from various sources, including groundwater, atmospheric deposition, point sources, and runoff, serve as input for this model, described by Burns and Cline (1985). The outputs are predicted concentrations for chronic and acute exposures, directly usable with ecological models or simulations.

SARAH2—This model works backwards compared to the others, to estimate the maximum allowable discharge of contaminants based on a preestablished surface water concentration protective of ecological resources (Vandergrift and Ambrose 1988).

Once models and other techniques have been used to estimate contaminant concentrations in the environmental media, the degree of contact or uptake by the organisms from the media must be assessed. This uptake by ecological receptors can occur by way of three pathways: inhalation, dermal contact, and ingestion. In terrestrial systems, the inhalation pathway is probably the most critical when contaminants volatilize from contaminated soils, and animals live in close contact with the soil. In such cases, contact or uptake must be estimated from literature values for animal respiration or other similar methods, as discussed in Chapter 6.

Dermal contact with contaminants, although generally difficult to assess, can be a critical pathway for terrestrial animals. Reptiles and amphibians can live in close contact with the soil, at least seasonally, and although direct contact with soils for most mammals and birds is generally limited by protective hair or feathers, hairless young mammals or female birds incubating eggs, however, have exposed areas. The important variables, such as duration and extent of direct contact with soils, is rarely documented in the literature because, outside of assessments at hazardous waste sites, there has been little interest in such topics. As discussed in Chapter 6, there are usually major assumptions that must be made to estimate the degree of contact and transfer of contaminants when considering dermal exposure in terrestrial animals.

Ingestion is probably the most commonly assessed and often the most important exposure pathway for animals. It is comparatively simple to estimate, because there is often relatively good documentation of ingestive rates, food sources are well researched, and the contaminant concentration of the food can be measured in the field. In comparison to uptake from dermal contact or inhalation, there has even been substantial research on the transfer of contaminants through the food web. This pathway for terrestrial animals is covered in detail in Chapter 6.

Contaminant exposure to terrestrial plants is generally restricted to passive uptake through the roots. However, exposure through areal deposition on leaves is possible and can be a concern. The degree of uptake through the soil and the resulting impacts have been considered in terms of soil concentration (Bodex et al. 1988). However, the site-specific soil type, groundwater regime, and plant species present can be important variables, and when plant uptake is identified as an issue, site-specific measurements are often necessary.

Exposure assessments in aquatic systems generally require a more holistic approach to exposure assessment then the evaluation of individual pathways described earlier for terrestrial systems. In aquatic ecosystems, the organisms are constantly exposed to a relatively uniform environment. The effects of contact, ingestion, and respiration are all a function of the contaminant concentration in the same environmental medium. There have been models developed, such as those summarized following, to integrate exposure in aquatic systems.

FGETS—The Food and Gill Exchange Model of Toxic Substances (Barber et al. 1988) calculates concentrations of selected organic chemicals in fish, considering both that from exposure to water and food intake. The model incorporates data bases for gill morphology and food exchange, which are used in combination with the properties of the contaminant to estimate toxicity and tissue concentration.

WASP—The Water Quality Analysis Simulation Program described by Connolly and Thomann (1985) uses contaminant concentrations input directly, as determined from measurements, or the output of models such as TOXI4, to predict the tissue concentrations in aquatic plants and animals. The model considers both uptake and loss functions including: the transfer across gills, surface sorption, ingestion, desorption, metabolism, growth, and excretion.

Thomann (1989) presents a simplified aquatic exposure model that highlights the important transfer and uptake variables. The model also illustrates the concepts that are calculated with precision in the more complicated computer-based models. A conceptual description of the model is presented below, to demonstrate the approach used to estimate aquatic exposure.

There are two primary mechanisms for the uptake of chemicals in the aquatic environment. These two mechanisms are reflected in the two terms bioconcentration factor (BCF) and bioaccumulation factor (BAF). The BCF reflects the organism's tissue concentration of the contaminant, resulting from the water concentration alone. The BAF includes tissue build-up from the water concentration, plus contaminants from feeding.

The BCF is simply defined by:

$$BCF = v/c$$

where v is the organism concentration and c is the water concentration. Most organic chemicals partition more readily to the lipid pool and the expression of BCF could be strongly dependent on the lipid content of the organism.

Hence, the BCF for organic chemicals is most commonly expressed on a lipid base. If the organism is in equilibrium with the water column, the lipid-based BCF is the octanol - water partitioning coefficient (K_{ow}). The K_{ow}s are compound-specific, and are empirically determined from the different solubilities of the compound in water and lipids. Thus, where uptake from the water is the only significant exposure route, the literature value for K_{ow} can be used to estimate the lipid-based organism concentration of the organism of concern. This concentration can, in turn, be converted to average whole body levels from literature values of typical lipid content. These figures can then be used to determine the consumption of contaminant by predators or organism effects expressed as a function of body concentration.

In most aquatic systems, exposure to the water column is not the only route of concern. Feeding is generally a potentially significant contaminant uptake mechanism, and thus the BAF (which includes concentration due to equilibrium with the water plus tissue build-up from contaminants in the food) is more applicable than the BCF. The tissue build-up from consumption can be expressed in a simplified manner as:

$$Pc = k*wc + aCRc - KPc$$

where: Pc = concentration in the organism of concern
   k = uptake rate from the water (L/day)
   wc = water concentration (g/L)
   a = chemical assimilation efficiency (g of chemical absorbed/g of chemical ingested)
   C = is the specific consumption rate of prey (g/day)
   K = is the chemical excretion plus metabolism rate (g/day)

There are many elaborations to make this model more accurate, and the variables can be standardized to a unit mass basis to make it more universally applicable. Variables can also be added to account for growth, which would dilute the tissue concentration. Thomann (1989) presents elaborations of the model, plus methods for estimating the chemical assimilation efficiency, metabolism, excretion, and other variables.

There are general rules of thumb as to when the BCF will suffice in estimating tissue concentrations, and when the more elaborate BAF is needed. Based on an evaluation of extensive research and field measurements, Thomann (1989) has concluded that if the log K_{ow} is below approximately 5, tissue concentrations are not increased due to feeding. This is because, at low octanol-water partitioning coefficients, any tissue concentrations above equilibrium due to feeding rapidly dissipate due to solution into the water, decreased uptake, and excretion. Thus below an octanol-water coefficient of about 5, the BAF approximately equals BCF, which approximately equals K_{ow}.

At a K_{ow} of approximately 5 to 6.5, food chain accumulation is significant and should be added to accumulation from partitioning with the water column, because dissolution to water is slow and can not keep up with food uptake (Thomann 1989). At values above 6.5, tissue accumulation from feeding generally accounts for all significant uptake, because dissolution is so slow it becomes insignificant in comparison to food uptake. At the higher coefficients, tissue concentration can be estimated by a method such as the formula given earlier, expanded to account for additional variables.

The exposure assessment is really the substance and dynamic portion of a contaminant's effects evaluation at a hazardous waste site. The assessment makes use of on-site physical and chemical measurements, combined with analytical tools such as mathematical models, to estimate the contaminant concentration in environmental media. The ecologist can then employ knowledge of the biological resources from professional experience, literature, and on-site investigations to determine the contact or interface of the organisms with the contaminated media. This information then forms the input for the toxicity assessment, which is a relatively simple, but critical, final element in the evaluation of contaminant effects.

Toxicity Assessment

If the exposure assessment is the test of existing or potential ecological effects resulting from the release of hazardous materials, then the toxicity assessment forms the template for grading the test. The objective of the toxicity assessment is to establish the quantitative relationship between ecological effects and the concentration, dose, or exposure of a contaminant of concern. In other words, the toxicity assessment determines the effects levels for the contaminants present on-site that do or might cause ecological risk or damage. Generally, the exercise results is a gradient of contaminant concentrations that produce a range of conditions from no effects, measurable impairment of a biological function, to mass mortality.

The toxicity assessment must be somewhat interactive with the exposure assessment, so that the form of the data from the exposure evaluation is comparable with the measure of effects resulting from the toxicity assessment. The evaluation must also be directed at the type and sensitivity of resources on-site. The effects levels considered should bracket the preestablished ecological assessment endpoints for the site. In preparing the toxicity assessment, the range of concentrations found on-site, or the levels and pathways predicted from the exposure assessment, can not be ignored. All of these factors must be considered, so that the toxicity assessment can be used directly as input for site-specific remediation decisions. As the toxicity assessment proceeds, there should be periodic checks with the exposure assessment and contaminant data from the site, to determine if a pathway or contaminant is not an issue, or if more precise toxicity information is necessary.

Where possible, the type of effects collected for the toxicity assessment should correspond to the critical exposure pathways at the site. For example, if ingestion is a primary pathway of concern, the biological effects from various doses via feeding should be documented in the toxicity assessment. Similar expressions of effects would be required for inhalation and dermal exposure pathways, if there are an issue at the site. The types of pathways dictate the form of the toxicity assessment. If the pathway is through total immersion, as in aquatic ecosystems, the assessment may document the relationship between water concentration and effects level. If the pathway is through ingestion or inhalation, the assessment may focus on dose per unit time. In the case of dermal contact, the assessment could result in an integrated function of soil concentration and duration of exposure.

Unfortunately, the available data often dictates how the effects level can be stated. If there is only information expressed on a food concentration basis, assumptions as to consumption rates and organism size may be necessary. Similarly, there may be no data on the biological species or function (e.g., fecundity or larval survival) of concern, and some interpolation, assumption, or surrogate measurement endpoint must be employed.

There are two different, but not exclusive, avenues for developing the relationship between contaminants and effects. The first avenue is published or otherwise available data on toxic effects related to the ecological endpoints. This approach should always be attempted first. If the available data, in comparison to on-site concentrations or exposure assessment predictions, indicates (with a sufficient margin of safety) that there are no potential effects, no additional toxicity assessment is warranted. Similarly, if there is existing toxicity data available for situations, compounds, endpoints, and species directly applicable to the specific site, further investigation is generally not necessary. These available levels then become the result of the toxicity assessment, and can be used as direct input for the risk characterization and remediation decisions.

The second avenue for establishing the contaminant-effect relationship is original toxicity testing. The decision to conduct site-specific toxicity testing should be preceded by a thorough review of existing data, not only to establish the need for additional testing, but also to precisely define the scope of the testing. Original testing is often needed if there is no directly comparable data available in the literature. Testing might also be required if, due to an uncertainty of prediction, the range of predicted exposure levels overlaps the range of concentrations reported to produce effects related to established endpoints. In such a case, site-specific original testing may be required to more precisely quantify the relationship between a contaminant and its effect. Sometimes, a quick assessment of relative toxicity is needed before data on the contaminant concentration of various on-site media is available. In such cases, screening-level site-specific toxicity testing using the potentially contaminated media may be appropriate. Site-specific complicating factors is another common justification for original testing as part of the toxicity assessment. Frequently, the combination of contaminants found at a hazardous waste site make the application of toxicity data for individual compounds inappropriate, and tests using the mixture of compounds present on-site are necessary for the toxicity assessment. Also, if community or ecosystem-level endpoints have been established, site-specific testing is generally required because there is little if any chance of finding toxicity information specific to the contaminants and community present on-site. Toxicity testing for individual hazardous waste sites is discussed in Chapter 8 of this book, and specific tests, approaches, and methods will not be addressed further here.

There are several primary sources of existing toxicity information for developing the relationship between contaminants and effects. EPA Quality Criteria for Water (U.S. EPA 1986) is an important reference, but it is specific to aquatic ecosystems. This document summarizes the available information on the toxicity of priority pollutants in the aquatic environment. Where sufficient information exists protection criteria are established for acute, chronic, and human health for both fresh water and marine systems. Generally, there is enough information presented so that modifications to the presented criteria can sometimes be manipulated to represent spite specific conditions, such as species or endpoints of concern. More compounds are being added to the Criteria as updates to the original document and some are revised as more toxilogical information becomes available. The treatment of each compound also provides a extensive reference list and thus is a good place to start a comprehensive search for toxicity information for individual compounds. Water quality criteria can also represent a place to start even for terrestrial receptors if no other references are immediately available.

TABLE 4-3.   Current U.S. Fish and Wildlife Publications in the
Hazards to Fish, Wildlife, and Invertebrates Series: A
Synoptic Review.

Compound or Topic Reviewed	U.S. Fish and Wildlife Service Publication No.
Mirex	85 (1.1)
Cadmium	85 (1.2)
Carbofuran	85 (1.3)
Toxaphene	85 (1.4)
Selenium	85 (1.5)
Chromium	85 (1.6)
Polychlorinated Biphenyl	85 (1.7)
Dioxin	85 (1.8)
Diazinon	85 (1.9)
Mercury	85 (1.10)
Polycyclic Aromatic Hydrocarbons	85 (1.11)
Arsenic	85 (1.12)
Chlorpyrifos	85 (1.13)
Lead	85 (1.14)
Tin	85 (1.15)
Index to Species	85 (1.16)
Pentachlorophenol	85 (1.17)
Atrazine	85 (1.18)
Molybdenum	85 (1.19)
Boron	85 (1.20)
Chlordane	85 (1.21)
Paraquat	85 (1.22)

Source for Publications: Librarian, Patuxent Wildlife Research Center, US. Fish and Wildlife Service, Laurel, MD 20708; or U.S. Fish and Wildlife Service Publication Unit, 1849 C Street, N.W., Mail Stop 130-ARLSQ, Washington, DC 20240

Beyer (1990) has assembled toxicity information applicable where soil is a potential contaminant pathway. The summary presents data from 29 sources and addresses 200 contaminants. Although much of the data do not relate to specific ecological effects, it is one of the few references currently available on soil contamination levels usable in an ecological assessment.

Another major source for ecological toxicity assessment information are the "Harzards" series publications developed by the U.S. Fish and Wildlife Service. The series consists of publications for individual contaminants and addresses the chemistry, typical background concentration, and toxicity. All publications have consistent titles ("..... Hazards to Fish, Wildlife, and Invertebrates: A Synoptic Review") preceded by the specific compound. The series is currently in development, with individual publications released as they are completed. Status and the availability of documents can be obtained from the Librarian, Patuxent Wildlife Research Center, U.S. Fish and Wildlife Service, Laurel, Maryland 20708. The compounds currently in the series are listed in Table 4-3.

A very abbreviated summary of the 71-page polychlorinated biphenyl review (Eisler 1986) follows as an example of the contents of these publications, and also to present toxicity assessment information for a common compound of concern, with potentially serious ecological implications. The publication, although somewhat dated now, gives extensive tables of background and toxicity data, which are herein only cited as ranges.

Polychlorinated biphenyl compounds (PCBs) have been manufactured in the United States and at least half a dozen other countries since 1930. Their most widespread use has been as dielectric agents in electrical equipment, but they have also been used in lubricants and flame retardants. Due to their extreme toxicity and environmental persistence, the manufacture and use of PCBs in the United States was essentially banned in 1979. However, an estimated 630 million kg were manufactured domestically before their production was halted.PCBs can be extremely slow to break down, with the rate of chemical and biologically mediated decay dependent on the degree and position of the chlorine on the benzene ring of the molecule. Their environmental persistence is paralleled by their insolubility in water. However, they do demonstrate a high degree of solubility in nonpolar organic solvents and biological lipids. Thus, they tend to remain in soils and sediments, and are highly susceptible to accumulation in fatty biological tissues. PCBS are easily taken up through the gut, skin, and respiratory system, and concentrate in virtually every tissue, with the highest levels generally in adipose tissue and skin. The less chlorinated forms of PCB that are readily metabolized can be excreted in urine and bile, but other isomers can accumulate almost indefinitely.

PCBs are ubiquitous in the environment and are extensively reported in algae, macrophytes, invertebrates, fish, reptiles, birds, and mammals. For example, all of the chinook salmon and half of the coho salmon collected by Rhorer et al. (1982) in 1980 from the Great Lakes exceeded the U. S. Food and Drug Administration's action level of 2.0 mg/kg fresh weight in fillets. Measured concentrations found in biological tissues in areas not necessarily associated with hazardous waste sites range over four to five orders of magnitude centered on the Food and Drug Administration's action level. There seems to be general agreement that concentrations of PCBs are declining in at least some areas of high contamination, and lower chlorinated isomers may be disappearing.

There is a relatively large data base on the toxicity of PCBs, covering a variety of organisms and end points. There are measurements of acute toxicity based on concentrations in environmental media and intake. A range of chronic endpoints, including growth and all aspects of reproduction, have been evaluated for both aquatic and terrestrial species, with the majority of the data focused on animals. Depuration is generally very slow to nonexistent, and effects tend to increase with duration of exposure, even up to 120 days. There is a high variability in the level of sensitivity to PCBs, with even closely related species demonstrating a one- to two-order of magnitude range in response.

Toxicity differs significantly among isomers, but because of the large number of isomers, the differing solubility, and the natural transformation among isomers, it has been difficult to establish ecological benchmarks for individual isomers with a high degree of certainty. Consequently, the trend has been to set criteria for total PCBs. The toxicity of PCBs can be expressed by concentration in environmental media, biological tissue levels, or as a dose rate. In freshwater and marine systems, protection criteria are most often expressed as environmental concentrations. The total PCB Quality Criteria for Water (U.S. EPA 1986) is 0.014 and 0.03 μ/l for fresh water and salt water, respectively. There are also protective criteria for dietary concentrations, which range from 0.3 to 0.5 mg/kg fresh weight.

There has been a reluctance to propose criteria for terrestrial environmental media such as soil. This is presumably due to the almost infinite complexities in transferring PCBs and other contaminants from soil to organisms. Consequently, the evaluation of transfer is relegated to site-specific assessment, and then the results can be compared to criteria established for diets or tissue concentrations. Birds generally seem to be more tolerant than mammals, and the dietary criteria for birds have been proposed at 3.0 mg/kg fresh weight. Mink seem to be the most sensitive mammalian species tested, and a dietary criterion of 0.64 mg/kg fresh weight has been proposed.

The ecological toxicity assessment for a hazardous waste site can range in complexity from looking up a number in a table to iterative bioassay tests gauging reproductive impairment. The degree of complexity is dependent on: the number and complexity of ecological endpoints; the contaminants of concern; the required degree of precision and accuracy; and the availability of existing data. One extreme would be at a site with a limited number of contaminants of concern, and water quality criteria for all of the subject compounds. At the site used in this example of a simple situation, all of the ecological endpoints would be extremely protective, direct, and aquatic based. The on-site aquatic habitat would also be reflective of the species used to formulate the criteria. In this situation, the toxicity assessment would simply be to cite the appropriate water quality criteria. A slightly more complicated, but still elementary, assessment might be a similar situation, but the local conditions would not support the species used to set the criteria, or the level of protection afforded by the criteria was in excess of the relevant endpoint. If appropriate toxicity data was available, the criteria could then be adjusted to reflect the species likely to be on-site and the designated endpoint.

With multiple contaminants, endpoints, and pathways, the toxicity assessment is necessarily more complicated. Even in an aquatic situation, if consumption of highly contaminated food is a significant pathway, water quality criteria alone are not always sufficient. Also, the possibility of additive and synergistic, or antagonistic, effects of multiple contaminants can be important. Although without detailed site-specific toxicity testing it is hard to treat these effects quantitatively, some consideration of such effects should be addressed in the toxicity assessment.

When there are important terrestrial endpoints, the situation is almost always complicated. The available toxicity data is generally not directly applicable. The data is frequently developed in laboratories for an evaluation of human exposure, and thus based on laboratory animals that might or might not be applicable to the species of concern on the site. Also, the mode of exposure and/or uptake used in the laboratory tests in distinctly different from natural situations. For example, dermal contact is typically evaluated by applying a uniform concentration of the contaminant directly to the same exposed area of an animal at the same time every day. This is obviously quite different from what would be expected on-site. Similarly, toxicity data for ingestion pathways is generally based on a diet of a uniform contaminant concentration. To be useful for a toxicity assessment, such data must frequently be converted to total dose per kg body mass. Such conversions necessitate assumptions and averaging over time. This approach can oversimplify the toxicity response, and introduce other potential errors. Also, the natural situation is not to consume a single food source with the same contaminant concentration, so additional sources of uncertainty are introduced when interpreting the laboratory data to the field situation. For example, it is often unclear if a diet with an average contaminant concentration similar to a constant laboratory diet, but with a high coefficient of variation, will have similar effects.

In summary, a toxicity assessment is a critical element in the environmental assessment for any hazardous waste site. The assessment could, in a theoretical sense, be conducted independent of all site-specific information save identification of the contaminants and ecological resources of concern. The product is a template or yardstick against which to gauge the implications of conditions observed or predicted onsite. The toxicity assessment alone does not lead to any decisions or understanding of the ecology of the site or the nature and extent of contamination. However, when combined with the site-specific information, the critical ecological decisions can be made with a good understanding of the implications. The assessment can produce widely accepted environmental media concentrations, demonstrated from existing data to be comparable with and protective of the ecological assessment endpoints. Alternatively, the template can be derived from concentrations in media that have been back-calculated from manipulations of data from a variety of species and exposure scenarios. The third possible method for developing the relationship between contaminants and effects is original toxicity testing, employing site-specific media, species, or other conditions. The most successful toxicity assessments tend to employ a mixture of these three methods of evaluating ecological toxicity.

Risk and Ecological Effects Levels

At least partially due to the public and regulatory attention to hazardous waste over the last ten years, risk assessment has evolved into a distinct science, or at least technology. The emphasis was originally on human health risk assessment, but the technology and approach are currently being applied to other areas. The assessment of ecological conditions and potential remediation at hazardous waste sites is one of the areas where risk assessment methods, conceptually similar to those used for human health, has been applied. The application of the approach to ecological issues is a separate topic and is discussed as such later under a separate heading. However, some form of risk characterization and input to remediation is an element of the ecological assessment at any hazardous waste site and is discussed in this section.

The foundation for an ecological risk assessment at a hazardous waste site is an expression of damage or the threat of damage to an ecological resource associated with the site. Although the damage could result from a variety of causes, this discussion of risk assessment is limited to the effects from contaminant release and exposure. The approach can be applied to other sources of ecological effects, such as damage from remediation activities, but contamination is perhaps the most straightforward source of risk, and thus represents an appropriate example. However, as ecological risk assessment methodology matures, it can be used not only for site-specific events or single sources of stress, but also to integrate risk from multiple sources, and even to prioritize ecological risk at the regional, or even global scale. In addition, when the methodology has been well developed, once the existing damage or threat of damage has been expressed, and the sources of damages discussed and possibly quantified, the future risk resulting from altering the sources can be evaluated.

The data base on the toxicity of contaminants of concern and a description of ecological resources associated with the site determine the precision, accuracy, and detail achievable for the expression of risk. Unfortunately, such data bases are generally limited and not always specific to situations of concern, so the ecological risk assessments is frequently not fully quantitative, and less than specific to individual endpoints. Even with these limitations, the ecological assessment must arrive at an evaluation of baseline risk at the site. Ecological risks are most easily expressed as some impairment of a biological function or condition, and there are two basic approaches to expressing baseline ecological risk. The evaluation can be a statement of the contaminant-related effects observed on-site and an interpretation of the associated ecological implications, in relation to appropriate endpoints. For example, elevated contaminant concentrations of on-site vegetation might be the observed effect, and the implications might be effects to herbivores feeding on-site. This approach of actually identifying and measuring an adverse effect differs significantly from the classic human health risk assessment, where the baseline risk is expressed as a probability of some undesirable effect in a hypothetical population.

The baseline risk can also be a comparison of on-site measured concentrations, to established benchmarks for the environmental media of concern, as identified during the toxicity assessment. The effects related to the benchmarks closest to on-site concentrations comprise the statement of baseline risk. For example, on-site concentrations might significantly exceed chronic water quality criteria, but not acute criteria. The baseline risk would then be characterized as a lack of acute effects, such as increased mortality in certain populations, but include the presence of chronic effects, such as reduced breeding success. The ecologist would then have to interpret these conditions in terms of site conditions and established endpoints. If there was a sufficient off-site recruitment supply, or there was no on-site breeding habitat for the species of concern, the exceedance of chronic criteria might pose only insignificant ecological risk. If, on the other hand, the site could potentially be an important breeding site for populating adjacent or even regional areas, the risk could be significant. This approach is more similar to the human health approach, but it generally lacks the probability element so prevalent in public health risk assessments. The lack of expression of probability is largely due to a limited data base. However, the emphasis on a population in ecology, as opposed to the individual in human health, also makes the probability of an event to an individual less appropriate for the ecological risk assessment.

To the extent possible, both of the above approaches, observed effects and comparison to benchmarks, should be employed to assess baseline risk. Often, the two approaches present different results. On-site concentrations might significantly exceed levels reported to be protective of a certain community or population, yet observations identify an apparently healthy assemblage. The reverse situation, that of observed effects but no identified contaminant concentration, is equally likely, and there is a wealth of possible explanations for such findings. The observed community could actually be transient, and not self-sustaining on-site. The community could be permanently present on-site, yet carry out critical life functions in specific areas of locally low contamination. Also, since benchmarks generally include a safety factor, the community could be existing on the margin between actual effects and the safety factor, thus appearing healthy but possibly without normal resilience and adaptability. A common explanation for healthy observations in the face of concentrations above benchmarks is the time or space scale of observation. It is possible, and perhaps likely, for a community to appear unaffected at a particular point in time and space. Yet, if the component populations could be viewed comprehensively, significant perturbations would be apparent. Observed effects at sites with contaminants below individual benchmark concentrations can result from synergistic or additive effects. The presence of additional compounds not considered in the assessment is another possibility. The lingering effects of historically high concentrations, or the occurrence of catastrophic episodic events (such as inundation with contaminated sediments during flooding) not detected in chemical sampling, could be other factors causing observed effects. Chance, and the often random nature of biological systems, is also a common and accurate explanation for the absence of expected ecological resources at a site.

The pivotal point in the ecological assessment is frequently the comparison between the risk characterization, or baseline risk assessment, and the ecological endpoints. If existing conditions are compatible with the ecological endpoints, the assessment of contaminant effects is complete, and remediation for ecological concerns would generally be deemed unnecessary. If there is a discrepancy between the baseline ecological risk and the endpoints, and the effects are attributable to a contaminant release or exposure, remediation for ecological purposes must be considered. In such cases, the contaminant concentrations associated with various relevant ecological effects must be established. These ecological effects levels then clearly demonstrate to decision makers the ecological conditions, and thus the level of ecological protection, associated with different cleanup levels. The ecological effects levels resulting from the risk characterization thus form the primary ecological input for developing remediation criteria.

The process of establishing ecological effects levels can be very straightforward. The concentrations can be taken directly from the toxicity assessment, if there are identified levels appropriate to the ecological endpoints. Even in such simple cases, achievement of the levels may not be practical or may be irrevocably destructive to the ecosystem, and thus a conscious and informed decision can be made to establish less protective remediation criteria. Also, because the effects levels are for individual contaminants, the possibilities of addictive or synergistic effects must be considered, and the criteria may need adjustment to accommodate such effects. If there are no ecological benchmarks for the contaminants, end points, and environmental media at issue, exposure and toxicity assessment results must be manipulated to establish re mediation criteria. In such cases, there may be greater uncertainty, and possibly a range of values for the remediation criteria would be required, but the process of establishing and evaluating remediation criteria for ecological concerns is the same.

Sometimes, the approach to establishing ecological effects levels can be simplified without heavy reliance on site-specific conditions. In certain situations, standards, and criteria are examples of accepted approaches for determining whether an area requires remediation. Accepted criteria can also be used to identify effects levels compatible with the applicable ecological endpoints.

Public health risk assessment is also an approach widely used for determining the need and extent of remediation. Compared to ecological risk methodology, there is extensive guidance and a track record for public health risk assessment (see Chapter 5), and the decisions resulting from the assessment are generally less subject to significant challenge. There are, however, uncertainties in a human health risk assessment, and assumptions are necessary. For example, the future land use, and thus potential exposure, at a site can be speculative. Also, the time on the site, potential for multiple pathways, and the appropriate "safety factors" are sometimes not well known.

Even with some uncertainties compared to ecological risk assessments, public health assessments are generally more precise, well documented, and acceptable methods for determining the need for remediation than an ecological risk assessment. Therefore, it is sometimes prudent to first establish if an area needs to be remediated based on public health concerns, and if it does, then an ecological risk assessment to determine the need for remediation for the specific area may not be necessary. Similarly, if a cleanup level has been established based on the human health risk assessment, it is often a better use of resources to evaluate the acceptability of the cleanup levels to the appropriate ecological endpoints, based on a mixture of quantitative and qualitative factors, rather than attempting to establish independent ecological-specific cleanup levels.

The ecologist is often faced with the opportunity and responsibility of establishing multiple effects levels for each contaminant of concern. As mentioned above, the most protective levels may not be achievable, and alternative criteria may be necessary. Also, different portions of the site may represent different exposure pathways, and thus distinct effects levels are appropriate. For example, at a single site there might be contaminated soils, wetland substrate, and lake sediments. Due to the physical differences in each of these media, the variety of organisms with diverse sensitivities associated with each media, and the distinct exposure pathways, individual criteria for each media may be appropriate. The development of effects levels for each media would require appropriate benchmarks, exposure assessments, and toxicity evaluations, to identify the concentrations protective of ecological endpoints associated with each environmental media.

Individual site characteristics may also have to be considered in establishing multiple effects levels for ecological protection. If a particular area within a site supports a distinct habitat or ecological function, a unique criterion may be appropriate for the area. For example, at a site with steep slopes or other features that are suitable for mammal dens, because of long exposure durations or inhalation in a confined space, remediation criteria lower than for the site as a whole may be appropriate. Similarly, at a single site, ecological remediation criteria for groundwater might be very different for the area where discharge is to a headwater stream, suitable for trout spawning and with little dilution, compared to the area discharging to a large river.

In some cases, evaluation of alternative remediation criteria must consider the effects of remediation activities. Either ecological, engineering, or economic effects can occur, and must be considered as part of finalizing remediation criteria. Achievement of the most protective criteria may be prohibitively expensive or not from an engineering standpoint feasible. Remediation to such levels may also be so destructive that the ecological feature or function targeted for restoration or protection is irrevocably destroyed. In such cases, alternatives that approach, but do not necessarily fully meet, the desired endpoint must be considered. In such cases, it is critical for the ecologist to develop and evaluate a range of effects levels, so that informed tradeoffs can be considered. For example, higher concentrations, perhaps protective of feeding and cover but not breeding, could be considered for the site if they substantially reduced the effects of remediation.

Achievement of the criteria in only certain portions of the site is another alternative remediation approach. For example, if a particular ecological effects level is based on a specific, extremely sensitive life stage or habitat requirement, the entire site may not need remediation to such a level for support of a self-sustaining population. A portion of the site may have low enough concentrations to support the sensitive function or a selected area may be designated for clean up to the lower remediation level. The remainder of the site could be left at existing levels or cleaned up to concentrations supportive of other endpoints. It may also be possible to establish a reduced, but acceptable, site population density as an alternative endpoint. It would then be necessary to determine what portion of the site must be remediated to the lower level, or what alternative levels reflecting some, but not full, breeding success would be required to achieve the alternative density. In such cases, the set of ecological effects levels can become large and complex, but ultimately result in a remediation plan that is technically and economically achievable, and provides an acceptable level of ecological protection.

The designation of partial site areas for the application of full remediation criteria or multiple criteria in various areas must be an iterative process. It can be useful to develop a graph or other display showing: the area of total remediation criteria compliance; degree of endpoint achievement; and the extent of the engineering, economic, or ecological effect. An evaluation of protection from contamination effects and achievement of other site remediation goals can then be made, and an acceptable balance considered. The evaluation of which specific areas need cleanup to the most stringent concentrations must include the ease, and thus cost, of remediation. The ecological considerations that should be incorporated in considering areas for the most extensive remediation include the existing resources in the area, the potential for habitat recreation, and alternative uses for the area following remediation.

A hypothetical example evaluating different areas for full remediation is shown in Figure 4–1. In this example, the cost of remediation and the population density that the site could support, given the area remediated, are compared. In the example, a model assuming an isolated population would be required to predict site density. The model would establish density based on area available for separate life functions, or with certain habitat characteristics. As more area is remediated, there is more area available for the most sensitive function and the population increases. At some point, about 15 individuals in this example, area requirements for functions other than the most sensitive becomes the limiting factor, and additional remediation produces only limited benefit. The changes in slope in population density can then be compared to changes in slope for cost, and an informed decision made as to the extent and degree of remediation. In the example, cost continues to rise significantly for remediation of more than about 14 acres, yet population density shows little increase. Therefore, it might not be productive to clean up more than 14 acres to the lowest remediation criterion. Remediation of eight acres, compared to four, adds relatively little cost, but population density is significantly higher when eight acres are cleaned up, so there

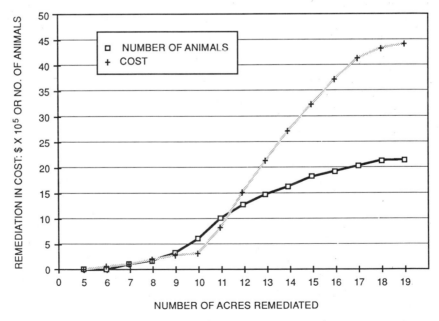

FIGURE 4-1.   Hypothetical comparison of population density achievable as a function of acres remediated and cost of remediation.

could be substantial benefit in applying the most stringent criteria to eight acres of the site.

A similar process can be followed at increasing levels of complexity. The benefits of various combinations of remediation criteria and areas treated can be compared in an effort to balance the cost to benefit ratio. Additional variables, such as minimum population size, can also be added to the model. There can also be numerous similar evaluations, such as population density of one species to another if soil remediation for species X destroys habitat for species Y. However, such comparisons are gross simplifications of complicated issues,and there is generally little assurance of the accuracy of predictions. The more complicated the comparisons and integrations of various areas and concentrations, the more assumptions necessary and the lower the degree of accuracy. Consequently, the simpler the comparison, the greater the likelihood of achieving the desired balance of ecological endpoints and other factors.

Evaluation of Remedial Alternatives and Site Restoration

Similar to any other construction project, such as the development of a shopping mall, site remediation can destroy or otherwise affect uncontaminated ecological resources. Because the RI/FS process is designed to be consistent with the National Environmental Policy Act (NEPA), the ecological impacts of the remediation must be understood by the decision makers before a remediation commitment is made. Also, where adverse effects are identified, alternatives with potentially less damaging impacts must be evaluated and considered, but not necessarily selected.

Construction of treatment or storage facilities, soil removal techniques, alterations of site hydrology, and site preparation are examples of remediation activities that can result in inadvertent damage to ecological resources. The ecological assessment should draw on the information developed as part of the preliminary site description and field investigation to identify the potential for such activities to alter ecological resources. Generally, the level of understanding of such resources developed to assess potential contaminant effects is more than adequate to identify possible conflicts between ecological resources and construction or operation activities. Identified infringement on the structure or function of resources on-site should be evaluated in relation to the ecological endpoints identified for the site. Although this is not normally done for nonhazardous waste projects (because ecological endpoints are rarely established), if they are available, a comparison to endpoints can be a meaningful exercise to establish the relative severity of the impact.

The identification and evaluation of ecological resources to be avoided is primarily for input into remediation decisions. However, the investigation phase of site remediation can sometimes cause inadvertent and unnecessary damage. The installation of wells can disrupt wetlands and site hydrology, but if such impacts are identified prior to the initiation of field investigations, a comparable well site without wetland infringement can be identified. Also, the location of investigation support activities, such as decontamination and field trailors, can usually be selected without ecological implications if the resources are generally identified prior to the major field investigation effort.

The identification of ecological impacts should be performed for each remediation alternative. This then allows the comparison of alternatives and their impacts, which is required by NEPA. It also provides the input for the overall evaluation of alternatives as part of the RI/FS process. Alternatives that destroy the environment they are designed to protect by remediating contaminant releases can thus be viewed more comprehensively, and informed judgments made.

The identification of impacts from alternatives also initiates the process of evaluating mitigating measures. The overall benefit of a particular remediation alternative can be enhanced, and thus the alternative made more attractive, if any inadvertent adverse impact can be reduced or eliminated. This is often not a difficult process, and with informed forethought can be achieved at little sacrifice to economics or remediation goals. A typical example is locating storage or treatment facilities well away from sensitive ecological resources. Designing access to avoid such resources or timing construction to avoid critical periods are other methods of eliminating impacts without substantially affecting beneficial aspects of remediation activities. Considering special excavation techniques, which minimize the need for large equipment, elaborate access, or extensive staging areas, is another way of minimizing inadvertent impacts.

Once the area to be remediated has been identified, and the removal and treatment techniques established, planning for site restoration is an opportunity to minimize impacts and achieve ecological endpoints. Critical habitats lost to remediation procedures, such as capping, or from past contaminant releases, can be replaced as a part of site restoration. The decisions as to what type of habitat to replace should draw from an understanding of the requirements for the established endpoints. It is often possible to significantly improve the overall ecological value of the site with no additional cost or sacrifice to other beneficial uses through ecologically aware site restoration.

Compliance With ARARs

As discussed in Chapter 3, remediation of hazardous waste sites is generally exempt from formal compliance with the Applicable or Relevant and Appropriate Regulations (ARARs). However, the remediation must comply with the substantive requirements of the permits and regulations applicable to the site conditions or individual contaminants. For ecologically related regulations, such as wetland permits, the endangered species act, and the migratory bird treaty, the assessment should provide documentation of substantive compliance. The documentation of compliance should really not be a substantial effort beyond the ecological evaluations performed as part of the field investigations or exposure assessment. If an ecological resource on-site is protected by an ARAR, some aspect of the resource should generally have been considered as an ecological endpoint. Thus, the requirements of the resource, and the potential effects of contaminant release and remediation, should have been fully evaluated. Documentation of ARAR substantive requirements is therefore only putting the information in the proper format for each potentially affected resource.

The requirements and jurisdiction of ARARs must be identified early in the ecological assessment process, so that the necessary information is collected as a part of the appropriate elements of the ecological assessment. The early identification of site-specific ARARs, and the collection of relevant site information, is also necessary so that the substantive requirements are not inadvertently violated. A common example involves the destruction of wetlands during site investigations. Frequently, temporary access roads or cleared paths are constructed to install wells, conduct seismic investigations, or dig test pits. If these access roads are through wetlands, or the debris from construction is deposited in wetlands, there can be a violation of the substantive requirements of Section 404 of the Clean Water Act and various state and local ARARs designed to protect wetlands. The situation can easily be prevented by early identification and delineation of the protected resource. If the season, timing, and coordination with other survey requirements permit, the wetlands should be fully delineated as in any other construction or development project, prior to any potentially destructive action. If full delineation is not possible prior to site investigation, at least the general locations of protected resources should be identified, with particular attention to areas subject to early site investigations, such as well or test pit installation. Once the location of the wetlands are mapped, and the wells or other sampling locations generally identified, it is usually a simple process to establish access or slightly modified sampling locations that minimize wetland impacts. Documentation of the process of minimizing impacts is generally sufficient demonstration of compliance with the substantive ARAR requirements. A similar process should be followed for other ARAR-protected resources, such as critical habitats for special-status species.

A remediation activity may be necessary for the protection of public health and the environment, even though it is not in conformance with the substantive requirements of one or more ARARs. In such cases, exemptions are possible under a specified set of conditions, as discussed in Chapter 3. The ecological assessment should provide the justification for the exemption when ecological resources or regulations are involved. In addition to showing compliance with the exemption conditions, the assessment must demonstrate that all reasonable actions have been taken to minimize the degree of noncompliance and the associated ecological effects.

The Endangered Species Act and wetland protection (Section 404 of the Clean Water Act and similar orders and regulations) are probably the two ecologically related ARARs most frequently requiring demonstration of substantive compliance. The process for the Endangered Species Act should closely follow the Biological Assessment for the species of concern, as described in Chapter 3. Documenting the 404(b)(1) demonstration should be included in the ecological assessment when jurisdictional wetlands are involved. The 404(b)(1) process is described in Chapter 3, and the Middle Marsh case study (Chapter 10) gives an example of the demonstration.

PHASING OF ECOLOGICAL ASSESSMENTS

As is evident from the above discussion of ecological assessment elements, these studies represent a broad range in their levels of investigation and complexity. The extent of the ecological assessment is dictated by the complexity of the resources onsite, the nature and extent of the contamination, the ecological effects, and the exposure pathways. Until there is some appreciation of the complexities of these site characteristics, the detail and extent of the ecological assessment can not be determined. Consequently, the assessment must proceed in phases, with the results of the initial investigations needed to define the objectives, methods, and extent of subsequent investigations. The objective of the early phases is simply to determine the need for an ecological assessment based on the resources present and the potential for damage. Subsequent phases are at increasing levels of detail, and are designed to address one or more of the overall ecological assessment objectives or, alternatively, identify the need for more detailed study.

Thus, a phased approach is necessary for determining the need and extent for subsequent ecological investigations and evaluations. The need is not only for the technical aspects, but also for the administrative and management aspects. The funding for studies must be justified and secured, and decision makers generally require the demonstration of need based or some initial indication of substantial ecological damage, risk, or resource on the site. The step-by-step approach also forces a need for periodic consensus, and provides the opportunity for refinement of the ecological endpoints and objectives based on more information than was available when the endpoints were originally designated.

Commonly, the first step or phase is a screening of on-site conditions to determine the potential presence of significant ecological resources and ecological damage due to measured contaminant concentrations. The potential for contaminant damage can often be assessed at the screening level without ecological field investigations, by comparing concentrations measured on-site to ecological benchmarks, as described earlier as part of the exposure assessment. A comparison to benchmarks can be supplemented, or followed if conditions dictate, by limited field investigations. Typical screening-level field studies consist of observations and some qualitative biological sampling. The initial field investigations not only give some indication of overt ecological damage due to a release of contaminants, but also identify sensitive resources associated with the site.

The Rapid Bioassessment Protocol (U.S. EPA 1989c) presents a model for a screening-level evaluation of ecological effects and resources at hazardous waste sites. The model, as summarized following, is specific for freshwater aquatic systems, but the principles and philosophy can easily be adapted to terrestrial and marine systems.

There are five separate Rapid Bioassessment Protocols (RBPs) addressing different aquatic resources and levels of investigation (designated RBPs I, II, III, IV, and V (U.S. EPA 1989c). Protocols I through III address the evaluation of benthic invertebrate communities, and the remaining RBPs deal with fish. The ascending number of the RBP indicates an increasing level of detail in the investigation, thus representing additional effort and description of the ecological resource or impairment. Although at different levels of detail, each protocol calls for an evaluation of the three basic ecosystem components: water quality/physical characterization; habitat assessment; and biosurvey.

RBPs I (benthic invertebrates) and IV (fish) are considered to be first-level screening evaluations, and can be used either exclusively in a simple situation, or as a preliminary step to aid in designing a more extensive program in a complex situation. The objective of the screening-level studies is to determine whether biological impairment or sensitive resources exist, and if further investigation is warranted. The benthic invertebrate protocol relies on systematic documentation of visual observations to evaluate the potential impairment or other characteristics of the community. Consequently, it is essential that an experienced biologist, who can easily recognize variations from expected conditions, perform the reconnaissance. The presence or absence of particularly tolerant or sensitive organisms, as well as their relative abundance, are important characteristics in judging the health and requirements of the system. Such an assessment is relatively straightforward when there are similar aquatic habitats upstream and downstream of a hazardous waste site. If similar habitats can not be found, or other potential sources of impairment affect upstream reaches, a reference site in an adjacent drainage basin may be necessary. When a reference site is required, the habitat type and stream flow are important variables to match with the potentially impaired water body at the hazardous waste site. The reference and hazardous waste site reconnaissance should be conducted under the same seasonal and flow conditions to minimize natural variability in comparing the sites.

EPA suggests a different approach for the screening level assessment of fish (U.S. EPA 1989c). Compared to invertebrates, fish are very difficult to observe, much less assess, in a "stream walk" type reconnaissance. At the other extreme, there is often extensive existing information and knowledge concerning fish, because of their importance to sportsmen and state agencies dealing with recreation and natural resources. Rarely can sport fishermen tell you if an area supports a healthy macroinvertebrate segment of the stream community, but they certainly know if it is a waste of time fishing a particular reach of the stream. They also know if the fish from a certain area are prone to abnormalities, or if the flesh has a tainted taste. Similarly, state agencies generally know which aquatic habitats are used for fishing. The agencies also often know of areas with particularly good or poor success in stocking efforts. EPA RBP IV takes advantage of the existing knowledge of fish and fisheries by using a questionnaire approach. The condition of a stream or stream reach potentially affected by a hazardous waste site is assessed based on the results of a questionnaire. In many ways, the questionnaire can provide more information than a field reconnaissance, because it covers conditions over a longer period of time and can include more information on the condition of the populations. The approach is limited, though, because there might not be information on the specific area of interest. However, if a section of stream or river appears even remotely suitable for fish, chances are a fisherman has made an attempt to fish the area, and thus could report some information. If the water body is not obviously suited for fishing, then an assessment of fish is most likely not warranted, and the screening of impairment should be based on the invertebrates in the system. Perhaps the greatest drawback of the questionnaire method is the accuracy and reliability of the data received. State agencies generally have accurate files and an interest in full disclosure. However, fishermen can exaggerate or withhold information on the location of favored fishing spots. Also "fishermen know a lot about fish, and some of it is true," so their reported information must be viewed accordingly and their assessment might not constitute a reliable scientific evaluation.

The protocol for benthic invertebrates beyond screening consists of two levels of investigation (RBP II and RBP III), while there is only one protocol beyond screening for fish assessment (RBP V). These procedures have the common objective of assessing the degree of impairment, as opposed to just determining whether impairment or significant resources exist, as in RBPs I and IV. Similar to the screening-level protocols, they also provide information for further assessment and prioritization of investigations. The two benthic invertebrate protocols differ in their level of detail in data collection and the quantification of results. Samples are returned to the laboratory for more thorough identification when RBP III is used. Also with the more detailed RBP III, additional quantitative data is developed from the samples, to allow distinction in the level of impairment. The protocol for fish assessment consists of electrofishing of 100 to 500 meters of stream, and documenting the number and biomass of each species collected. The data can then be used to calculate a variety of indices and employ other quantitative techniques to assess the degree of impairment.

In terrestrial systems, a parallel to the Phase I Rapid assessment is a qualitative survey of vegetation. This can be accomplished by first reviewing any available aerial photography of the site. If alteration of vegetation is significant, it can generally be detected, even from high-altitude photography. An alteration of species composition or density indicated from photographs can then be compared to natural variations in soils, topography, or other natural differences. This information is frequently available from U.S. Soil Conservation Service maps, National Wetland Inventory maps, and U.S. Geological Survey topographic maps.

Even in a terrestrial screening-level identification of effects from releases, field reconnaissance is necessary if areas of altered vegetation are indicated from photographs or other information sources. The field inspection can verify the correlation of the altered vegetation with soil, topographic, or hydrological information. The reconnaissance is often the only reliable means to identify noncontaminant-caused alterations in vegetation, such as fire, man-made physical disturbances, or historic land use.

As in the Rapid Bioassessment protocol for aquatic systems, a screening-level evaluation of terrestrial resources can be expanded beyond review or maps, photographs, and field reconnaissance. Academic resources and regulatory personnel can be contacted to identify historically noted unusual ecological characteristics of the site. Information of this type is often associated with hunting resources. There can also be information from environmental or natural resource organizations, such as data from the Audobon Society Christmas Bird Counts. Often, such surveys will identify the presence or absence of a species in areas of suitable habitat. Also, in some programs with a continuous data base, the disappearance of a species from an area could be correlated with releases from a hazardous waste site.

Screening-level identification of contaminant effects in the terrestrial environment can include qualitative sampling and detailed observations. Often, small mammal traps can be easily used for overnight capture. If there are no captures reported, a concern could be raised. At the other extreme, if the survey is conducted in the appropriate time of year, capture of lactating females or immature animals generally indicates an unaffected population. Based on such information, impacts on small mammals from contaminant releases could possibly be eliminated as a concern. Similarly, observations during the spring by a trained biologist can often identify successful bird breeding, and thus provide critical information to the remedial investigation. If the reconnaissance identifies the area as a suitable habitat for animals that leave obvious signs, limited additional field observations can often confirm the presence of an active population. For example, if the area is suitable for amphibians, and observations do not confirm frog songs at dusk in the spring, there is reason for concern. Also, racoon sign (scat and tracks) are generally ubiquitous and obvious in suitable habitat areas. If a few hours on a site do not reveal such sign, additional investigation may be called for.

The field investigations described above, both for terrestrial and aquatic systems, can be followed by more detailed and quantitative investigations. Such detailed investigations approach those used for research projects, and generally should not be entered into lightly as a part of hazardous waste site ecological assessments. All chemical and available field ecological data should be thoroughly reviewed, and some exposure modeling and toxicity assessment performed to confirm a need, before more detailed studies are initiated. If more studies are needed, they can generally be narrowly focused on specific problems and locations. Frequently, bioassays, either in situ or in the laboratory, can provide the needed information after limited field investigations have provided a general understanding of the ecological features and stresses on the site. When viewed in the light of the overall site ecology, the bioassays can provide invaluable information.

A removal action, as discussed in Chapter 3, is a special case of hazardous waste site action, where screening-level ecological investigation is critical. A removal action can be initiated prior to any RI/FS action, when there is imminent threat to human health or the environment. A reconnaissance-level ecological assessment for a removal action is often needed to determine if any action is necessary to protect ecological resources associated with the site. The evaluation is also sometimes needed to identify potential effects from the removal action and to evaluate measures to mitigate any impacts. The removal can produce impacts either from a release of contaminants to off-site resources, or by physical disruption of on-site critical habitats. The preliminary ecological evaluation conducted prior to or during the removal action can also supply important information in deciding the need for remedial investigations and, if needed, the scope for such investigations.

Often, removal actions are conducted with a narrow focus on eliminating an immediate threat to human health, and little attention is given to ecological considerations. Thus, important resources can inadvertently sustain damage with long-term implications that are difficult and costly to rectify. Common inadvertent actions that affect resources are often not location-specific, and can be avoided. Examples include creating a temporary access road that fills wetlands or disrupts surface drainage to a stream during a critical period. Storage of materials or other removal action staging activities sometimes occur within or adjacent to critical habitats. Diversion of surface drainage or stream flow, frequently necessary to prevent migration, can inadvertently deprive downstream reaches of water needed to support breeding or other sensitive aquatic life process. A simple identification of ecological characteristics potentially at risk, early in the evaluation of removal actions, can usually provide the information necessary to select comparable alternatives with little or no significant ecological damage.

The phased approach is not unique to the ecological assessment in the investigation and remediation of hazardous waste sites. The RI/FS, the subsequent design, and even the construction is generally phased. In fact, implementation of the designated action often includes monitoring of contaminants following construction as an acknowledgement that additional action may ultimately be needed. Phasing was originally incorporated into Superfund because much of the investigation and remediation technology was new without a proven history, and there was a desire to try the simplest solutions first and see how well they worked. The situation with ecological assessments at a hazardous waste site is analogous to the whole Superfund process, and thus phasing is similarly an appropriate approach.

ECOLOGICAL RISK ASSESSMENT

An ecological risk assessment can be viewed as organizing and expressing much of the critical information collected and evaluated in the various elements of the ecological assessment discussed earlier. An ecological risk assessment at a hazardous waste site combines the exposure assessment, observation of effects, endpoints, toxicity assessment, and risk characterization to comment on the relationship of ecological resources and potential or realized damage. Suter (1990) has categorized ecological risk assessments as either retrospective or predictive. At a hazardous waste site, both types of risk assessment are appropriate and necessary, as a retrospective assessment generally establishes the assumptions, methods, and numerical relationships to make predictions. These two types of risk assessment represent the two purposes of ecological risk assessment and two of the basic objectives of ecological assessments at hazardous waste sites: a determination of the effects from past releases of hazardous substances; and defining future conditions acceptable to established ecological endpoints, or ecological effects levels to be used as input in establishing cleanup levels.

The need for incorporating a formalized expression of risk into hazardous waste decisions was recognized early in Superfund. By 1986, after years of development, the EPA produced a series of guidances on human health risk assessment. The approach was based on the paradigm established by the National Academy of Science, which simply stated defines risk assessment as four processes: hazard identification; dose response determination; exposure evaluation; and risk characterization. The approach and methodologies have been modified since 1986, but the basic paradigm has not changed. With the recent concern of ecological resources at hazardous waste sites, the EPA has recognized the need for on ecological risk assessment approach to parallel the human health assessment. Early guidance and methods and approaches were issued so that ecological risk would be considered as part of the RI/FS, and some experience could be developed in the field of ecological risk assessment (U.S. EPA 1989d and 1989e, U.S. EPA 1990). However, the regulatory and scientific communities recognized that an open and comprehensive examination of ecological risk was needed. In 1990 and 1991, the EPA started this process of open examination for input to formulating comprehensive and accepted policy and guidance related to ecological risk assessment methodology. Toward this end, they have held a number of symposia and published initial thoughts on the issue, but they have not yet developed an official approach, or even finalized the philosophy of ecological risk assessment (U.S. EPA 1991).

Even without published regulatory guidance, ecological risk assessments are often required, and almost always useful as a part of the hazardous waste site RI/FS process. The lack of formal guidance from the primary regulatory agency can be a hindrance, but until the guidance is developed, it can also represent an opportunity for ecologists to apply a variety of approaches. The approaches can be structured around the specific site, resources, and contaminants. Thus, the degree of detail can be matched to the extent of the resource or the problem, and not be constrained or unnecessarily elaborate in order to meet formal guidance. The lessons learned from experimental approaches in this early era of ecological risk assessment can hopefully be incorporated as the regulatory agencies and scientific community formulate ecological risk assessment methodology.

A common approach in ecological risk assessment is to begin with the paradigm and model used for human health risk assessment. This is generally appropriate. However, the differences, complications, and uncertainties must be recognized. The first and foremost difference is in the endpoints considered. Human health assessments have a well-established and accepted endpoint, whereas ecological endpoints are site-specific, not universally accepted, and there are often multiple endpoints at a single site. There are also significant uncertainties and difficulties in ecological risk assessments related to: different sensitivities for various species and endpoints, and the need to extrapolate toxicity data and predictions among endpoints, taxonomic groups, and from laboratory to field situations (U.S. EPA 1991).

The evaluation of retrospective or baseline ecological risk is also significantly different from the classical human health risk assessment. In a human health evaluation of a hazardous waste site, the calculation of baseline risk is probably the most critical and telling exercise. The contaminants present on-site are mathematically combined with the exposure pathways and the human use of the site, to determine the chance or risk of harm or disease to which the human population has been exposed. This formalized and documented method is necessary to evaluate human health effects, because actual effects can not be measure due to the almost infinite number of variables not related to site conditions and, of course, the unacceptability of direct toxicity testing. This process for the human health evaluation then provides a single number, integrating all contaminants and pathways, representing the increased chance of some undesirable situation (e.g., increased cancer rate), which is well suited for summarizing the site conditions for decision makers.

The evaluation of baseline ecological risk is a very different process. There is not a universally accepted endpoint, and ecological endpoints have to be developed on a site-by-site basis. Also, there is insufficient data and a lack of an accepted method to estimate risk on a numerical basis or other frequency distribution associated with an ecological endpoint. The evaluation generally tends to be bimodal: the ecological effect is expected to occur or it is not expected. Even if it was possible to calculate the increased risk of damage to an individual, it generally would not be appropriate for ecological resources, because individuals are not ecologically important. Changes to individuals only have ecological implications if they collectively result in changes to the population. Therefore, in contrast to the product of the human health assessment, the ecological baseline risk assessment yields a statement about the achievement of an endpoint. Often, the significance or importance of the endpoint is not readily apparent to non-ecologist decision makers. When pressed to interpret the meaning and significance, the ecologist usually must confess that the conclusion is based on several assumptions, interpolations of data, and sometimes on a wealth of speculations. It is not unusual for the number of opinions concerning the significance and uncertainty of an observed effect to equal the number of ecologists involved in the assessment.

As an important step in establishing ecological risk assessment philosophy, the Society of Environmental Toxicology and Chemistry (SETAC) has made a distinction between true risk assessment and hazard assessment (cited in U.S. EPA 1991). As defined by SETAC, a risk assessment is "...the process of assigning magnitudes and probabilities to an adverse effect resulting from human activities or natural catastrophes." Consequently, a risk assessment focuses on probabilities of clearly defined effects employing scientific methods. In contrast, SETAC (cited in U.S. EPA 1991) defined hazard assessment as "...a quotient or margin of safety comparing the toxicological endpoint of interest to an estimate of exposure concentration. A judgment is then made on the adequacy of the margin of safety." Others have more generally defined hazard assessment as simply the determination of an effect from some form of stress (U.S. EPA 1991). Thus, in contrast to a true risk assessment, a hazard assessment relies more on the margin of safety and the expert judgment of the assessor than on formal techniques, such as mathematical and statistical models, in evaluating an existing or potential threat to a resource.

A true expression of risk is generally the end product of an evaluation of human health issues at hazardous waste sites. The probability of health or well being of a single individual is well established, and toxicological data is generally complete enough for a consensus on the expression of probability. Consequently, it has been common practice to use risk as an expression of probability to protect the vast majority of individuals. This has generally been a successful and accepted approach in human health, because the probability of adverse effect can be compared to a risk an individual willingly takes, such as smoking or driving on the freeway. Society can then determine what level of risk is acceptable, in comparison to the cost or inconvenience required to achieve a greater level of protection.

Although a complete risk assessment is an admirable goal for ecological consensus, in comparison to human health, there is a very different objective, approach, and philosophy in the protection of ecological resources. Preservation of the individual is really inconsequential. Protection of habitats, self-sustaining populations, and ecological function are the critical ecological endpoints. Consequently, the ecological risk assessment should be structured around the maintenance of these resources. Unfortunately, we can not currently express the probability of achieving these endpoints as a function of a contaminant concentration. This is due either to insufficient research, natural variability, the extreme complexity of ecosystems compared to human health, or most likely it is a combination of all of these.

The current approach to protecting ecological resources has been to establish conditions deemed suitable for the continued existence of the most sensitive population or function in the ecosystem. This has often resulted in identifying the concentration of a contaminant that has no effect on any life function of any species tested. This could well be much more stringent than the maximum concentration that the ecosystem could tolerate and continue to exist under in an unaltered fashion. Thus, under this scenario, if there is no reduction in any individual functional or structural component of the community, there will be no alteration in the community as an entity. Most ecologists would agree that this is a very conservative approach, and in most cases provides a significant margin of safety. However, because of the scarcity of data, and variability in the data that does exist, ecologists and regulators have historically been reluctant to set limits that allow for some effects on individuals or other subunits but protect the ecosystem integrity. The result has been somewhat of an irony, because the objective is to disregard individual survival, yet the limits set to protect the ecosystem are generally much more stringent than those needed to protect the individual.

This conservative approach has been dominant in establishing criteria for ecological protection, such as the EPA's Quality Criteria for Water (U.S. EPA 1986). Laboratory test results have been reviewed, and criteria developed for individual compounds, based on the results for the most sensitive species. Safety factors are frequently employed, and significant success of the most sensitive life functions are considered. There is some consideration of level of protection in Water Quality Criteria by establishing both acute and chronic levels of protection. These levels of protection are generally related to duration or area of exposure. They are rarely employed based on some predetermined, acceptable, less-than-full protection of all functions that can sustain the ecosystem.

Based on this history of ecological assessment and protection, it is probably unrealistic to expect ecological risk assessments at individual hazardous waste sites to define the probability of a sustained ecosystem as a function of contaminant concentration. However, a form of ecological risk assessment can certainly provide critical input to remediation evaluation and decision making at the site. An ecological risk assessment at a hazardous waste site is thus likely to be closer to a hazard assessment. A hazard assessment, as defined by the Society of Environmental Toxicology and Chemistry (SETAC), consists of calculating a margin of safety by comparing a toxologically determined safe concentration to an estimate of exposure, and then using professional judgment to assess the ability of the level to provide the desired protection (U.S. EPA 1991). This is in contrast to strictly employing scientific methods to establish probabilities of specific effects, which is what SEATAC considers a true ecological risk assessment.

A complete ecological risk assessment at a hazardous waste site is an admirable goal. It is possible to perform such an assessment for some specific parameters and endpoints. For example, for some well-studied compounds, it may be possible to establish the relationship between media concentration and reproductive success for an individual and perhaps even for a population. If both the importance of the population to the ecosystem and the minimum acceptable reproduction rate could be established, risk and remediation criteria could be established in a manner similar to a public health risk assessment. In other cases, due to a lack of data or consensus on ecological endpoints and their importance, a hazard assessment may be all that is possible. The utility of a hazard assessment for ecological resources should not be dismissed. If the conclusion is that there is no existing risk warranting remediation, the hazard assessment achieves everything that a true risk assessment would for the baseline risk. If there is existing risk, and the protective hazard assessment demonstrates that remediation criteria established for public health are protective of the ecological endpoints, the hazard assessment is an adequate tool. The hazard assessment can fall short when specific ecological remediation criteria are required. Without the probability and documented relation to broadly understood and accepted goals provided by a true risk assessment, establishing remediation criteria based on ecological protection is difficult and can be subject to challenge.

TECHNICAL APPROACH SUMMARY

An ecological assessment at a hazardous waste site consists of a number of separate but highly interdependent investigations, which are directed toward the common objectives of: determining existing ecological impacts; establishing ecological effects levels; and evaluating ecological site restoration. The individual investigations, or elements of the ecological assessment, are frequently conducted in a phased and iterative approach, with the specific aspects and level of detail dependent on the complexity of the ecological resource or contaminant effects revealed in previous investigations. Ecological risk assessment methodology, which can be a convenient method to integrate the results of all elements, is an emerging science or technology and experience gained at hazardous waste sites is proving invaluable in establishing techniques to evaluate and manage ecological risk.

There is no mandated or even widely accepted standardized technical approach for ecological assessments at hazardous waste sites. This applies to the type of individual investigations needed, the composition of each individual investigation, and the sequencing and integration of the separate elements of the assessment. There are a number of topics that must be addressed in some level of detail at every hazardous waste site with potential ecological effects or resources, in order to meet the Superfund objectives.

The first of these elements is the preliminary site description. This task forms the basic understanding of the site, so that an appropriate investigation methodology can be established. The description is usually based on a review of existing documents, a site reconnaissance, and discussions with people familiar with various aspects of the site.

The preliminary description also forms the basis for the next two elements: establishing ecological endpoints and developing an ecological work plan. Ecological endpoints must relate to the overall site objectives and future use, and must have an assessment and measurement component. The assessment component establishes the ecological use goal for the site and adjacent area, whereas the measurement endpoints permit evaluation of relative success in achieving the goal. Both the assessment and the measurement endpoints must have a target component (i.e., species, population, or community) and an action or effect (i.e., survival or reproduction). The endpoints must have broad relevance and support, because they will ultimately form the basis for all decisions affecting ecological resources. The ecological work plan outlines the approach for evaluating and predicting the relationship between site conditions and ecological endpoints.

The majority of the ecological investigation is conducted as the field investigation and bioassay elements of the assessment. These are adaptations of classical ecological studies, but must be focused on the objectives of the assessment and the site-specific ecological endpoints. The focusing of effort and a phased approach are most critical for these elements, because they are where the majority of effort is expended, and they must form the scientific basis for any remediation or restoration decision. These studies are usually the exciting and challenging elements for any ecologist, because they involve the hands-on investigation of the relationship between organisms and their environment.

The next series of elements are directed at establishing ecological effects levels that relate various contaminant concentrations to resulting ecological conditions. These investigations incorporate measured site conditions and toxicological data (both site-specific and literature values) to establish the ecological/contaminant relationship. Sometimes the relationships can be set by simply relating concentrations to established contaminant criteria or other ecological benchmarks. The process can also be as complicated as detailed site and media-specific investigations involving laboratory bioassay, in situ exposure of organisms to on-site conditions, and research projects directed at indigenous populations. The first individual element in this series is the hazard investigation, where the contaminants potentially causing a hazard and the ecological resources at risk are identified. The next element is the exposure assessment, where the interface between the contaminant and the ecological target is identified and sometimes quantified. As part of the toxicity assessment, the relationship between the contaminants of concern and resources at risk is quantified. The final element in the series is the integration of all of the above steps, to establish numerical ecological effects levels.

The final two elements of the ecological assessment are not unique to hazardous waste sites. One is documenting compliance with pertinent ecological laws, regulations, and permits or ARARs. This must be done for all types of projects and hazardous waste remediation is no exception, but only the substantive components of the ARARs must be adhered to at Superfund sites. The final element might seem trivial after the more sophisticated investigations. However, ecological restoration following remediation must be completed to achieve real ecological protection. This is true for all types of projects, but it can have special importance for hazardous waste situations, because the site is often left with no structures or other anthropogenic conditions that preclude exceptional ecological potential.

The approach to ecological assessments in this chapter is presented as food for thought. Not all elements are needed at every site, and there will undoubtedly be sites where elements not described here are critical to achieving the objectives of the ecological assessments. However, the basic philosophy of describing the biotic and abiotic conditions in order to define existing ecological damage, the potential for future exposure and damage, and identifying opportunities to protect and enhance critical resources is applicable to every site. The approaches to accomplish these objectives presented in this chapter have proven successful at many sites, and should be useful for adaptation to new sites. The case studies presented in Chapters 9 and 10 vary from the specific approach outlined in this chapter, but the common philosophy is adhered to, and the reasons for the variations are apparent and logical.

REFERENCES

Ambrose, R. B., 1987. WASP4, A General Water Quality Model for Toxic and Conventional Pollutants. EPA/600/3–87/039.

Ambrose, R. B., and T. O. Barnwell, 1989. Environmental software at the U.S. Environmental Protection Agency's center for exposure assessment modeling. *Environmental Software*. 4(2): 76–93.

Barber, M. C., L. A. Suarez, and R. R. Lassiter, 1988. FGETS (Food and Gill Exchange of Toxic Substances): A simulation model for predicting bioaccumulation of nonpolar organic pollutants by fish. EPA/600/3–87/038.

Beyer, N. W., 1990. Evaluating soil contamination. U.S. Department of the Interior, Fish and Wildlife Service. *Biological Report* 90(2).

Bodex, Itamar, W. J. Lyman, W. F. Reehi, and D. H. Rosenblatt, 1988. *Environmental Inorganic Chemistry*. SETAC Special Publications Series. New York: Pergamon Press.

Burns, L. A., and D. M. Cline, 1985. Exposure analysis modeling system: Reference manual for EXAMS II. EPA/600/3–85/038.

Cairns, J. Jr., and J. R. Pratt, 1987. Ecotoxicological effect indices: A rapidly evolving system. *Water Sciences and Technology*, 19(11):1–12.

Connolly, J. P., and R. V. Thomann, 1985. WASTOX, a framework for modeling the fate of toxic chemicals in aquatic environments, Part 2: Food chain. EPA/600/4–85/040.

Eisler, R., 1986. Polychlorinated biphenyl hazards to fish, wildlife, and invertebrates: A synoptic review. U.S. Fish and Wildlife Service. *Biological Report* 85(1.7). 72 pp.

Hynes, H. B. N., 1970. *The Ecology of Running Waters*. Liverpool, UK: Liverpool University Press.

Keeton William T., 1972. Biological Science, Second Edition. New York: W.W. Norton & Company, Inc.

McCarthy, J. F., and L. R. Shugart, 1990. *Biomarkeres of environmental contamination*. Boca Raton, FL: Lewis Publishers, CRC Press, Inc.

Oviatt, C. A., J. G. Quinn, I. T. Maughan, J. T. Ellis, B. K. Sullivan, J. N. Gearing, P. J. Gearing, C. D. Hunt, P. A. Sampou, and J. A. Latimer, 1987. Fate and effects of sewage sludge in the coastal marine environment: A mesocosm experiment. *Marine Ecology Progress Series*, 41:187–203.

Rohrer, T. K., J. C. Forney, and J. G. Hartig, 1982. Organochlorine and heavy metal residues in standard fillets of coho and chinook salmon of the Great Lakes 1980. *J. Great Lakes Res.* 8:623–634.

Suter, F. W. II, 1990. Endpoints for regional ecological risk assessments. *Environ. Man.* 14(1):9–23. Thomann, R. V., 1989. Bioaccumulation model of organic chemical distribution in aquatic food chains. *Environmental Science and Technology* 23(6):699–707.

U.S. EPA, 1991. Summary Report on issues in ecological risk assessment. EPA/625/3–91/018. PB91–172122.

U.S. EPA, 1990. Quantifying Effects in Ecological Site Assessments: Biological and Statistical Considerations. EPA/600√/–90/152.

U.S. EPA, 1989a. Ecological Assessment of Hazardous Waste Sites: A field an laboratory reference. EPA/600/3–89/013. Edited by: W. Warren-Hicks, B. R. Parkhust, and S. S. Baker, Jr.

U.S. EPA, 1989b. Terrestrial Ecosystem Exposure Assessment Model (TEEAM). EPA/600/3–88/038.

U.S. EPA, 1989c. Rapid Bioassessment Protocols for Use in Streams and Rivers: Benthic Macroinvertebrates and Fish. EPA/444/4–89–001. Edited by James A. Plafkin.

U.S. EPA, 1989d. Risk Assessment Guidance for Superfund: Environmental Evaluation Manual. EPA/540/1–69/001A. OSWER Directive 9285.7–01.

U.S. EPA, Region I, 1989e. Supplemental Risk Assessment Guidance for the Superfund Program. EPA/901/5–89/001.

U.S. EPA, 1988a. Guidance for Conducting Remedial Investigations and Feasibility Studies Under CERCLA. EPA/540/G–89/004.

U.S. EPA, 1988b. Superfund Exposure Assessment. EPA/540/1–88/001. OSWER Directive 9285.5–1.

U.S. EPA, 1986. Quality Criteria for Water. Office of Water Regulation and Standards. EPA/440/5–86/001.

Vandergrift, S. B., and R. B. Ambrose, 1988. SARAH2, A Near Field Exposure Assessment Model for Surface Waters. EPA/600/3–88/020.

Waddell, T. E. 1989. Superfund Exposure Assessment Manual Technical Appendix, Exposure Analysis of Ecological Receptors. Environmental Research Laboratory, U.S. EPA. Athens, Georgia.

Warren-Hicks, W., and B. R. Parkhurst, 1991. Ecological risk assessment methods in water quality standards and regulations: Case study. Paper presented and manuscript available: Water Pollution Control Federation, 64th Annual Conference, Session 19. Toronto, 7–10 October 1991.

5

Human Health Risk Assessments

Mary E. Doyle
John C. Young

Metcalf & Eddy, Inc.

INTRODUCTION

A human health risk assessment is a procedure used to evaluate the health implications of past or proposed actions. The intent of the assessment is to inform the public and decision makers of possible public health benefits and repercussions of the actions. Risk assessments have been used by many regulatory agencies in making a wide variety of decisions. Evaluations of hazardous waste sites, comparisons of remedial technologies, siting of landfills and incinerators, and the setting of occupational, drinking water, and food standards are some examples of where risk assessment has been used. A public health risk assessment is one part of the evaluation conducted to make decisions involving government funds or approvals. Other evaluations considered together with the health risk assessment to make these decisions include ecological risk assessments, economic analyses, and technical feasibility studies.

This chapter focuses on risk assessments conducted as part of the remedial investigation and feasibility study process required for evaluating hazardous waste sites under the Comprehensive Environmental Response, Compensation, and Liability Act of 1980 (CERCLA, 42 U.S.C. 9605), as amended by the Superfund Amendments and Reauthorization Act (SARA, Pub L. 99–499). The original act and the amendments are herein referred to as CERCLA or Superfund. The practices described here can also be applied to risk assessments conducted under other legislative directives. This information is directed toward an ecologist or biologist conducting ecological risk assessments, to provide some understanding of the health assessment often conducted concurrently to evaluate human health risks.

Health risks are estimated based on the contaminants detected at the site, the toxicity of the contaminants, and the expected characteristics of human exposure. The size of the exposure, estimated as contaminant intake or dose, is based on an evaluation of the ways in which people may be exposed, and the frequency and length of contact with contaminants. Examples of exposures that may be estimated are: ingestion of contaminated groundwater or surface water; direct contact with contaminated media,

113

such as water or soil, with absorption of contaminants through the skin; and inhalation of contaminants in the air, as vapors or associated with dusts.

The requirement to conduct a risk assessment, and the standard approach used to evaluate human health impacts, are presented in this chapter. The risk assessment's place within the feasibility study, to identify and evaluate remedial objectives, as well as additional information on analytical data needs and dose-response information, are presented in the last subsections of this chapter. Since many CERCLA risk assessments are conducted with both federal and state input, specific observations and caveats are noted at some of the points where consideration of state regulations is useful.

REGULATORY AUTHORITY AND GUIDANCE

Congress laid the basis for the use of risk assessments to evaluate hazardous waste sites in its passage of CERCLA. The act requires that remedial actions at hazardous waste sites be selected to protect human health and the environment. The U.S. Environmental Protection Agency (EPA) further established the requirements for risk analysis in the National Oil and Hazardous Substances Pollution Contingency Plan (NCP), which provides the regulatory structure and procedures for responding to discharges of oil or releases of hazardous substances (U.S. EPA 1990a). Section 300.430(d) of the NCP specifies that data collected as part of the remedial investigation be used to:

> Conduct a site-specific baseline risk assessment to characterize the current and potential threats to human health and the environment that may be posed by contaminants migrating into groundwater or surface water, releasing to air, leaching through soil, remaining in the soil, and bioaccumulating in the food chain.

The NCP further specifies that the results of the baseline assessment will be used to assist in developing "acceptable exposure levels" for use in the feasibility study. Acceptable exposure levels are defined for systemic toxicants (i.e., noncarcinogens) as the level to which the general population can be exposed for a lifetime without adverse effects. For carcinogens, an acceptable exposure level is defined as the concentration corresponding to an excess upper-bound lifetime cancer risk to an individual of 10^{-4} to 10^{-6} (U.S. EPA 1990a). This can also be stated as: the exposure of ten thousand to a million people will result in approximately one additional cancer case.

Until recently, despite the requirement to protect human health and the environment, the primary focus of risk assessments has been on public health. Following a report by the Science Advisory Board in 1990, stating that more attention should be given to ecological threats, a greater emphasis has been given to evaluating ecological risks.

All new orders or decrees for RI/FSs under CERCLA, since a directive promulgated on August 28, 1990, have risk assessments conducted by the EPA or a state, if the state oversight is federally funded (U.S. EPA, 1990b). Any orders in existence at the time of the directive, which specified that the risk assessment would be performed by the entities that at some time had some control over the waste or the Potentially Responsible Parties (PRPs), remained in place.

The EPA has provided numerous guidance documents and guidelines for conducting human health risk assessments. Some of the most useful include:

Risk Assessment Guidance for Superfund, Volume 1: Human Health Evaluation Manual (Part A) (U.S. EPA 1989a).
Superfund Exposure Assessment Manual, (U.S. EPA 1988a).

There are also five guidelines for assessing the health risks of environmental pollutants (U.S. EPA 1986a–e).

1. Guidelines for carcinogenic risk assessment (U.S. EPA 1986a).
2. Guidelines for estimating exposures (U.S. EPA 1986b).
3. Guidelines for mutagenicity risk assessment (U.S. EPA 1986c).
4. Guidelines for health assessments of suspect development toxicants (U.S. EPA 1986d).
5. Guidelines for health risk assessment of chemical mixtures (U.S. EPA 1986e).

This chapter draws from each of the above references, but is based primarily on the Human Health Evaluation Manual, which incorporates information from the other documents.

SUMMARY OF THE BASELINE HUMAN HEALTH RISK ASSESSMENT

The objective of a baseline human health risk assessment is to estimate the risk to human health should no action be taken at a contaminated site. This information is then used by a risk manager or decision maker to determine whether remedial action is necessary.

The baseline risk assessment has four steps (U.S. EPA 1989a):

• Data collection and evaluation
• Exposure assessment
• Toxicity assessment
• Risk characterization

Data are collected to characterize chemical contamination and the site, and to provide the necessary information to understand the contaminant fate and transport, including input parameters for environmental modeling. In the exposure assessment, potential exposure pathways are evaluated, which describes the ways in which a contaminant migrates from a source to a point where human contact is made with the contaminant. The manner in which humans contact contaminants (e.g., ingestion, inhalation, or dermal contact) is referred to as an exposure route. The magnitude, frequency, and duration of specific routes of human exposure are assessed to estimate the contaminant intake or dose. The toxicity assessment includes both an assessment of the potential for contaminants detected to produce health effects, and the evaluation and presentation of dose-response data. The final step characterizes risks based on each of the earlier steps, by integrating contaminant concentrations, estimated intake, and dose-response data, to estimate the potential carcinogenic and nonncarcinogenic risks from current and potential future exposures to the environmental contamination. Critical to any risk characterization is the discussion of uncertainties, to put the estimate of risk into proper perspective. The risk assessment provides the information necessary to decision makers as they work through the process of determining the overall most suitable site cleanup approaches and methods. Each of the risk assessment steps are discussed in greater detail following.

Data Collection and Evaluation

Data collection and evaluation starts with a review of the historical information to determine the site history, to evaluate what data are available, and to identify gaps in the information. Data reviewed includes information on environmental conditions, such as geology, hydrogeology, and soil characteristics, and contaminant data available in previous site or area studies. Site-specific historical reports, as well as information available in area-wide data bases, such as geological and water resources reports, are often available for review.

Risk assessors involved in scoping the data collection portion of the study should ensure that data are collected to define contaminants present on-site, to measure concentrations of the contaminants, to identify source areas and the extent of contamination, and to evaluate the potential and pathways for contaminant migration. A preliminary evaluation of potential exposure pathways will help to select the media and locations to sample (see the later section on Analytical Data Needs for a Human Health Risk Assessment).

The data collected must be of adequate quality to decide whether the estimated risks are acceptable or unacceptable, and to decide what remedial actions are necessary. The quality of the data are specified by the method used to collect the data and the required data validation. For example, soil gas data collected using a field gas chromatograph may be useful for evaluating the potential for contaminant migration into the air, but may not be considered of adequate quality to assess risks from the inhalation of airborne contaminants or to make decisions on remedial actions. Data validation is classified by the amount of documentation and the scrutiny that the data receives after laboratory analysis. Additional guidance for evaluating data quality is available in:

Risk Assessment Guidance for Superfund, Volume 1, Human Health Evaluation Manual (Part A) (U.S. EPA 1989a)
Guidance for Data Useability in Risk Assessment (U.S. EPA 1990c)

Unfortunately, information on background concentrations of contaminants is often overlooked, or only a minimal background sampling program is proposed. Background concentrations of contaminants in the different media sampled are critical for an understanding of site-related risks. For example, in Massachusetts, naturally occurring arsenic in soil is often present at concentrations resulting in risks above the EPA-defined acceptable risk ranges. Adequate background data will allow an evaluation of whether arsenic detected at a site is related to site activities or merely present at a background concentration.

The data collected should provide adequate information to define background and on-site concentrations of contaminants, as well as to provide input parameters for fate and transport modeling. Analytical data collected to estimate exposure should focus on contaminants that are known or expected to be at the site, and that are toxic to humans. Information on site conditions needed for fate and transport modeling should include data identified for specific migration pathways at the site. For example, porosity, hydraulic conductivity, and the total organic content of the soil may be useful in assessing the migration of contaminants from the soil into and through groundwater.

Once the analytical data from environmental sampling have been collected, they are evaluated to determine whether the data are suitable for use in the risk assessment. An assessment of the suitability of the data includes both an evaluation of the methods used to collect the data (as discussed previously), as well as data validation criteria. Sample quantitative limits, data qualifiers, blank contamination, and tentatively identified compounds are reviewed to determine whether the data will be included in further evaluations.

Contaminants of concern are included in the risk evaluation. These are typically identified for each medium as site-related contaminants, with data of adequate quality to be included in the risk assessment (U.S. EPA 1989a). In some instances, not all contaminants detected at the site are used in the risk assessment. This is often the case where the number of contaminants is large and a review of the data indicates that certain contaminants can be dropped from subsequent analysis with little or no impact on the final estimate of risk. Narrowing the list of contaminants may be based on historical information, concentration, frequency of detection, toxicity, mobility, persistence, and bioaccumulation, as well as other criteria. Some examples of contaminants often excluded from the list of contaminants of concern include metals that are essential minerals, and chemicals of low toxicity that are also detected infrequently and at low concentrations. In practice, regulatory agencies often specify that all site-related contaminants with analytical data of adequate quality and at concentrations above background levels be retained in the assessment. Where detected contaminants are eliminated from the evaluation, substantial justification is often required.

It is important to review the distribution and concentration of contaminants in each medium when summarizing the data to maximize its use in estimating exposures. Figures presenting sample locations, source areas and plumes, location of sample maximums, and/or important site characteristics are useful when presenting the data, and may provide information on how to summarize the data. This visual representation may indicate two different areas of soil contamination that, due to accessibility or proximity to a residential area, may result in exposures of different frequencies. The soil could then be categorized into different subgroups for the two locations, for separate exposure assessments. Some recommended information to present is the contaminant name, the ratio of frequency of detection to the number of samples analyzed (and retained in the assessment), the range of sample quantitation limits, the range of concentrations detected, and background levels (U.S. EPA 1989a).

Exposure Assessment

The objective of the exposure assessment is to evaluate the potential for human contact with contaminants in the environment, and to estimate the intake (or dose) of a contaminant by a human following contact with the contaminants. The calculated dose integrates the contaminant concentration, the potential for contact, the expected frequency and duration of the exposure, and the body weight of the exposed person.

Site conditions, such as the physical setting, land use, and the presence of residential or business areas in the vicinity, will affect the final estimate of the dose by, defining the potential for contact with the contaminants, as well as the expected duration and frequency of exposure. An exposure pathway describes the path of a contaminant from the source to a route of contact (e.g., ingestion, inhalation, or dermal contact). A risk assessor identifies possible exposure pathways and evaluates how likely an exposure is to occur. The information collected on the conditions at the site is used to evaluate whether people are likely to come into contact with contaminants at the site, given the current and potential future land use in the vicinity of the site. If it is possible that exposure will occur, the concentration of the contaminants contacted, and the frequency and duration of the exposures are estimated, so that the human dose can be calculated. Each of these steps, characterization of the site conditions, identification of the exposure pathways, and quantification of exposure, is discussed in the following subsubsections.

Characterization of Site Conditions

The evaluation of site conditions—including physical characteristics, land use, human populations, and local activities—assists in evaluating the potential for exposure. Present and future conditions on and in the vicinity of the site should be evaluated, so that the potential for both current and foreseeable exposure can be assessed. Land use and activities on nearby properties are important, since there may be current or future exposures from the migration of contaminants to off-site properties, and since adjacent property use may affect the potential for frequency of access to the site.

Characterizing the physical setting assists in evaluating whether people can contact contaminants on the site, or if contaminants can migrate off-site to areas where human contact can occur. Physical characteristics of interest include the topography; climate; vegetative cover; surface water locations, size, depth, and use; and access to the site area. Physical characteristics of the site affect the potential routes of migration and the potential for exposure to contaminants at a specific site. For example, topography defined by steep inclines may indicate migration of contaminants by surface runoff. Warm temperatures can result in greater volatilization of chemicals, while substantial rainfall will reduce exposures to dusts. Dense vegetation can also minimize the potential for dust exposures and may, in fact, limit direct contact with contaminants in surface soils by providing a cover. The proximity of a surface water body to areas of contamination and the ability of the surface water body to support fish, or regular recreational activities, will impact the potential for exposures to contaminants in surface water. Distances to residential areas, the presence or absence of access roads, the density of forest or undergrowth, as well as the presence of physical barriers, such as cliffs or fences, provide an indication of the ease and likelihood of access to the site.

A thorough evaluation of the area's land use is critical to subsequent steps of the exposure assessment. Land use in the vicinity of the site is presented in order to evaluate human populations who may come in contact with contaminants at or in the vicinity of the site. Typical land use categories are: residential, industrial, commercial, agricultural, and recreational. Local water resources (surface water or groundwater) should be checked to see whether they are used as a potable water supply or, in the case of surface water, to support significant fisheries. Children and adults may come in contact with contaminants at the site in residential, and possibly in recreational, areas. A more limited subgroup of the population, working adults, may be the only people who come in contact with contaminants at a site in an industrial area. Children can have very different rates of exposure to contaminants than adults, due in part to the amount of time they may spend playing outside. As a result, accessibility to children is an important consideration when evaluating land use. Specific land use information is the most useful. For example, knowing that a recreational area is used frequently by residents living nearby for fishing can factor into the exposure assessment by suggesting the use of a higher than average fish ingestion rate, to estimate intake.

Defining land use will also assist in evaluating exposure frequency and duration. Exposures on residential properties are often assumed to be daily and, in the case of inhalation exposures, may occur 24 hours each day. Industrial exposures are more limited, based on a 5-day, 40- to 50-hour, work week. Recreational exposures may be even less frequent. For example, exposures in a recreational area in a remote location may be anticipated as a couple of exposures each year. On the other hand, exposures to contaminants in a recreational area in an urban setting may result in daily exposures to some individuals.

Both present and reasonably foreseeable land use must be discussed. The steps described above should be addressed for both types of land use. A change in current land use might result in different people coming in contact with the contaminants at the site under different conditions, and may change the frequency and duration of contaminant contact.

The understanding of different land uses is to the human health evaluation what the delineation of habitat is to the ecological assessment. By understanding what activities can potentially occur within different areas of the site or affected area, the existing risk and protective contaminant concentrations can be evaluated, based on expected pathways. Just as a breeding habitat represents one type of sensitivity to contaminants and one set of pathways for ecological resources, residential land use is different from open space for humans.

Information on land use is available from local residents, town officials, planning boards, local land use and zoning maps, road and topographic maps, and through visual observations. Present land use is easily compiled from the above sources. Potential future land use is not always as easily determined. Zoning maps, local planning boards, town engineers, and local developers may provide some indication of future land use plans.

Such information can be used to define human receptors, both in terms of the number of people who are or could be exposed, and the characteristics of the population that may affect exposure. Census data provides some indication of populations located nearby. However, information collected to evaluate land use is often more useful in defining potential human receptors. Human receptors at industrial properties may include employees, contractors, visitors, and trespassers from nearby residential populations. Recreational land may be used in a variety of ways, resulting in exposures to many different groups of people. If people use recreational areas for fishing or hunting, for example, people exposed to contaminants from eating fish and game, not merely the people who fish or hunt, should be considered. Local sporting goods stores, state fish and game agencies, and sporting clubs may have information on whether hunting or fishing is conducted on or in the vicinity of the site.

Activities conducted at the site can impact the potential for exposure by providing easy access to contaminants or by increasing the potential for contaminant migration. Sites where regular activities involving digging, or substantial heavy truck or other vehicle traffic on dirt roads or paths will produce dust, and may expose previously buried wastes. Careful consideration of activities on-site will assist in defining populations with the potential for the largest exposures. Again, both activities presently conducted at the site, as well as the potential for future activities, are included in this evaluation.

Sensitive populations with the potential to come in contact with site contamination should be considered carefully. People may be considered sensitive receptors because their age or activities are expected to result in higher exposure rates. For example, children often have a higher rate of exposure to contaminated soil by inadvertent ingestion, due to the amount of time they spend playing outdoors. Other sensitive receptors are identified due to compromised health, such as the chronically ill. The risk assessment method used accounts for expected variability in human sensitivity in the dose-response data; therefore, it is conducted to protect all people. Proximity to a chronic care hospital may not impact calculations of exposure or risk. However, identification of a local chronic care hospital may factor into the risk management decisions made about the acceptability of the risks. In addition, the community may be concerned about chronically ill people living adjacent to a site, so it is important to address this concern in the assessment. Sensitive receptors can be identified by observing the proximity of the site to schools, day-care centers, parks, and hospitals or other health-care facilities, including nursing homes.

Identification of Potential Exposure Pathways

An exposure pathway consists of: the path from the source of the contamination, the process of the contaminant getting from the source to where an exposure can take place, a point of contact where exposure occurs, and intake by a human receptor. When evaluating potential exposure pathways, consider sources of contaminants and release mechanisms, retention or transport media, exposure points, and routes of exposure. An exposure pathway is complete if contaminant intake can occur. For example, dermal contact with contaminants in soil is a complete pathway having the contaminated soil as a source and retention medium, an exposure point at the source, and dermal absorption as the route of exposure. If there is a cover over the contaminated soil, there is no exposure point and the pathway is not complete. Figure 5–1 presents various exposure pathways that may occur at a landfill site. The risk assessor evaluates whether the pathways are likely to be complete, including determining whether there is a point of exposure. Other less direct pathways may also be considered such as

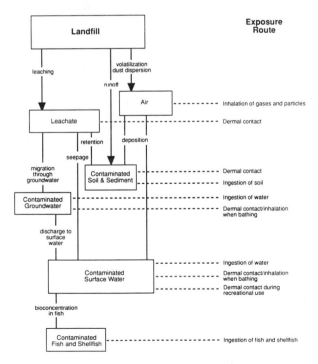

FIGURE 5-1.   Potential exposure pathways for landfills.

uptake of contaminants in soil or irrigation water by plants, which are then eaten by people, or bioconcentration of contaminants in animals eating and drinking contaminated materials, which are subsequently eaten by people.

When evaluating potential pathways, consider the exposure in a sequential manner, starting at the source of the contamination. The source may be identified as physical waste piles, compromised containers, contaminated soils, groundwater, surface water, sediments, or biota. An exposure pathway is not complete where a source is contained with no potential for release. However, even with a contained source, a future exposure may occur should the containment fail, as in the case of a deteriorating drum or tank.

For each source, the potential release mechanisms, the transport media, and the potentially impacted medium need to be delineated. For example, surface sources such as lagoons, soil, and waste piles may release contaminants to the air by volatilization or dust dispersion. Contaminants could reach groundwater by leaching, and then migrating to surface water by groundwater discharge and surface runoff. The soil could be contaminated by leaching, deposition of airborne contaminants, surface runoff, or by tracking, where soil is carried off on the feet of humans and animals passing through the site.

An exposure point is the location or point at which a person may come in contact with contaminants. Exposure points may be located at the source, or at a location some distance from the source, where contact is to media impacted by transport. Where there is no point of exposure, the pathway is not complete. For example, where groundwater is contaminated, but the groundwater is not used as a potable water supply or for any other use in the vicinity of the site, there is no exposure point, and the groundwater exposure pathway is not complete. In this instance, potential future uses of the groundwater and the potential for contaminant migration from groundwater to other media, such as air or surface water would be considered. All of the earlier information, including human access, should be evaluated when identifying exposure points.

An exposure will occur at a point through a specific route or routes. Exposure routes often considered as part of risk assessments are ingestion, dermal absorption, and inhalation. Exposure to contaminated soil may occur through all three routes. Inadvertent ingestion of soil may occur when eating or smoking, following hand contact with soil. For children, ingestion of soil can be an important exposure route if they play outside in areas with contaminated soil. Contaminants in soil which adheres to hands, feet, or other part of a body may be absorbed through the skin. Soil particles can become airborne through actions of wind, vehicle traffic, or digging. Exposure to the soil particles in the air may occur by inhalation.

Quantification of Exposure

All complete pathways identified should be evaluated quantitatively, unless substantial information is available to eliminate them. The EPA identifies four reasons why a complete pathway might not be evaluated: 1) exposure from one pathway will be lower than the exposure from another pathway, in the same medium at the same exposure point, 2) the magnitude of the exposure is expected to be low, 3) both the probability of occurrence and the magnitude of the risk are low, 4) a lack of data makes quantification impossible (U.S. EPA 1989a). Where a lack of data is the reason that a pathway is not quantified, the pathway should be included in the assessment and discussed as part of the risk and uncertainty evaluations (U.S. EPA 1989a). The presentation of exposure pathways should clearly identify those pathways chosen to quantify, those to eliminate, and those to discuss qualitatively, as well as the basis for the decisions.

There are many unknowns and uncertainties associated with the risk assessment process. In general, conservative assumptions are used when estimating exposure and risk, to ensure that risks are not underestimated. One conservative step in the process is estimating the exposure for a reasonably maximally exposed individual. This concept of estimating a reasonable maximum exposure (RME) involves focusing on estimations of exposure representative of high contact rates, large surface areas of skin contacting contaminants, exposure to areas of high contamination, etc., while considering what is reasonable for the circumstances at the site. Professional judgment may be used when defining exposure parameters to estimate dose. However, regulatory agencies often specify the exposure assumptions to be used when estimating the dose for the RME, in an attempt to ensure an even-handed application of the risk assessment process to the regulated community.

For those pathways identified for quantitative assessment, the dose or intake of a contaminant for dermal and ingestion exposures is estimated. Inhalation exposures are an exception, as they are typically evaluated based on air concentrations, so dose calculations are not required. Otherwise, contact with contaminants per unit body weight each day is expressed in units of mg/kg/day, meaning milligrams of contaminant per kilogram of body weight per day. Intake is estimated from the exposure point concentration, contact rate, frequency, and duration of the exposure, body weight, and averaging time. Exposure point concentrations are estimated from both measured analytical data and from environmental models. Numerous guidance documents are available for environmental models, including:

- Superfund Exposure Assessment Manual (U.S. EPA 1988a)
- Air/Superfund National Technical Guidance Study Series, Volumes I, II, III, IV (U.S. EPA 1989d)

The basic equation used to estimate intake is (U.S. EPA 1989a):

$$I = \frac{C \times CR \times EF \times ED}{BW \times AT}$$

Where:

    I—intake
    C—chemical concentration in the exposure medium
    CR—contact rate
    EF—exposure frequency
    ED—exposure duration
    BW—body weight
    AT—averaging time

For each of the parameters needed to quantify exposure, site-specific information, the populations with the potential to be exposed, and agency guidance should be reviewed prior to selecting the values. For example, a different contact rate for soil ingestion is appropriate if a child, rather than an adult, is exposed, and state agencies may provide soil ingestion rates different from those used by EPA. Where possible, site-specific information should be evaluated to assess whether site-specific exposure parameters are more appropriate. Standard default values are available for many exposure parameters, in the absence of site-specific information (U.S. EPA 1991b).

Under some circumstances, the calculation of contact rate requires two or three variables. Where contamination is in water, the dermal contact rate includes two variables: the surface area of the skin making contact and the permeability of the contaminant through the skin. In this instance, the dose calculated is an absorbed dose.

An applied dose is the dose calculated from the exposure concentration only, while an absorbed dose takes into account not only the concentration contacted, but the amount absorbed into the body. An absorbed dose may be calculated based on the percent absorbed through the skin or intestinal tract. Most toxicity data are developed from applied doses. A toxicity value developed as an applied dose is not appropriate to use to evaluate risk from an exposure estimated as an absorbed dose. As a result, either the dose or the toxicity criteria must be adjusted (see the section on Dose-Response Data). A contact rate used for ingestion (such as drinking two liters of water a day) will provide an estimate of the applied dose, while an absorbed dose is estimated only after multiplying the chemical-specific applied dose by the proportion absorbed following ingestion, and a dermal dose is estimated as an absorbed dose.

Exposure frequency is often a site-specific consideration based on site conditions, although the default frequency is assumed to be daily for some pathways. For example, for the ingestion of water from a potable water supply, a daily exposure is most often assumed. Less than daily frequencies are used for other exposures, such as contact with surface water when swimming or contact with soil. Exposure frequencies may differ depending on the area of the country where the site is located. The frequency of swimming may be higher in the warmer regions of the country; where the ground is frozen, and/or snow covered for a portion of the year, contact with soil may be limited. Where statistical data are available for potential frequencies of exposure, the 95th percentile value is often used as a conservative estimate of exposure. There are multiple sources of information on exposure frequency, some of which are listed below:

- Superfund Exposure Assessment Manual (U.S. EPA 1988a)
- Exposure Factors Handbook (U.S. EPA 1989b)
- Risk Assessment Guidance for Superfund, Volume I: Human Health Evaluation Manual, Supplemental Guidance: "Standard Default Exposure Factors" Interim Final (U.S. EPA 1991b)

Exposure duration is the length of time that an exposure is expected to occur. It is estimated based on the pathway and the population evaluated. Typical durations for residential exposures may be 9 years (national median time at one residence—50th percentile) 30 years (national upper-bound time at one residence—90th percentile) or 70 to 75 years (representing a typical lifetime) (U.S. EPA 1989a,b). Information available from a local town hall may be useful when evaluating whether the site is in a location likely to be similar to the national average. An actual location history could show that people living in the vicinity of a site tend to stay in the same area for a lifetime (perhaps families live in the same area for generations).

Where worker or child exposures are anticipated, other exposure duration values would be used. As was discussed for the frequency of exposure, and consistent with the conservative nature of the risk assessment, the 95th percentile value for exposure durations is used where statistical data are available. The average body weight of 70 kg is used to represent a typical American adult. Age- and gender-specific values for body weight are available in various guidance documents (U.S. EPA 1985, 1989a, 1989b, 1991a).

The averaging time is defined depending on the toxic effect of the chemical being evaluated. To evaluate acute and subchronic effects, shorter averaging times would be used than those used to evaluate chronic effects. Acute effects are evaluated for exposure durations sometimes as short as one exposure event. The duration for assessing subchronic effects is often defined as two weeks to seven years, and chronic effects as seven years to a lifetime (U.S. EPA 1989a). Noncarcinogenic effects are evaluated by averaging over the duration of the exposure, while the exposures for carcinogenic effects are averaged across a lifetime.

Toxicity Assessment

The toxicity assessment includes both an evaluation of whether contaminants attributable to the site or site activities can cause adverse health effects in humans, and an evaluation of dose-response relationship between the dose of a particular contaminant and the possibility of an adverse health effect. The EPA has assessed the toxicity of numerous contaminants, and has provided the information in the Integrated Risk Information System (IRIS), an on-line data base. A secondary source of toxicity data is the Health Effects Assessment Summary Tables (HEAST), provided quarterly by the EPA's Environmental Criteria and Assessment Office. A summary of dose-response information is presented here to provide information on how it is used in the risk assessment. A more in-depth discussion of dose-response data, how it is developed, and sources of information is in the later section on Dose-Response Data.

Reference dose (RfDs) or reference concentration (RfC) values and critical effects are required to evaluate substances with noncarcinogenic effects. A chronic RfD or RfC is considered to be the average daily intake level or airborne concentration unlikely to cause significant adverse threshold health effects in humans (including sensitive populations) exposed for a lifetime. The RfD is a dose estimated in units of milligrams of chemical per kilogram of body weight per day (mg/kg/day). An RfC is an air concentration estimated in micrograms of chemical per cubic meter of air ($\mu g/m^3$). The RfD is compared to an estimated dose resulting from an exposure to contaminants, to assess the likelihood of health impacts from the estimated exposure. It is often referred to as an oral RfD for evaluating oral exposures. An RfC (often referred to as an inhalation RfC) is used to evaluate the potential health impacts from an estimated or measured concentration in air. Oral RfDs are often used to compare to dermal doses, after an adjustment is made to ensure that both represent absorbed dose, as necessary.

The values for RfDs and RfCs are generated by the EPA based on the assumption that threshold levels exist for noncarcinogenic health effects (U.S. EPA 1989a). The critical effect is the health effect observed in the study used to calculate the RfD or RfC. The toxicity criteria are set to prevent the critical effect from occurring, following an exposure. Calculated intake levels are compared with RfDs, as discussed in the risk characterization section, where a calculated intake level less than the RfD indicates that adverse health impacts, from the exposure, are not expected to occur. Similarly, where measured or calculated air concentrations are less than RfCs, adverse impacts from an inhaled exposure are not expected. A smaller RfD indicates that a compound may result in a health effect at a lower dose than one with a larger RfD. The severity of the critical health effects used to generate RfDs and RfCs varies between compounds, but they do provide a gauge of relative toxicity for analytes having noncarcinogenic health effects.

To evaluate carcinogenic effects, the carcinogenic weight of evidence and carcinogenic slope factors, or unit risks, are compiled. The EPA has developed a carcinogen classification scheme using a weight-of-evidence approach to classify the likelihood of a chemical to be a human carcinogen. In this system, contaminants are classified as known (Class A), probable (Class B1 or B2), or possible (Class C) human carcinogens. Chemicals not classified as human carcinogens bear a Class D designation, and Class E contaminants indicate evidence of noncarcinogenicity in humans. Information considered in developing the classification includes human studies of the association between cancer incidence and exposure, as well as long-term animal studies conducted under controlled laboratory conditions. A slope factor is an upper-bound probability estimate of developing cancer for a given intake of a contaminant. The slope factor has units of the reciprocal of milligrams per kilogram per day $[(mg/kg/day)^{-1}]$. An inhalation unit risk is the upper-bound probability estimate of developing cancer for exposure to one microgram of contaminant per cubic meter of air $[(\mu g/m^3)^{-1}]$. Both are used to estimate the upper-bound cancer risk from an estimated exposure to a specific contaminant. The slope factor and inhalation unit risk for various carcinogens provides a relative measure of the strength of a carcinogen. A larger slope factor or unit risk indicates a more potent carcinogen.

RfDs and RfCs are developed using uncertainty factors to account for intraspecies extrapolation, interspecies extrapolation, extrapolating subchronic exposure to evaluate chronic conditions, the use of a lowest observed adverse effect level instead of a no observable adverse effect level from an exposure, and uncertainty in the experimental data. The slope factor and unit risks are upper-bound estimates of the risks based on an upper 95th percentile confidence limit on the slope of the dose-response curve (see the later section on Dose-Response Data). As a result, the actual hazard from exposure to noncarcinogens or the actual cancer risk is unlikely to be higher than the estimated value using the toxicity criteria, and may be less than the estimated values.

Dose-response data are not available for all chemicals; some methods for developing RfDs, RfCs, slope factors, and unit risks are discussed in the later section on dose-response. Where neither dose-response nor data for developing toxicity values available, a chemical may be retained for discussion in the risk assessment uncertainty section without quantifying risk.

Risk Characterization

Information from intake calculations and the toxicity data are integrated, to characterize carcinogenic and noncarcinogenic risks. Initially, a risk assessor estimates contaminant-specific risks for one exposure pathway. Since exposure at hazardous waste sites is seldom limited to one contaminant or one exposure route, the method for estimating exposures to multiple contaminants from a variety of exposure routes is also described.

Risks From Exposure to One Contaminant From One Exposure Route

Contaminant-specific carcinogenic risks are estimated by multiplying the chronic daily intake by the slope factors (SF), or the estimated air concentration by the inhalation unit risk for each contaminant. For example, exposure to 2 $\mu g/m^3$ of a contaminant with a unit risk of 3×10^{-6} $(\mu g/m)^{-1}$ results in an estimated upper-bound lifetime cancer risk of 6×10^{-6}. The estimate of risk is considered an upper-bound estimate of risk, and it is likely that the true risk is less than the value predicted.

The excess individual lifetime cancer risk associated with a given exposure is expressed as a proportion (e.g., 1×10^{-6} or one in a million). It represents the incremental increase in an individual's lifetime risk or chance of developing cancer that is attributable to the exposure. Another way to view a one-in-a-million risk is that, given an exposure to a million persons, one additional case of cancer may occur from the exposure. This estimate of carcinogenic risk is valid at low risk levels, where risks are estimated at less than 0.01. The level of increased cancer risk considered negligible is still widely debated, but among scientific and regulatory communities, the range of 10^{-4} to 10^{-6} is often considered the acceptable risk range. This risk range has been adopted by the EPA for use with Superfund sites, as is defined in the NCP (U.S. EPA 1990a).

Contaminant-specific noncarcinogenic effects are estimated by a calculated hazard quotient. The quotient is calculated by dividing the estimated intake by the reference dose (RfD) or, in the case of air exposures, the air concentration by the RfC, making sure that intake and toxicity criteria are for the same exposure duration. A hazard quotient of less than one means that the estimated intake is less than the RfD or RfC, and adverse health effects are not expected. A hazard quotient of greater than one indicates that adverse health impacts may occur, but does not indicate that adverse effects will occur.

When calculating the hazard quotient or increased lifetime cancer risk, it is important to ensure that the intake values and toxicity data are in the same (or inverse) units, and that both represent either the applied dose or absorbed dose, as was discussed in the exposure assessment. Dermal intake is calculated as an absorbed dose. Where the toxicity value for a chemical is expressed as an administered dose, either the toxicity value or the estimated absorbed dose should be adjusted. For example, where the toxicity value is an administered dose, it can be adjusted for the percent of the chemical absorbed for the route of administration used in the study from which the toxicity value was estimated. In this way, the absorbed dermal dose is evaluated using an absorbed dose for the toxicity value. In other instances, certain toxicity values have been calculated as an absorbed dose. In those cases, the exposure intake for each exposure route should be presented as an absorbed dose.

Risks From Exposures to Mixtures and Multiple Routes
Most often, people are exposed to a mixture of contaminants through a variety of different routes or pathways. Little scientific information is available on the toxicity of contaminant mixtures, or how the potential health impacts from a contaminant are affected by interactions with other chemicals. A contaminant in solution with other contaminants may be absorbed into the body at a higher or lower rate than in its pure form, and natural detoxification mechanisms may be affected differently by the contaminant mixture. Without additional contaminant or mixture-specific information, estimates of both carcinogenic and noncarcinogenic risks from exposure to contaminant mixtures are most frequently based on an assumption of additivity. To estimate the total risk or hazard from an exposure pathway, contaminant-specific cancer risks are summed and hazard quotients are summed. Where synergistic effects occur, the assumption of additivity will underestimate risks, while should the mixture cause a decrease in toxicity (antidotal effects), the assumption will overestimate risk.

The sum of contaminant-specific cancer risks for a certain pathway, is an estimate of upper-bound excess lifetime cancer risk from exposure to multiple contaminants by one exposure pathway. The simple summing of risks is considered appropriate for most Superfund risk assessments where the total cancer risks do not exceed 0.01, though certain limitations do exist. The limitations include: giving equal weight, when summing risks, to contaminants with different weights of evidence, and to contaminants with slope factors derived from animal and human data; and summing risks calculated based on an upper 95th percentile confidence limit of the probability for response, which may result in an overestimate of risk, where numerous contaminants contribute large portions of the risk (U.S. EPA 1989a).

The sum of the hazard quotients results in what the EPA refers to as a hazard index. Only hazard quotients for contaminant exposures for similar exposure durations are summed. Where an initial summing of hazard quotients results in a hazard index close to or greater than one, a second step is taken, where major effects from contaminant exposures are considered, and only those contaminants causing effects on the same organ or system are summed to determine more than one hazard index for the exposure pathway. It is important to note that major effects includes both the critical effect used to develop the RfD or RfC, as well as other effects seen at higher doses.

In some instances, it may be appropriate to sum pathway-specific risks to estimate the total risks for a population likely to be exposed by more than one pathway in other cases, pathway-specific risks may be of greatest interest. Whether or not to sum pathway-specific risks may be a function of how likely it is that a person or group of people will be exposed by more than one pathway. In some instances, summing risks from all exposure pathways may result in an unreasonable overestimate of risk, where it is unlikely (or impossible) that a person would be exposed to contaminants by each pathway evaluated in the risk assessment. On the other hand, where exposure to surface water during recreational use is estimated for both dermal contact with and ingestion of water, risks from these two exposure pathways are summed to assess the risk from total contact with the surface water.

Uncertainty of Assessments

The human health risk assessment is not complete without a description of the uncertainty associated with the risk assessment process and, where possible, a conclusion on whether the uncertainty is likely to over- or underrepresent risk. Risk assessments are conducted to ensure that public health is protected. As a result, uncertainties are often addressed by using conservative assumptions, where the actual values are not known. Conservative assumption may be reflected in each step of the assessment and, while protective of public health, should be identified to define the uncertainties in the final estimation of risk. Information on uncertainty is valuable to the risk manager when interpreting the results of risk characterization. It needs to be clearly expressed so that it can be used when making risk management decisions. For example, where the calculated risk level is above the acceptable risk range, but the major component of risk is based on a Group C carcinogen or where the major component of a hazard index is based on an RfD with a large uncertainty factor, the risk manager may decide that no remedial action is necessary.

The uncertainty associated with the risk evaluation is the result of the uncertainty associated with the data, with the assumptions used in developing the exposure scenarios, and with models used to evaluate the exposures. General sources of uncertainty include:

• Environmental sampling
• Analytical chemistry
• Selection of substances used to calculate risk

- Modeling, fate, and transport assumptions
- Exposure scenario development
- Toxicological data
- Characterizing risks to multiple chemicals from multiple exposure pathways
- Complex interactions of the above

Each of these uncertainties is discussed below. In a risk assessment, it is important to identify site-specific uncertainties for each of these areas, as appropriate.

The first two sources of uncertainty are common to any sampling and measurement routine, and are associated with the representativeness of the sampling, as well as the analytical capabilities of the instrumentation, such as the ability to detect low concentrations. Uncertainties arising from the contaminants selected to characterize risks come from the limits of the analysis, limitations of toxicity data, and the understanding of the background data, to name a few examples. Contaminants analyzed are often limited to groups of contaminants, such as priority pollutants, which may exclude substances at the site. In some instances, there are no toxicity data presently available for the contaminants detected. Where no toxicity data are available for a contaminant, risks from exposure to the contaminant cannot be quantified, and thus the overall risks may be underestimated. An inadequate understanding of background concentrations may result in the inclusion or exclusion of site-related contaminants.

When environmental models are used to estimate exposure point concentrations, environmental conditions at the site, or the chemical and physical properties of contaminants, may not be fully characterized. For example, the organic content of the soil, groundwater flow rate, or the partitioning of a contaminant between an organic and aqueous phase may not be fully understood, and each may be estimated in a groundwater model; these estimations introduce uncertainty in the calculation of exposure point concentration and, ultimately, into the estimation of risk. Where exposure point concentrations are estimated directly from analytical data, the handling of samples where contaminants are not detected should be specified, since the method used will effect the estimate of exposure.

In developing a scenario for an exposure pathway, simplifying assumptions are used to calculate the dose. When evaluating exposures, actual exposure point concentrations, contact rates, body weights, and exposure frequency and duration are often not known. The risk assessor often relies on literature values to estimate exposure to an actual or, in the case of future exposures, hypothetical population. The magnitude of the uncertainties can vary tremendously, from a smaller unknown associated with the estimate of body weight (where there is a large amount of data) to larger unknowns associated with the absorption of contaminants in soil or water across the skin. The tendency for each assumption to over- or underestimate risk should be identified. As is consistent with the overall protectiveness of the risk assessment, conservative assumptions are often used to avoid underestimation of risk.

Some of the unknowns associated with toxicological data include uncertainties associated with the animal experimentation, extrapolating high experimental doses to the lower dose generally of concern, given environmental conditions, and extrapolating human health effects from animal data. Toxicity data are not available to evaluate dermal exposures. The use of oral RfDs and slope factors to estimate the risks from dermal exposures contributes to uncertainty. An oral RfD or slope factor, based on an administered dose, may be adjusted to reflect an absorbed dose, to be consistent with the estimated absorbed dermal dose. This practice assumes that the toxicity or carcinogenicity is unaffected by the route of exposure. This is not the case if the toxic endpoint differs depending on how the exposure occurs. Where information is available indicating that dermal exposures result in different health impacts than oral exposures, the use of oral toxicity values to evaluate dermal exposures is not appropriate. For this reason, and the limited information often available on absorption, the practice of using oral toxicity data to evaluate impacts from dermal exposure presents additional uncertainties in the risk assessment.

The risk from exposure to multiple contaminants acting on the same target organ or system are assumed to be additive. In addition, risks from exposures to contaminants by different exposure routes are often assumed to be additive. Scientific information on risks from exposure to contaminant mixtures and multiple routes or pathways is minimal. Assuming additivity may underestimate the risks where a contaminant in a mixture is absorbed into the body more readily than in its pure form, or where the toxicity of one of the contaminants is activated by the presence of another one.

Given the variety of uncertainties associated with each step of the risk assessment process, no numerical estimate of uncertainty is generally made. The evaluation should not be considered to be a predictor of absolute risks. However, it is a procedure used to identify the areas of greatest concern in evaluating remediation objectives.

HUMAN HEALTH RISK ASSESSMENT IN THE FEASIBILITY STUDY

The first questions that the risk assessment is used to answer are what is the level of risk, and is remediation necessary to protect public health? Where risk levels are assessed as unacceptable, a feasibility study is initiated and the baseline risk assessment results are used, together with other information, such as standards and criteria to set remedial action objectives and to evaluate different remedial actions. A variety of possible remedial actions are evaluated as part of the feasibility study; the ability of action to protect public health is one of the evaluation criteria. Health impacts should continue to be assessed during the implementation and review of the adequacy of the alternative remedial actions.

Remedial Action Objectives

Remedial action objectives are established to clearly define why remediation is required, and to establish goals to reach with remedial action. The establishment of remedial action objectives is the point where risk management is introduced, as opposed to simply calculating risk, as in the assessment phase. Risk management includes developing objectives that are set based on regulatory policies, analytical capability, as well as other considerations, such as environmental background concentrations.

Remedial action objectives are developed to be specific to contaminants, media of concern, and exposure pathways. An example of a remedial action objective is the prevention of excess risk of cancer greater than 10^{-4} to 10^{-6} from exposure to polycyclic aromatic hydrocarbons in soils in a residential area. An allowable contaminant concentration or concentration range specific to an environmental media, referred to as a preliminary remediation goal, is also specified.

Initially, preliminary remediation goals are developed prior to the completion of the risk assessment, based on contaminant-specific applicable or relevant and appropriate requirements (ARARs) such as drinking water standards or ambient water quality criteria. These initial goals are modified at the completion of the risk assessment, to provide goals that reflect site-specific conditions and considerations, and to insure the protection of human health and the environment. The goals are established as necessary for exposure pathways and contaminant of concern identified in the baseline risk assessment as resulting in hazards or risk above acceptable levels. They include both promulgated standards and concentrations calculated to prevent unacceptable exposures to contaminants without existing standards.

The NCP states that the baseline risk assessment is used to assist in developing "acceptable exposure levels," which are defined for noncarcinogens as the concentration to which the general population can be exposed for a lifetime without adverse effects (e.g., a hazard index of less than one) and, for carcinogens, as concentration levels corresponding to an excess upper-bound lifetime cancer risk to an individual of 10^{-4} to 10^{-6}. The risk assessment input in the development of site remedial objectives is discussed in *Risk Assessment Guidance for Superfund, Volume 1:Human Health Evaluation Manual Part B, Development of Risk-based Preliminary Remediation Goals* (U.S. EPA 1991a).

Screening and Detailed Analysis of Alternatives

Health impacts must be considered when screening site remediation technologies, and again during the detailed analysis of cleanup alternatives. Often, the health impact assessments are qualitative, where the ability of each technology to meet ARARs or cleanup goals, and any risk associated with the implementation of the technology, are discussed. The risk evaluations of the technologies and alternatives includes an assessment of both short-term and long-term risks. The evaluation must also consider any new risk presented by the action.

These new risks may result from dusts produced during excavation or soil handling, releases to the air from treatment, such as soil venting or air stripping of groundwater (short-term risks), or creating new chemicals during the treatment process. For example, larger molecular-weight chlorinated solvents may break down into vinyl chloride during treatment by processes such as bioremediation (potential long-term risks). Where the treatment alternative provides a method to minimize each of the existing or new risks, it should be described.

In the screening steps, the risk assessment is used primarily to identify technologies or processes that present acceptable or unacceptable levels or risk. In the detailed analysis of alternatives, additional information is required, so that the levels of risk associated with each of the alternatives can be compared. Aside from presenting the results of the baseline risk assessment, qualitative assessments are often adequate for each of the steps of the FS. However, there are circumstances where a quantitative assessment may be warranted. For example, where the decision to select one of two or more alternatives is not clear, the risk reduction for the different alternatives appears to be similar when evaluated qualitatively, or where there are residential populations nearby and new risks introduced by remediation may be unacceptable, a quantitative assessment may be warranted (U.S. EPA 1991b). These estimates of risk can also be used to assess the need for including additional controls to minimize new risks. For example, quantitative evaluations of risks from airborne contaminants during excavation may indicate the need to control dusts or volatile emissions during remedial activities.

The detailed analysis of alternatives required under the NCP includes nine evaluation criteria. Risk assessment factors into several of the nine criteria, listed below:

1. Overall protection of human health and the environment
2. Compliance with ARARs (unless waiver-applicable)
3. Long-term effectiveness and permanence
4. Reduction of toxicity, mobility, or volume through the use of treatment
5. Short-term effectiveness
6. Implementability
7. Cost
8. State acceptance
9. Community acceptance

The first two criteria are defined in the NCP as threshold criteria. These are thresholds, to be met before a remedy may be selected. The risk information in the remedial action objectives is critical in the evaluation of the first criterion, where the evaluation of protectiveness of a remedial alternative looks at how the alternative achieves and maintains overall protection of human health and the environment, and how health risks are reduced (U.S. EPA 1989a, U.S. EPA 1988a).

The third through seventh criteria are considered primary balancing criteria. Risk information factors into both the evaluation of long-term and short-term effectiveness and the permanence of the alternative (criteria 3 and 5). Long-term effectiveness and performance focuses on residual risks associated with any contaminants left on-site at the completion of the remedial action, and how the alternative maintains the protection of human health when the objectives are met. This may include an evaluation of the adequacy of controls that are in place at the completion of the remedial action. The risk evaluation portion of short-term effectiveness focuses on any risks possible during the implementation of the remedial action. For example, new risks may be created by exposing subsurface contamination after surface soils are excavated. There could also be new risks associated with the treatment process, such as transport, storage, and the use of chemicals in the treatment process, or emissions of chemicals, during treatment. This evaluation should look both at protection of workers and the public, and the time necessary to meet remedial objectives (U.S. EPA 1988b).

Human health risks do not generally factor into the evaluation of the last two modifying criteria, although the perception of human health protection often factors significantly into community acceptance. The perception of risks is important to consider in each step of the risk assessment process. Addressing specific areas of community concern early in the remedial process may assist in obtaining greater community support, thus hastening the cleanup, and achieving the ultimate goal of protecting public health. Risk assessments may also be used following the feasibility study, for example, to evaluate risks after data has been collected during pilot studies, or to assess final compliance with remedial objectives.

The primary guidance document for remedial alternatives evaluation is *Risk Assessment Guidance for Superfund, Volume I:Human Health Evaluation Manual Part C Risk Evaluation of Remedial Alternatives* (U.S. EPA 1991b). Additional guidance on how risk assessment is used to evaluate alternatives is also available in Part A of the Superfund Guidance (U.S. EPA 1989a), and in the *Guidance for Conducting Remedial Investigations and Feasibility Studies Under CERCLA* (U.S. EPA 1988b).

ANALYTICAL DATA NEEDS FOR A
HUMAN HEALTH RISK ASSESSMENT

Analytical data collected as part of the RI/FS processes are collected for various reasons: to evaluate the nature and extent of the contamination, to estimate exposure point concentrations, and to evaluate the feasibility of remedial technologies. Data quality required for Superfund risk assessments, including approaches to plan for collecting useable data, is provided in the *Guidance for Data Useability is Risk Assessment* (U.S. EPA 1990c). The risk assessor provides input when developing the scope of the fieldwork and other investigations as part of the RI, to ensure adequate data collection to estimate exposure. The amount of data collected is often a reflection of the information required, as well as the project schedule and the budget. With this in mind, a phased approach to data collection is often preferable.

For human health assessment, information is needed on contaminant concentrations in each of the media where exposure may occur. It is important to consider potential current and future exposure pathways when selecting media and locations to sample. For example, surface soil sampling may be required to evaluate current, infrequent exposures to high levels of contamination at source areas, as well as periphery areas to evaluate more frequent exposures to areas of lower levels of contamination. Subsurface soil sampling would provide information on exposure point concentrations to construction workers, should digging occur on the site, and may be needed to estimate future exposure to contaminants detected in subsurface soils that are on the ground surface following excavation activities on the site. The data collected must be adequate to provide measured concentrations at specific exposure points or the required input information for contaminant fate and transport modeling. A list of media often sampled as a first phase of study is given here, along with examples of reasons why the specific medium is sampled.

Medium	Examples of Reasons to Sample
Soil Gas	Evaluate the extent of soil and/or groundwater contamination and the potential sources of contaminants migrating into air
Surface Soil	Evaluate contaminant concentrations where exposures to surface soils may occur and evaluate the potential for contaminant migration into air
Subsurface Soils	Evaluate contaminant concentrations where exposures to subsurface soil may occur presently, or where it is contacted as surface soil, should the soil be excavated (e.g., following construction), or where these data are needed to evaluate the source of contaminants leaching into groundwater
Groundwater	Evaluate contaminant concentrations where exposure to groundwater may occur from use of the water as a potable water source, irrigation water source, or where groundwater may be a source of surface water contamination
Surface Water	Evaluate contaminant concentrations where exposure to surface water or ingestion of fish caught in the surface water may occur
Sediment	Evaluate contaminant concentration where dermal contact with sediment may occur
A second phase of study may include:	
Air	Evaluate contaminant concentrations where surface soil, surface water, and/or soil gas results indicate a potential for contaminant migration into air
Fish Tissue	Evaluate contaminant concentrations where surface water contamination indicates the potential for bioconcentration of contaminants in fish

The field approach at each site is unique; it is dependent on the site history, contaminants discharged or disposed of at the site, and the local receptors or populations at risk of exposure. The above listing is merely an example of an approach that can be used.

When selecting locations for sampling, the risk assessor should evaluate whether the locations are representative of likely exposure points, and whether they will fulfill the goal of the exposure assessment, for site-wide exposures, as well as exposure to hot spots. Adequate background data are needed to determine the site-related origin of contaminants, compared to background levels. Natural and anthropogenic background levels may be present, complicating evaluation of the contaminant source. Natural background is limited to naturally occurring levels of certain metals in environmental media. Anthropogenic background evaluations may extend to polycyclic aromatic hydrocarbons related to pavement, railroads, or other off-site asphalt and tarring activities; and pesticides, which may be related to historical uses of pesticides both on and off of the site. When evaluating anthropogenic background concentrations, it is useful to consider the site history. Low area background concentrations found on a site with a history of DDT disposal may be useful in determining incremental risks over background levels, and may also be useful in the FS when establishing cleanup goals. Adequate numbers of background samples should be collected, so that a statistical evaluation of whether site concentrations are consistent with or above background concentrations can be made. Background sample locations should be selected carefully, to ensure that the selected locations have not been impacted by the site.

There are differing approaches as to how anthropogenic background evaluations can be handled in the estimate of risk. It is useful to understand whether the contamination is likely to be site-related, or if contamination ubiquitous in the environment due to other than site-related activities. Contaminants present in sampled media, even at concentrations consistent with anthropogenic background levels can be retained in the risk assessment and considered separately when characterizing risk, to avoid omitting contaminants that may be important contributors to overall risks; or contaminants present at background concentrations can be eliminated from the assessment during the preliminary data evaluation phase, to focus the risk assessment on site-specific contamination and risk.

DOSE-RESPONSE DATA

Risk assessment combines information and assumptions specific to a site (contaminants present, site accessibility, etc.) with toxicological information that is not specific to the site. Toxicological information is presented in a human health risk assessment primarily in the dose-response section, where the relationship of quantified dose to human health effects is described or predicted. Once site-specific exposures are estimated, they are then compared to the assembled dose-response data in the risk characterization section of the assessment, in order to estimate site risk. Dose-response data used in human health risk assessments are derived to estimate the risk to people from the exposure to contamination at a site. Dose-response data used in a risk assessment must be relevant to the exposure routes identified in the exposure assessment, such as ingestion, inhalation, or dermal contact.

Quantitative risk assessments present one or more numerical indices that describe the potential for toxic effects at a site. Each index is calculated as a function of expected dose and the estimated degree of toxicity of each contaminant to humans.

Numerical estimates of toxicity for many environmental contaminants have been derived by the U.S. EPA from available research data; these estimates include reference doses (RfDs) and reference concentrations (RfCs), which are used to predict noncancer risks, and slope factors (SFs) and unit risks, which are used to predict carcinogenic risk.

Reference doses are doses at which any adverse noncarcinogenic effect is considered highly unlikely. Slope factors indicate the potential carcinogenic potency of a chemical thought to cause cancer. In some circumstances, when concentrations rather then doses are evaluated, reference concentrations and unit risks are used instead of RfDs and slope factors. Reference concentrations are concentrations at which any adverse noncarcinogenic effect is considered unlikely. Unit risks indicate the potential risk of a carcinogen per unit of concentration.

The dose-response section of a risk assessment presents toxicity information, and gives RfDs and/or SFs as quantitative indices of that toxicity. Tables showing RfDs and SFs are readily available. The information in this section is presented to assist ecologists in developing approaches to evaluate toxicity for ecological risk assessment. This section summarizes the process that is normally used to reduce available toxicological data to numerical dose-response values for human health risk assessments, following EPA methodology.

General Findings of Toxicological Studies

Toxicologic and pharmacologic studies show a few very consistent results. The data generally show that a sufficiently small dose of a chemical either has no effect, or the effect is too small to be measurable. On the other hand, for most chemicals, a large enough dose is fatal.

The relationship of response to dose can be graphed, as illustrated by Figure 5–2. Dose, as the independent variable, is usually represented by the X-axis, and response as the Y-axis. The figure illustrates how experimental dose-response data from a human or animal population can be plotted. In nearly all cases, the percentage of animals responding increases as doses increases. Depending on the chemical, the dose-response curve may be linear, curve up, or curve down within a particular range of doses.

Studies show that toxic effects may vary, depending on the route of exposure. Both the effects observed and the sensitivity to each effect may differ between routes of exposure. As an example, the principal toxic effect of cadmium inhalation is a reduction in lung capacity, while the primary effect of excessive cadmium ingestion involves kidney damage (Hammond and Beliles, 1980). The type and degree of the effect also depends on the medium in which the chemical is applied. There is often a difference in the magnitude of effects between administration in oil versus water, in studies of oral exposures (NAS 1977). The metabolism and distribution of toxins may also vary by exposure route. This information is typically difficult to acquire. The severity of effects may also depend on the duration of exposure. Some chemicals are essential nutrients, or otherwise beneficial to the animal, at low doses and toxic at high doses. As a result, simple classification of chemicals as toxic or nontoxic are not always possible.

Humans and their evolutionary ancestors have been exposed to potentially toxic chemicals throughout evolution. These include plant and fungal toxins, animal venom, and minerals. Animals have evolved some mechanisms to detoxify and/or excrete these chemicals. Animals have also evolved some repair mechanisms, including some repair of the damage caused by toxins. Animals may use the same mechanisms to detoxify, excrete, and/or repair the damage from naturally occurring and synthetic toxins.

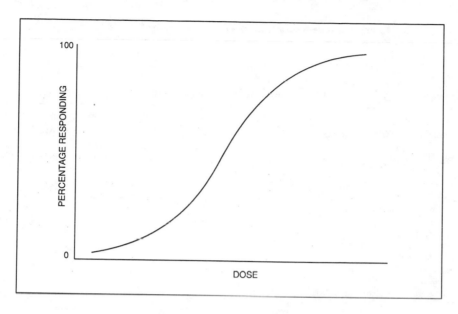

FIGURE 5-2.   Dose-response curve.

There is a concept of threshold dose, which is that a healthy organism can tolerate a quantifiable dose of a toxin with no adverse effects. The animal tolerates the toxin either by utilizing reserve capacity in the damaged organ or system, by detoxification or excretion, or by repairing the damage. Each of these strategies would generally be favored by evolution. The individual is either not adversely affected by the toxin, or else is sufficiently resilient so that no lasting damage occurs. However, at some threshold level, the ability to tolerate the toxin is exceeded, leading to a toxic effect.

Another hypothesis is that damage from some chemicals can occur at the smallest doses, and that some types of damage may not be repairable. If this applies, then there would be no threshold dose below which there are no adverse effects. The commonly held theory is that mutagenic and carcinogenic effects have no dose threshold. On the other hand, regulators almost always assume that all other manifestations of toxicity (e.g., neurotoxicity, kidney damage, reproductive toxicity) have a threshold below which there is no adverse effect.

The truth of either of these basic assumptions cannot be determined purely by experimentation. To test for the presence of small effects as the dose approaches zero, one of two problems is presented. Either the effect is so mild that the experimenter cannot measure any effect, or it is so infrequent that an impractically high number of test animals would be required to statistically confirm the existence of any effect.

Whether the toxicity of a chemical is best described by a threshold or nonthreshold model can be judged by the shape of the dose-respond curve at the lowest dose ranges measured, and on postulated mechanisms of metabolism, toxic action, excretion, and repair. One case where the assumptions of threshold levels have been challenged is lead, for which there may be no practical dose threshold that would be safe (U.S. EPA 1991c).

Once it has been determined whether the toxicity of a contaminant will be considered threshold or nonthreshold, data are reviewed to determine a reference dose or a slope factor, as appropriate. Reference doses (or reference concentrations, for inhalation) for a considerable number of contaminants, and slope factors, have been derived for

TABLE 5-1.   Example of Dose-Response Data Available From the U.S. EPA.

Chemical	Oral RfD mg/kg/ day	Critical Effect (oral)	Inhalation RfC ug/cu.m.	Carcinogen Class	Oral SF per mg/kg/ day	Inhalation Unit Risk per ug/cu.m	Reference
Benzene	NA	NA	NA	A	0.029	0.0000083	IRIS 1/22/92
Cadmium	0.0005*	Significant proteinuria	NA	B1	NA	0.0018	IRIS 1/22/92
DDT	0.0005	Liver lesions	NA	B2	0.34	0.000097	IRIS 2/12/92
Phenol	0.6	Reduced fetal body weight	NA	D	—	—	IRIS 1/20/92

*RfD for cadmium in water only
NA Not available in IRIS
— Not applicable (not carcinogenic)
Critical effects not available for inhalation
RfD Reference Dose
RfC Reference Concentration
SF Slope Factor
IRIS Integrated Risk Information System (U.S. EPA 1992)

evaluating the ingestion and/or inhalation of many environmental contaminants that the EPA has classified as Group A (carcinogens) or as Group B (probable carcinogens), and for some of those classified as Group C (possible carcinogens). Table 5–1 presents recent EPA dose-response data for four representative compounds. Most carcinogens also have observable threshold effects; some of these have published reference doses as well as slope factors. Benzene is an example of a commonly encountered carcinogen for which there are known noncarcinogenic effects, but for which no reference dose is currently published (U.S. EPA 1991c).

Derivation of Dose-Response Data

The risk assessor should investigate the derivation of contaminant-specific dose-response data, to ensure that the data are applicable to the condition being assessed. For example, the chemical being assessed may not be exactly the same as that for which dose-response data were derived, or the exposure route being assessed may not be the same as that for which toxicity data were derived. In addition, most dose-response data are derived from animal experimentation. Dose-response data from animal studies are converted for use in evaluating human health risks by a set of extrapolations. Two different methodologies are use by the EPA, depending on whether the effect is considered threshold or nonthreshold, as described in this section.

Contaminant identification would seem to be straightforward. However, to ensure that appropriate dose-response data are used, the chemical form of the contaminants must be clearly distinguished. The identification of the chemical is primarily an issue for inorganic contaminants. Excepting elemental mercury, exposures to elemental metals are rare. Likewise, laboratory studies generally involve exposure to salts, oxides, or other metal compounds, rather than to elemental metals. Absorption, metabolism, toxicity, and excretion of metals depend on the compound present and its physical form, yet many RfDs and slope factors for metals, as presented by EPA, do not specify the metal compounds or forms to which the number applies. An exception to this rule, which illustrates the problem (potentially for all metals), is the inhalation slope factor for nickel, which is applicable specifically only to nickel refinery dust and nickel subsulfide (U.S. EPA 1991d). The correct identification of organic contaminants is an issue in the few cases where the RfD or slope factor is listed for a designated mixture of compounds, such as mixed xylenes, or PCB Aroclor 1260. Since, for example, PCBs would not be present in the environment in a combination exactly matching Aroclor 1260, the risk assessor must interpret the dose-response data carefully.

Using dose-response data for a variety of exposure routes is the second problem in applicability. There are three basic chronic exposure routes for chemicals: inhalation, ingestion, and dermal contact. Direct damage to the eyes, which can be so critical in accidental exposures, is generally not considered for chronic effects. When data for either inhalation, ingestion, or dermal contact are not available, the best estimate of toxicity would incorporate data from the exposure route for which there are available data. In some cases, the EPA does present dose-response data that has been converted for use in evaluating exposures by routes other than that used in the study from which it was derived. In other cases, the risk assessor must make the extrapolation.

For example, risk assessors typically use an oral reference dose for dermal exposures, since dermal reference doses are not available. Typically, a one-to-one correspondence is assumed between the toxicity of dermal exposure and ingestion exposures, on an absorbed-dose basis. This correspondence is appropriate for systemic toxins, but not for toxins that have localized effects. Absorption efficiency is considered when assessing toxicity from dermal exposure. Dermal exposures are often described as absorbed dose, while most RfDs are described in terms of administered dose. A reference dose applicable to dermal exposures can be calculated by adjusting the administered oral reference dose by the oral absorption efficiency. Alternatively, a factor called a relative absorption factor (RAF), can be included in the exposure equation to calculate a dose to compare with the administered oral reference dose. The RAF is the ratio of the dermal absorption efficiency to the oral absorption efficiency.

Dose-response information generated and published by the EPA does include data that has been converted between ingestion and inhalation exposures. The EPA generally corrects for known or expected differences in absorption between ingestion and inhalation. The data are presented on the basis of the amount of chemical that is inhaled or ingested; the extrapolation can incorporate the expected difference in absorption across lung and gut membranes.

Some contaminant exposures are better described by exposure concentrations than by estimating dose. When the adverse effect is acute and results from direct contact with the target organ, then concentration is a better description than dose. Examples are direct skin irritants, eye irritants, stomach irritants, and acids and bases whose toxicity is mostly function of pH. A comparison of concentrations, rather than doses, may be warranted when inhalation of contaminants is under evaluation.

The EPA has tended to convert between the RfD and the inhalation RfC, using a 70 kilogram adult with a ventilation rate on 20 cubic meters per day. Guidance is not currently established throughout the agency as to whether the RfD or the RfC should be used consistently (i.e., for receptors other than the 70 kilogram adult breathing 20 cubic meters of air per day). The EPA appears to be moving towards the direct use of RfCs and inhalation unit risks for air exposures to all contaminants, instead of using RfDs and slope factors; inhalation unit risks are highlighted in recent summary tables

(U.S. EPA 1991d). By comparison, occupational limits for contaminants in air have historically been set as concentrations only.

Ecological assessors should make a conscious choice of whether dose or concentration is the most appropriate measure. The amount of PCB-contaminated fish that a mink consumes affects its risks, so dose is an appropriate comparison; whereas the concentration of copper, rather than the amount of water, that passes through a trout's gills is the major determinant of absorption of copper, so concentration is a more appropriate measure. Often, the decision must be based on the form of the data available.

Sources of Toxicity Data

There are at least three types of toxicological data available from which to derive estimates of toxicity; actual human experience, laboratory animal experimentation, and data adapted from structurally similar contaminants. Dose-response estimates rely heavily on laboratory studies, rather than on actual human experience, due to the difficulty in collecting useable data for humans. In animal data, the exact doses are known, but uncertainty in extrapolation to humans is high. In data from humans, the uncertainty in extrapolation is much lower. Ethical considerations preclude administration of measured doses of most harmful chemicals to humans, so that when a toxic effect is observed in humans, as in occupational exposures, the dose often must be estimated indirectly. Among humans, the most sensitive manifestions of chemical toxicity are often subjective. Headaches and feelings of fatigue are difficult to measure objectively. In contrast, observable toxic effects in animals can be measured.

Laboratory studies can be run consistently for a fairly large number of animals at a variety of doses, for a variety of contaminants. This allows a comparison of the toxicity of multiple contaminants under comparable conditions. The data are developed under highly controlled conditions. While interspecies extrapolation appears to be more precise and objective than direct evaluation of human experience, it is questionable whether the extrapolated data are as accurate for humans as data derived from human experience.

For some contaminants, neither animal nor human toxicity data are available. In some such cases, data may be adapted from structurally similar chemicals. In a few classes of organic chemicals, such as polycyclic aromatic hydrocarbons, dose-response values from one compound are sometimes used to evaluate several related compounds.

Most EPA dose-response estimates use animal data. The most sensitive species, sex, strain, and effect are selected in the derivation of RfDs. To derive an RfD, the highest dose showing no effect is preferred, provided that the experimental data are of good quality. Since many of the original trials were set up as range-finding experiments, the highest of the 'no observed adverse effects levels' (NOAELs) may be three to ten times lower than the 'lowest observed adverse effects level' (LOAEL). If there is no NOAEL, then the LOAEL is used to derive an RfD.

Generally, toxicological studies are designed to evaluate the toxicity of one contaminant, not interactions between contaminants. The validity of these data in application to humans is often questioned, since humans are exposed to a mixture of different toxins. A simple addition of effects is the most common method used to evaluate exposure to mixtures of chemical.

In some instances, tests used to derive dose-response data have been performed on animals deprived of certain nutrients. The toxicologist should consider whether data derived from such a test is useful for assessing public health risks. Likewise, some cancer tests are performed on strains of mice that are predisposed to high rates of tumor formation. The applicability of these data to humans should also be evaluated before applying it to critical site remediation decisions.

Extrapolation to Humans From Animal Data

Human dose-response estimates are often extrapolated from experimental animal data. Extrapolations may address differences in the time span of exposure, variable sensitivity by species, variable sensitivity by individuals, and differences in the severity of effects. The degree of variability in sensitivity, and the effect of duration of exposure on toxicity, are typically unknown, so these are addressed explicitly with uncertainty factors. The uncertainty factors introduce a degree of conservatism into the human dose-response estimates.

Threshold Effects

When deriving a safe threshold level of exposure for humans, scaling assumptions for animal size and lifetime are used, and the resultant dose is divided by one or more uncertainty factors. For threshold effects, the EPA scales daily doses by body weight (Calabrese et al. 1992), and the duration of a mouse lifetime is considered to be equivalent to a human lifetime. These assumptions are valid if lifetime dose rates between species have equivalent effects when the daily intake rate is directly proportional to body weight.

The rate or capacity of many metabolic processes is better predicted by body surface area than by the body weight of an animal (Calabrese 1983). As a result, the human response to many drugs is predicted well by animal response, when dose is scaled by body surface area (Calabrese 1983). Similarly, the toxic effects of a contaminant may be similar among species when the dose is measured on the basis of surface area. Therefore, a surface-area scaling factor might be more appropriate than the body weight scaling factor used by the EPA. It would be more conservative, by roughly an order of magnitude, to use a surface-area scaling factor when converting toxicity values from a mouse to a human.

The calculated dose from this basic extrapolation is then modified by the uncertainty factors. Typically, one order of magnitude is used for each factor that applies (i.e., the dose is divided by 10). Any or all of the following may contribute to a total uncertainty factor (U.S. EPA 1989b):

1. The experiment may have been run for less than the total time span of interest (e.g., if a chronic RfD must be derived from a subchronic study).
2. Humans may be more sensitive than the test animal (i.e., the scaling assumption may not account properly for interspecies variability).
3. Some humans may be more sensitive than the average. Since sets of laboratory animals lack genetic and environmental diversity, the test data may not account for intraspecies variability.
4. The test may not have identified a nonzero dose at which there was no effect. Extrapolation from a dose that caused an effect requires an additional safety factor.

As an example, the EPA has established an RfD for toluene based on a calculated NOAEL of 223 mg/kg/day in rats exposed by gavage to toluene over 13 weeks. Using 10-fold uncertainty factors for each of three extrapolations—from the subchronic test to a chronic RfD, from rat to human, and from an average human to a sensitive individual—the EPA derived an RfD that was 1,000 times lower than the animal NOAEL, or 0.2 mg/kg/day (U.S. EPA 1991c).

It is not clear that the factor of ten for interspecies differences is conservative. On the other hand, assuming that each safety factor is conservative, the combination of several safety factors is likely to underestimate a safe dose rate. Combined factors as high as 10,000 are reported, by the U.S. EPA in the derivation of an RfD (e.g., strychnine) (U.S. EPA 1991d). Use of such a high combined uncertainty factor is likely to result in an overestimate of chemical toxicity.

The EPA is beginning to require that assessments report a qualitative description of the level of confidence in the RfD (i.e., high, moderate, or low, as determined by the EPA). EPA guidance is not explicit about the relationship between the overall quantitative uncertainty factor and the qualitative description of confidence in the RfD.

Carcinogenic Effects

When deriving slope factors to describe the carcinogenic potency of a contaminant, the EPA extrapolates from existing data to potential effects in humans without using the ten-fold uncertainty factors used for noncarginogenic risks. Instead, the derivations include some conservative steps. Among these steps is scaling by body surface area when extrapolating from a small experimental animal to a human. Lifetime average daily doses are considered to have equivalent risks for different species if the daily dose rates are directly proportional to the surface areas of the animals; the surface area is considered to be adequately described as the two-thirds power of body weight (U.S. EPA 1989a). This surface area scaling factor is more conservative than an extrapolation based on body weight. In this extrapolation, it is implicitly assumed that the risk of cancer in a human lifetime is the same as the risk of cancer in a (much shorter) animal lifetime, provided the average daily doses are equivalent.

The approach used for cancer effects includes two additional conservative steps. A straight-line extrapolation is used to predict the effects of low doses. This approach is more conservative than most of the biologically plausible alternatives (U.S. EPA 1989e). Secondly, instead of using the best-fitting estimate of the slope of the dose-response line, the EPA uses the upper 95% confidence limit on the slope, based on the study data. The EPA will often calculate, as a summary figure for all acceptable experimental slope factors, the geometric mean of the 95% upper confidence limit estimates. This model assumes that there is one slope factor shared by all species, to be derived for humans from the study animals. There are sometimes differences in scaling between animals; body-weight scaling may be used between rat and mouse, surface area scaling used to extrapolate from rodents to an adult human, and body-weight scaling used again between a human adult and child.

Specific information showing interspecies differences in metabolism or toxicity is usually not available, and no additional allowance for differences in sensitivity between species in normally included in the estimate of a slope factor. The potential for individual variation in humans does not enter into the equation. Also, while the age of exposure tends to greatly affect lifetime cancer risks (Weisburger 1975), slope factors are calculated and presented for use with average lifetime exposures. Carcinogenicity tests are designed to last the lifetime of the animals, but the dose rate may change as the animal grows. A more sophisticated risk prediction model than the use of slope factors would be needed to account for latency periods and other determinants of cancer risk.

Interspecies Extrapolation for Ecological Assessments

In most ecological assessments, dose-response data derived from tests of the species of interest will not be readily available, so interspecies extrapolation will be required. In some cases, there will be data available for a closely related species; in others, no data for the same phylum will be available. In human risk assessments, animal data are almost always from the same class (i.e. mammals) and occasionally from the same order (i.e., primates). Toxicity data is usually available for rodents (or rabbits), so the taxonomic distances and size differentials in interspecies extrapolations to humans tend to be stable. No distinction is generally made in the uncertainty factor used to extrapolate monkey data to humans, compared with that used to extrapolate rat or mouse data to humans.

Since most interspecies extrapolations to humans are from rodents or other small mammals, regulators have not been challenged to extrapolate between closely related species or between phyla to estimate toxicity to humans. In ecological assessments, the differences in taxonomy and size between test species and target species will sometimes be narrow (e.g., trout to salmon), and sometimes be wide (e.g., water flea to salmon). The uncertainty of an extrapolation depends on the degree of taxonomic or ecologic difference between species. The need to extrapolate over both narrow and wide taxonomic or ecologic differences poses additional complexity to the ecologic risk assessor in using interspecies extrapolations.

Sources and Presentation of Dose-Response Data

Reference doses, reference concentrations, and slope factors for many environmental contaminants have been compiled by the EPA. There are two sources of current approved EPA dose-response data. Each is updated according to a different schedule, differences between these sources reflect changes in progress. The Health Effects Assessment Summary Tables (HEAST) are updated quarterly, and present the numerical values in a convenient format (U.S. EPA 1991d). HEAST tabulates RfDs, RfCs, SFs and unit risks, critical effects, the total uncertainty factor, carcinogenicity group, the species and target organ from which SFs were calculated, and references. The Integrated Risk Information System (IRIS) is a data base available for computer access. IRIS provides much of the same information, plus a narrative description of the studies from which the dose-response data were derived, a review of the quality of the studies for purposes of establishing dose-response values, and the overall uncertainty associated with the data. Information such as drinking water standards and water quality criteria is also included.

IRIS provides more information about each chemical than HEAST. In addition, IRIS is updated monthly. Therefore, the EPA's preference is that IRIS be cited as the primary source of dose-response data. To be assured of having the most complete information, the health risk assessor should search both data sets.

Even if no U.S. EPA dose-response data are available in the published sources, the risk assessor does have some options. Data for a closely related chemical may be appropriate for substitution. There may be sufficient experimental data available for derivation by the assessor of a provisional RfD or slope factor. Finally, the contaminant exposure in question may simply be omitted from quantitative assessment and described qualitatively.

A complete risk assessment report should contain RfDs, RfCs, slope factors, and unit risks for the contaminants of concern as appropriate for exposure routes anticipated at the site. These numbers are normally tabulated along with the carcinogenicity classification of each contaminant, and the effects of concern. In more detailed reports, the assessor tabulates the species tested, the route of application, and uncertainty factors.

Risk assessments also typically include, for each contaminant of concern, a toxicity summary in the text, a toxicological profile in an appendix, or both. These sections should include information on the absorption and metabolism of the contaminant, threshold toxicity, carcinogenicity, and special concerns. A brief description of the derivation of the dose-response data is normally included.

REFERENCES

Calabrese, E. J., 1983. Principles of Animal Extrapolation. New York: John Wiley and Sons.

Calabrese, E. J., B. D. Beck, and W. R. Chappell, 1992. Does the animal-to-human uncertainty factor incorporate interspecies differences in surface area? *Regulatory toxicology and pharmacology* 15:172–179.

Hammond, P. B., and R. P. Beliles, 1980. Metals, *in* Casarett and Doull's *Toxicology*, 2nd edition. New York: MacMillan.

NAS, 1977. Principles and Procedures for Evaluating the Toxicity of Household Substances, National Academy of Sciences. National Academy Press, Washington, D.C.

U.S. EPA, 1991a. Risk Assessment Guidance for Superfund: Volume I—Human Health Evaluation Manual (Part B, Development of Risk-based Preliminary Remediation Goals) Interim, Office of Emergency and Remedial Response. Publication 9285.7–01B, December 1991.

U.S. EPA, 1991b. Risk Assessment Guidance for Superfund: Volume I—Human Health Evaluation Manual (Part C, Risk Evaluation of Remedial Alternatives) Interim, Office of Emergency and Remedial Response. Publication 9285.7–01C, December 1991.

U.S. EPA, 1991c. Integrated Risk Information System (IRIS) online database. Office of Health and Environmental Assessment, Environmental Criteria and Assessment Office, Cincinnati, OH.

U.S. EPA, 1991d. Health Effects Assessment Summary Tables, FY1991. OERR 9200.6-303(91-1).

U.S. EPA, 1990a. National Oil and Hazardous Substances Pollution Contingency Plan (NCP) 40 CFR Part 300. Federal Register Volume 55, No. 46, March 8, 1990.

U.S. EPA, 1990b. EPA memorandum of performance on risk assessments in Remedial Investigations/Feasibility Studies (RI/FSs) by Potentially Responsible Parties (PRPs). OSWER Directive No. 9835.15, August 28, 1990.

U.S. EPA, 1990c. Guidance for Data Useability in Risk Assessment. Interim Final. Office of Emergency Remedial Response, Washington, D.C. EPA/540/G–90/008.

U.S. EPA, 1989a. Risk Assessment Guidance for Superfund, Volume I: Human Health Evaluation Manual (Part A). Interim Final. Office of Emergency Response. EPA/540/1–89/002.

U.S. EPA, 1989b. Exposure Factors Handbook. Environmental Assessment Group, Office of Health and Environmental Assessment, Washington, D.C. EPA/600/1–89/043, May 1899.

U.S. EPA, 1989c. Supplemental Risk Assessment Guidance for the Superfund Program. Draft Final. U.S. EPA Region I Risk Assessment Work Group. EPA 901/5–89–001, June 1989.

U.S. EPA, 1989d. Air Superfund National Technical Guidance Study Series, Volumes I, II, III, IV. EPA 450/1–89–001, 002, 003, 004, July 1989.

U.S. EPA, 1988a. Superfund Exposure Assessment Manual. Office of Emergency and Remedial Response. EPA/540/1–1–88/001 (OSWER Directive 9285.5–1).

U.S. EPA, 1988b. Guidance for Conducting Remedial Investigations and Feasibility Studies Under CERCLA. Interim Final. Office of Emergency and Remedial Response (OSWER Directive 9355.3–01).

U.S. EPA, 1986a. Guidelines for Carcinogenic Risk Assessment, 51 Fed. Reg., 33992 (September 24, 1986).

U.S. EPA, 1986b. Guidelines for Estimating Exposures, 51 Fed. Reg., 34042 (September 24, 1986).

U.S. EPA, 1986c. Guidelines for Mutagenicity Risk Assessment, 51 Fed. Reg., 34006 (September 24, 1986).

U.S. EPA, 1986d. Guidelines for Health Assessments of Suspect Development Toxicants, 51 Fed. Reg., 34028 (September 24, 1986).

U.S. EPA, 1986e. Guidelines for Health Assessment of Chemical Mixtures, 51 Fed. Reg., 34014 (September 24, 1986).

U.S. EPA, 1985. Development of Statistical Distributions or Ranges of Standard Factors Used in Exposure Assessments. Office of Health and Environmental Assessment, Office of Research and Development. EPA/600/8–85/010. August, 1985.

Weisburger, E. K., 1975. Industrial Cancer Risks in Dangerous Properties of Industrial Materials. 4th Edition. New York: Van Nostrand.

6

Biological Transfer of Contaminants in Terrestrial Ecosystems

Stephen E. Petron

Metcalf & Eddy, Inc.

INTRODUCTION

The transfer of contaminants through an ecosystem, and subsequent degradation of the ecosystem, presents a difficult challenge for the proving of cause and effect. Often, when attempting to show cause and effect in ecological contamination events, it is impossible because of the difficulty in measuring and showing an effect (Harris, Regier, and Francis 1990)! This probably stems from an inability to sort out the necessary information from the multitude of possible scenarios of contamination within one system. The effects of catastrophic contamination events are often easy to identify. However, these easily identifiable effects are often only of the acute type. Acute effects were readily apparent to millions of viewers as television presented graphic pictures of the oil-soaked animals from the Exxon Valdez disaster in Prince William's Sound off the coast of Alaska in 1989. Likewise, a release of toxic gas in 1984 from a Union Carbide Corporation chemical plant in Bhopal, India resulted in readily identifiable acute effects. The corpses of livestock and humans were easily observed and counted. In both of these cases, it was relatively simple to demonstrate an effect of the environmental contamination. One simply went around and tallied bodies. After correcting for search effectiveness, the world learned that up to 350,000 birds and more than 1,000 sea otters immediately died in the Exxon Valdez incident (Davidson 1990) and 2,500 people immediately died in the Bhopal chemical release (Weir 1987).

But, do these tallies really represent the totality of the effect of these disasters? The obvious answer is no, but then what was the effect? The answer to this question remains unknown today. If, in fact, one removes the anthropocentric viewpoint, the above-mentioned acute effects from these disasters probably represent a minor part of the actual effect on the whole ecosystem. Who knows how many microorganisms, small-sized birds, mammals, amphibians, reptiles, insects, and fish immediately died as a result of these incidents, and are continuing to die. Although there are studies underway, scientists can still really only guess. Such organisms, by their nature and small size, are less obvious and more difficult to observe. Contributing to the difficulty are the complex interactions between the varied organisms of an ecosystem, which may mask or dilute effects to the point where they are not readily discernable.

Furthermore, such acute, immediate effects of the two disasters may only represent a small amount of the total effect, if chronic, long-term effects are considered. It is possible that whole life structures have been destroyed or reconstructed in new ways, greatly altering the ecosystem. Chronic effects, as opposed to acute effects, such as those mentioned earlier, do not show up on TV or in the results of simple tallies. This is because the effects take longer, sometimes considerably longer, to occur. Furthermore, the effects may not, and probably do not, result in dramatic or massive deaths. Rather, what happens is probably a more subtle, sublethal disruption of the ecosystem. To the casual observer, and cursory investigation, there will apparently be nothing amiss.

Presently, one can only speculate about the total ecologic effects of the two disasters discussed earlier. Studies have been undertaken, but demonstrative results will take years to obtain. All the while, the ecosystem continually melds and dilutes the contaminant through its systems, continually rendering the effects more and more subtle.

The difficulty in showing chronic effects of long-term exposure to contamination can not be understated. An example of the long-term insidious nature of environmental contamination is the Love Canal situation. In the 1940s to the early 1950s, Hooker Electrochemical Corporation dumped chemical waste in an unused abandoned canal (Levine 1982). The dump was closed and covered, and in the late 1950s was developed. It took until the 1970s, nearly 20 years later, before people began to publicly complain about health problems (Whelan 1985). As of 1985, there was no "proven" health effects that could be solely attributed to the chemical contamination. This means that cause and effect could not be established for humans, one of the easiest animals to study, living on top of a huge toxic dump. That the effect went so long before even being suspected, and then never actually being attributed to contamination, is testimony to the insidious nature of environmental contamination.

One thing that the Love Canal and Bhopal examples have in common is that the focus of the effect was on humans. Little mention and effort was made to document the effects on other animal life. While the emphasis on humans makes a certain sense, society also has a responsibility to the health of other animals, and the ecosystem as a whole. However, if 30 years is not enough time to prove an effect on humans at Love Canal, one can only imagine how long it would take to show effects on ecosystems. We can only guess what effects were incurred by the ecosystem during these same years, and what changes were the result. Humans make excellent study subjects, as compared to other life. Humans can tell you what they are feeling and are easily measured, tested, and counted. Wild animals, on the other hand, seldom possess these traits, at least to this degree. Where would the investigator start in considering the multitude of possibilities?

One of the primary goals of ecological assessments is to ascertain whether there is an ecological effect of the contamination. Considerable amounts have been written on the difficulties and uncertainties associated with establishing cause and effect at the community or ecosystem level (U.S. EPA 1989a; Harris et al. 1990). Harris et al. (1990) contended that toxicity at the ecosystem level must be studied by investigations of particular ecosystem properties, such as trophic structure, sedimentation, and carbon transfer efficiencies. The problems with studying these aspects, from an ecosystem perspective, are troublesome in their own right, because of their complexity and the requirement of considerable effort and technical sophistication. Hirsch (1980) discussed the problems with showing cause and effect for changes in ecosystems, these problems can be summarized as:

1. Ecosystems are complex and dynamic. They are in a constant state of change, and express a wide variability of time and space, making only very obvious signals observable:
2. Warning signs are often masked by the background noise of natural variability:
3. Natural stresses and variability may be greater in magnitude than contamination-induced stress.

The study of ecological transfer is one method that may enable the resolution of the interwoven mesh that represents ecosystem contamination. That the contaminant moves along defined, albeit incongruous, pathways in the ecosystem, allows it to be traced, isolated, and measured. This evaluation of pathways, or pathway analysis, as it will be called in this chapter, represents a powerful tool in the study, documentation, and characterization of contaminated ecosystems. However, the description of the complete contaminated system is a monumental task and, in even moderately complex systems, is probably not practicable. Therefore, as suggested by Harris et al. (1990), the analysis usually must be organism based. Harris et al. (1990) felt that because toxicity is expressed and most frequently measured at the organism level, despite its ultimate manifestation at the community or ecosystem level, it may be reasonable to use an organism-based model for analysis. They cautioned that, to have any value in establishing validity or shedding light on the ecosystem level, the investigator must have a clear understanding of the organism's position in the ecosystem. The biological transfer of contaminants through organisms is the mechanism through which contaminants cause ecosystem effects. As such, it presents an opportunity to understand and document contamination and its effects at a hazardous waste site.

For pragmatic reasons, managers responsible for the documentation of site contamination focus the collection of data on single organisms, rather than higher levels of biological organization. The information currently available on toxicity is generally only available for individual organisms. Considerable research is required before toxic effects at the ecosystem level will be available. Such research should be encouraged, but is typically out of the scope of an exposure assessment.

Objectives

One objective of this chapter is to provide a basic understanding of the biological transfer of environmental contaminants in terrestrial organisms, and to describe the proper application of this knowledge in exposure analysis, risk assessments, and the setting of critical contamination thresholds. It is also the objective here to show the reader how to use his or her knowledge of ecology and contamination transfer to identify susceptible species and primary exposure pathways, estimate coefficients of transfer, estimate exposure, and determine toxicity risks. This chapter also presents information so that the reader can calculate contaminant levels that are protective of a variety of ecosystem attributes or effects levels, which will, in turn, provide informed input to the complex decision making process of establishing cleanup levels.

Approach

To accomplish the above objectives, the chapter is separated into two major sections, transfer dynamics and transfer pathways analysis. The first section, on transfer dynamics, provides an introductory, general understanding of biological transfer. Biological transfer is defined here as the process that redistributes contaminants through normal biological actions.

To provide the reader with a general understanding, the chapter will briefly describe biological transfer. It will introduce some fundamental aspects of biological transfer, such as intake and fate. These two concepts are theoretically definable for all levels of biological organization (cell, tissue, organism, community, or ecosystem), but will only be presented for individual organisms in this chapter. The relationship of intake and fate to exposure assessment will be discussed, as will the quantification of various compartments of intake and fate and their application in determining exposure and the evaluation of effects levels.

Contaminant transfer behaviors, such as bioconcentration, bioaccumulation, and biomagnification, will be introduced. Their description through ratios is often essential to understanding the magnitude of contamination and the consideration of cleanup levels. Their use allows the direct reconciliation of exposure levels in organisms to environmental concentrations and, therefore, allows the reconciliation of critical toxicity levels to critical effects levels.

The second section, on transfer pathways analysis, shows how this understanding of biological transfer can be coupled with an understanding of ecology to conduct quantitatively based exposure assessments. The essence of the quantitative approach is to establish the relationship of exposure to known critical toxicity levels. This quantified relationship then becomes the critical input to the ecological evaluation of cleanup levels. It will be demonstrated why quantified exposures and effects relationships are essential to the defensible and meaningful evaluation of effects limits.

The chapter shows how using contaminant transfer knowledge and the ecology of the site will allow for a logical, defensible estimation of exposure. Contaminant characteristics, coupled with ecological characteristics, including life history information, will allow for the selection of key species. These species need to have established relationships to the contamination and the ecosystem, making them ideal choices for the determination of exposure. The key species should represent the most "important" (sociologically and ecologically) species on the site that have a potential for exposure. Using the knowledge of transfer dynamics and mechanics, it is often possible to quantify the potential.

This chapter focuses on the application of biological transfer to ecological assessments, through an analysis of the exposure of primarily individual species of birds and mammals. This is consistent with the approaches used elsewhere (Harris et al. 1990). Communities and ecosystems are generally too complex and intractable to adequately and fully address toxic contamination within the scientific and time constraints in place at most hazardous waste sites. Finally, from both pragmatic and efficiency standpoints, individual species are easily identified and delimited, are most easily measured, and have the most quality information available. Therefore, they are the easiest to study and demonstrate exposure and effects levels. Individual species are also most easily recognized and understood by both decision makers and a public untrained in ecology. The chapter contains many real and hypothetical examples using individual species, identified by their common names in the text. The Latin names of the example species are provided in Appendix A of this book.

TRANSFER DYNAMICS

Budgets and Pathways

A complete understanding of the transfer of contaminants between terrestrial organisms and either their environment or other organisms requires a basic understanding of the two fundamental aspects of transfer—intake and fate. Intake, also often called uptake or biouptake, is the act or process of the contaminant being taken into an organism from a source outside of the organism. Fate is the disposition of the contaminant once it has been taken in by an organism. These concepts, and the keeping track of contaminant intake and fate by budgets or balances, are not new (Clark et al. 1988). They have been applied, often in the form of input-output budgets using mass balances, for all levels of biological organization. Borman and Likens (1979) used budget balancing, quantification of input, and the output of energy, minerals, and nutrients in their long-term extensive study of the Hubbard Brook ecosystem in New Hampshire. Others have applied the method to lower levels of organization. Harris et al. (1990) reported on the use of mass balance in the prediction of PCB levels in fish based upon intake into the Great Lakes ecosystem. The practice is equally applicable to a single organism. Furthermore, the use of budget balancing to understand transfer in single organisms is more easily done than for larger systems, such as communities or ecosystems. This is because the boundaries of the organism are readily identifiable, and more closely meet the definition of a closed system, which is an essential assumption in all budget balance efforts. Furthermore, the reaction of an individual organism is better ascertained than are those of more complex biological organizations. The focus is on the individual organism and not on the surrounding environment. In fact, it can be assumed that the environment is limitless.

In a closed system such as a single organism, contaminants can be taken up by the organism and then suffer some fate. This relationship must represent an equality in a closed system.

$$INTAKE = FATE$$

This equation governs the transfer of contaminants, must always be true, and the understanding and knowledge of the component pathways of intake and fate allows its use in exposure assessments. Contaminant intake by organisms is composed of three possible pathways, ingestion, inhalation, and dermal absorption. Contaminant fate in organisms is also composed of three pathways, excretion, detoxification, and retention. Using the above relationship of intake and fate, the relationship of these six pathways is:

$$Ingestion + Inhalation + Dermal\ Absorption = Excretion + Detoxification + Retention$$

This equation governs the transport of contaminants, and knowledge of the individual intake and fate pathways, and their interrelationships, is critical for use in exposure assessments. The intakes (ingestion, inhalation, and dermal absorption) all represent processes or means by which a terrestrial organism obtains contamination from its environment. If the environment has no contamination, then intake must equal zero. As soon as one species' environment is contaminated, whether in the soil, water, air, or other organisms, then the opportunity is available for the animals to intake the contaminant. The equation provides a multitude of opportunities to evaluate potential or actual exposure to contaminants.

Intake

An important aspect of intake is that much of the available contaminant toxicity information is in the form of intake. This is useful when attempting to determine the critical dosage (intake) for an organism. The critical dosages are usually given in terms of amount of contaminant per unit of food, air, or contact medium. For example, Platonow and Karstad (1973) found that dietary levels of PCBs as low as 0.64 ppm fed to female minks for 160 days resulted in reduced reproduction.

Of the three intake pathways, the ingestion of contaminants can probably be considered to be the most common, in terms of both frequency and magnitude. Ingestion can include secondary contamination, where contaminated food is consumed, or primary contamination, where contaminated water, sediments, or soil are ingested.

Some of the factors that influence the amount of contamination that an organism will obtain from ingestion include the contaminant concentration in food (trophic situations) and water, sediments, and soil (nontrophic situations), food and water requirements, foraging habits, propensity to ingest soil, and the amount of time spent at the contaminated site. For example, FWS (1988) found that, after ingestion, assimilation of lead from the gastrointestinal tract varied depending on the age, sex, and diet of the organism. Of these, they found diet to be the major factor determining lead absorption. The Natural Resource Council (NRC) (1978) found that mercury burdens in terrestrial mammals are usually directly related to their diets. Herbivores have the lowest burden, and carnivores preying on aquatic animals have the highest.

Exposure from inhalation requires that the contaminant volatilizes at ambient temperatures. Once volatilized, the contaminant may become available to inhalation by air-breathing organisms. This exposure is generally not as common as ingestion. For significant exposure to occur, the organism has to either inhale air with a very high concentration of the compound or have prolonged inhalation of lower concentration. High concentrations are probably uncommon, but could occur in an area where large amounts of the contaminant is present. The more frequent type of exposure is a protracted exposure through a confined air space, where the volatile contaminant is able to equilibrate with its source. In this situation, air mixing is minimal, as compared to the open air situation. Such confined air spaces exist in burrows and dens. Increasing the possibilities for this scenario is the fact that the burrow could be dug down into higher concentrations of contaminated soils. Finally, in the case of dens, the young are born in the burrow. Young animals may be more susceptible to effects than adults, because of the smaller body weight and greater ventilation and metabolism (Robbins 1983). The soil concentration, and therefore burrow air levels, coupled with the amount of time that the animal spends in the burrow, are the primary factors determining the severity of the exposure.

Dermal absorption can occur when contaminants come in direct contact with unprotected skin. The extent of absorption depends on the physicochemical properties of the contaminant, such as molecular size, water and lipid solubility, extent of ionization, and hydrolysis of the contaminant at the pH in the epidermis and dermis. Smaller-sized lipid-soluble contaminants are more readily absorbed than are larger-sized water-soluble ones. Local factors, such as temperature and blood circulation to the skin, also influence the rate of absorption (Gilbertson et al. 1991). As with the other intake pathways, the time that the organism is in contact with the contaminated medium, and the concentration of the contaminant, are the primary factors influencing the potential for adverse effects from dermal absorption. Also, another factor is the amount of exposed skin of an animal. Some animals may have only small portions of unprotected skin. Dermal exposure could be important for animals that are in frequent contact with the water, sediments, or soil. Examples include birds, whose feathers can serve as excellent conduits to the skin (U.S. EPA 1991), and for those who, during incubation, have a featherless brood patch on their underside. A brood patch is a highly vascularized feather-free area on the underside, and facilitates heat transfer to eggs during incubation. A duck could feasibly get contaminants on its brood patch while swimming. It could also pick up contaminants from sediments, for those that build their nest on the ground. Similarly, a bird with contaminants on its underside could transfer it to the eggs. Eggs are known to be very sensitive to some contaminants. For example, small amounts of petroleum compounds will seriously affect developing embryos (Eisler 1987).

Other animals that could potentially be exposed through this pathway include burrowing and denning animals, and amphibians and reptiles. The amphibians and reptiles often hibernate in sediments and conduct their metabolic exchange through their skin surfaces. Animals that nest in burrows are particularly susceptible if their young are born naked, whereas the fur on furred young would serve as a barrier to contaminant absorption.

Fates
There are three possible fates, excretion, detoxification, and retention. Fates are generally less documented than are intakes. The exception is retention levels, as represented by tissue concentrations. The use of fate when attempting to relate the media concentration of contaminants to ecological conditions requires at least one more transformation than does the use of intakes. The only exception is if site-specific empirical data for paired media concentrations and levels of retention are available for the location and media being evaluated.

Excretion refers to the elimination of the contaminant in its original or toxic form from the body. Contaminants can be excreted through urine, sweat, feces, or regurgitation. Contaminants can be excreted three ways. The first is the excretion of unabsorbed contaminants from ingested material. In this case, the contaminant has no ability to do systemic harm. The second way in which contaminants are excreted is if they are ingested and absorbed into the intestinal cell lining, and then the cell is sloughed off (Luckey and Venugopal 1977). The third way is if the contaminant is in the animal's system, and is expelled through any of the excretory mechanisms.

There is currently limited opportunity for the use of wild animals' excretion to gain insight into exposure to contamination. One may be able to collect excreta and test it for evidence of contamination or detoxified byproducts. This would allow some insight into the animal's exposure. This has the tremendous advantage of not requiring handling or sacrifice of the subject species. Frenzel and Anthony (1989) used this technique to study the contamination of bald eagles in the Klamath Basin, Oregon. They collected expelled castings (undigested material expelled from the gizzard through the mouth) at roosts and were able to evaluate relative exposures in different subgroups of eagles. There remains much potential in the documentation of contaminant evidence in excreta, and the subsequent relation of the concentration to both intake and, ultimately, environmental concentration. The primary drawback is that there is little information available on the relationships of excreta concentrations to toxicity levels and environmental concentrations, or intake levels. Its use, therefore, would require considerable extrapolation.

Detoxification refers to the chemical alteration of the contaminant into a nontoxic form. As such, detoxification includes a special case of metabolism (e.g. oxidation, reduction, or hydrolysis), where the end-products are nontoxic (Luckey and Venugopal 1977). Detoxification does not include metabolism that produces toxic byproducts, because the animal would still be subject to adverse effects. It also does not include chemical alteration into toxic forms, such as that common with PAHs. Some contaminants are readily detoxified into unharmful metabolites, whereas others are very resistant to detoxification. For example, many PAHs are very readily metabolized into nontoxic forms (Eisler 1987). In fact, detoxification occurs so rapidly in some PAHs that they do not persist in the body in toxic form long enough to cause harm. This does not mean that rapid chemical alteration automatically means detoxification. For example, mercury, selenium, and many of the PAHs are not overly toxic in their pure form, but it is rather the metabolites that are toxic. In contrast, some compounds undergo little chemical transformation, and are not readily detoxified. These compounds, such as PCBs, are said to be persistent. Persistent contaminants are particularly of concern because their availability greatly exceeds the life spans of many wildlife species.

Knowledge of detoxification characteristics has some limited use in exposure assessments. It is extremely valuable in the screening of contaminants to determine the contaminants of concern. Persistent contaminants are of particular concern because they offer the opportunity for significant accumulation and transfer to other organisms, by longer-lived organisms if the compound is not excreted. Indicators of detoxification, metabolic byproducts, can be used in determining exposure levels, if the rates of detoxification are known. Detoxification information is also difficult to use in evaluating the environmental concentrations, because of the many transformations and extrapolations required. Its best use is simply as an indicator that an organism is exposed. Quantitative assessment should generally use other approaches.

Retention refers to the storage of the contaminant, or its harmful metabolic byproducts, while it is in the organism and before it is detoxified or excreted. Of the possible contaminant fates, detoxification, excretion, and retention, there is more available information, by far, on retention. Since retention results in the actual body burden of the contaminant, it is what provides the primary opportunity for adverse systemic effects. After the contaminant is taken in, its ability to have deleterious effects is gone as soon as it is excreted or detoxified. Retention represents the point during biological transfer that organisms incur effects. Contaminants only have deleterious effects during the time they are in a toxic form in the organism, this represents the time of retention. Because of the close relationship between retention and deleterious effects, an understanding of retention is critical when conducting an exposure assessment. It is further useful because much toxicity information is provided in the form of tissue concentrations. This information base is extremely valuable to the evaluation of critical exposure levels.

The amount of contaminant retained in the organism is a function of the rate of intake, detoxification, and excretion. Generally, the higher the intake, the higher the retention. High intake, however, can be mitigated by high rates of excretion or detoxification. For example, high intakes of some PAHs may not result in high body burdens, because of their rapid metabolism or other detoxification. Extremely low excretion and detoxification also can result, given prolonged exposures, in high retention levels. The classic examples of this case are the extremely persistent (resistant to detoxification) PCBs. PCBs have been shown to reach surprising levels in areas that have comparatively low environmental contamination (Gilbertson et al. 1991).

Anderson et al. (1991) demonstrated that PCBs retained their biological activity in mice for nearly a year, which probably far exceeds the average life span of wild mice.

Depending on the contaminant, retention levels can reach equilibrium between intakes and other fates. In such cases, the body burden represents some constant fraction of the intake. In other cases, the body will seem to have an infinite affinity for the contaminant, steadily increasing its concentration as a function of time and intake rate. The aspects are dependent upon the behavior of the contaminant. Some important behaviors to understand when evaluating retention levels are persistence, as already discussed, and the solubility of the compound. Contaminants are generally either soluble in water (called hydrophilic) or soluble in lipids (lipophilic). Generally, water-soluble compounds are easily excreted in the urine and have relatively short retention times. Lipophilic contaminants, on the other hand, are often not excreted in urine at all, and exhibit overall low excretion. It is the lipophilic, persistent compounds that often reach the highest retention levels. These contaminants often become sequestered in adipose tissue, or in various organs with high lipid contents. Some organs that are often targeted by lipophilic contaminants include the brain, liver, or kidney. These organs and adipose tissue can have quite high levels, while the blood or muscle tissue exhibit low concentrations, and therefore represent important biomarkers for exposures. For example, FWS (1988) showed that amounts of lead in bone are indicative of estimated lead exposure and metabolism.

Because of the direct relationhip to intake, retention levels can be used as an indicator of exposure, and thus of environmental concentration. This is accomplished through the evaluation of tissue concentrations. This has the significant advantage of allowing the sampling of either dead or live animals. However, as Clark et al. (1988) points out, the use of contaminant concentration in wildlife tissue as an indicator of environmental concentration is predicated on an assumption that there exists some fairly constant ratio of tissue to environmental concentration for each chemical. As mentioned earlier, that is not always the case, but it is true often enough to be of value. As Clark et al. (1988) further points out when discussing concentrations in some birds, due to the fact that tissue concentration is a function of the exposure history over a determinable time period, and of growth, lipid content, and metabolism, it should be possible to interpret tissue concentration changes by back-calculating the exposure that caused these levels.

Finally, when monitoring tissue levels for evidence of contamination, one must sample the target organ. Selection of the wrong tissue can be misleading. Anderson et al. (1991) found a ten-fold difference in total PCB concentration between liver tissue and blood in mice contaminated with a single dose. As a general rule, hydrophilic contaminants usually locate in blood, muscle, and other tissues with a high water content. Lipophilic contaminants usually locate in brain, liver, kidney, adipose tissue, or other tissues with a high lipid content.

Contaminant Behaviors

Certain contaminants exhibit identifiable behaviors in their biological transfer. The behavior is a result of the individual contaminant's chemical characteristics, its solubility, and persistence. These behaviors are bioaccumulation, biomagnification, and bioconcentration. They can often be expressed by ratios, which allow their use in evaluating exposure and environmental contamination. Understanding contaminant transfer behaviors is necessary to conduct an efficient exposure analysis. They provide the clues to what to look for and where.

Bioconcentration

Bioconcentration of contaminants occurs when organisms intake and retain contaminants through nontrophic means (U.S. EPA 1989a; U.S. EPA 1989c). These means can include uptake by plants, diffusion in aquatic organisms, and dermal absorption and inhalation in animals. Bioconcentration is most frequently associated with aquatic systems, where both plants and small-sized animals absorb the contaminant directly from the water or sediments. It is of importance to terrestrial systems in that the organisms concentrating the contaminant are often an important food source for terrestrial animals, leading to possible bioaccumulation. Bioconcentration is often expressed through a ratio of organism concentration to environmental media concentration.

Organism concentration/environmental media concentration.

This ratio can be useful in determining critical environmental concentrations, particularly where aquatic organisms are involved. Often, bioconcentration is the first step in ecosystem exposure. For example, aquatic plants have been shown to bioconcentrate selenium 500 times or more, while terrestrial plants, with a few notable exceptions, seldom bioconcentrate it at all (Hoffman et al. 1990). Terrestrial animals eating aquatic plants with high levels of selenium can intake critical amounts. Bioaccumulation of mercury and lead in many animals can often be attributed to bioconcentration of the elements in aquatic systems first (Wren 1987; FWS 1988).

Bioaccumulation

Bioaccumulation occurs when contaminants are passed between organisms through trophic as well as nontrophic means. Organism contamination is therefore strongly influenced by the consumption of contaminated food. In fact, accumulation via feeding is often the governing factor in the amount of contaminant retained. As in bioconcentration, bioaccumulation is often expressed as a ratio of the organism's tissue contamination to that of the abiotic environment, soil, water, or sediment (U.S. EPA 1989c). This is a reasonable expression of contaminant concentration in the tissue from both trophic and nonthropic pathways, as it assumes that the organism's food, as well as the actual organism, has some quantifiable relationship with the concentration in the environmental media. This factor, called the bioaccumulation factor (BAF) is calculated as follows:

Tissue concentration/Abiotic environmental concentration

Knowledge of a BAF allows the use of tissue concentration (retention level) to back-calculate to environmental concentration. Usually, tissue concentrations are lower than the environmental concentration (BAF is less than 1). In some cases, however, if intake from the food source is very high and/or retention is high, accumulation levels can be higher than that of the environment.

Biomagnification

A special case of bioaccumulation is biomagnification. Biomagnification occurs when each successive trophic level has increased contaminant concentrations, relative to their food source. As a result, top predators often exhibit extremely high concentrations of persistent contaminants, such as PCBs and mercury (Wren 1987). For example, biomagnification of a variety of organcholorines has been demonstrated to occur in many of the fish-eating birds of the Great Lakes, resulting in chronic impairment of reproduction (Gilbertson et al. 1991; Hoffman et al. 1990). The significant potential for negative effects resulting from biomagnification probably first became widely recognized through Rachel Carson's book *Silent Spring* (1962), in which she documented numerous cases, including increased mortality in robins.

TRANSFER PATHWAY ANALYSIS

Transfer pathway analysis couples ecological knowledge to knowledge of the toxic behavior of the contaminant. Pathway analysis provides a logical sequential methodology based on an animal's life history, which helps to filter through the myriad of possibilities, finally concentrating on one or a few that are truly the most relevant or important. This process will improve data resolution and reduce uncertainty in conclusions, because it will allow for the concentration of resources. Rather than diluting resources across the entire ecosystem and hoping to see a trend or to turn up some sort of statistically important result, investigative resources should be concentrated on a particular aspect of the ecosystem most likely to exhibit a effect. This aspect is typically individually organisms. Individual organisms (called key species in this chapter) chosen for intensive study must be selected through a logical, scientifically based, and defensible process. The result is high-quality information upon which definitive conclusions can be based, rather than weak, diluted anecdotal information, from which only conjecture is possible. Key to this approach is not only the collection of quality information on selected species, but also information on the relationships between species, contaminant pathways, and the ecosystem. Proper pathway analysis allows for the solid establishment of such relationships.

The EPA (1989a, 1989b) also puts emphasis on the analysis of pathways while characterizing exposure, but does not discuss how the investigator is to manage the infinite number of pathways possible at a site. It also calls for assurances that "all likely exposure pathways have been considered." This is probably impossible and not necessary. More critical is that all ***important*** pathways are considered. The investigations as to which pathways are important and deserve further examination must be conducted carefully though.

Transfer pathway analysis provides answers to the following questions: What important animals are using a site, and do they have an opportunity to come into contact with the contaminant? What is the exposure of these animals? How does the exposure relate to known toxic levels? If an animal is observed or expected to be incurring unacceptable adverse effects, then to what level does the environmental contaminant concentration need to be lowered in order to reduce the effects on the animal to an acceptable level?

The approach of transfer pathway analysis is a logical sequence of increasing detailed investigation (Figure 6–1). Some key aspects that must be kept in mind include:

1. The determination of whether any animals are exposed requires looking for something amiss, and then setting up tests to determine if it is attributable to contamination.
2. The fact that nothing is overtly amiss does not mean that there is no effect (wildlife just are not that apparent). One must look to see whether there are any particularly susceptible species and set up tests for those species. Determining species that are susceptible requires integration of the animal's behavior, ecology, and physiology with the contamination and its media.

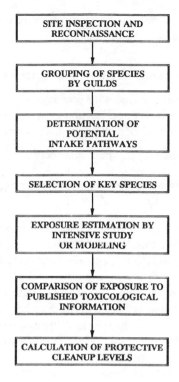

FIGURE 6-1.   Approach used in conducting contaminant pathway analysis.

3. In the case of discovered carcasses, one can test them for retention body levels, and then proceed to reconciling these to environmental levels with published BAFs or BCFS (if nontrophic).

4. In the case where no carcasses are discovered, but species absences are conspicuous, a second choice is to estimate exposure using pathway analysis. This can include estimation of intakes, retention, or (in some cases) excretion.

5. Once exposure is quantified, it must be compared to known toxic levels to determine if there is potential for harm.

6. The environmental conditions responsible for the toxic exposure must be calculated.

7. The final step is to calculate the minimal environmental concentrations that will result in minimal acceptable effect, by calculating effects levels for selected animals determined to be subject to potential or adverse effects.

Consideration of site cleanup levels goes beyond a concentration protective of every possible animal on site. The remediation decision is more than simply a biological (no effect allowed) decision. Sociological, political, and fiscal aspects come into play. For example, it would be unwise to set protection limits for Norway rats or mice, unless a pathway was established to another important species requiring protection. Furthermore, one must remember that, if there is no need to have the animal on-site, then don't proceed (that's why "socially or ecologically important" or ecologically representative animals are tested). If the only requirement is for the animal to be on-site there is no need to worry about reproduction if it can move on to and off of (except for very small ranged animals) the site. If the requirement is to have reproduction on-site, then look at on-site reproductive effects.

Exposure Screening of Species

Transfer pathway analysis should begin with a screening of potentially exposed species. This screening process begins with the collection of site biological and contamination history. Some useful source of biological information for sites include:

- Local resource agencies (municipal, state, and federal)
- Literature
- Private natural resource groups (Audubon, Wildlife Federation, etc.)
- Site records

Next, a general on-site resource inventory should be accomplished by a walk-over of the site (Jones 1986). Data should be collected on vegetation types and relative acreage, wetland cover types and relative acreage, mammals and their relative apparent abundance based upon sign or observation, reptiles and amphibians and their relative abundance based upon sign or observation, and topography. This information should be quantified through professional estimates as much as possible (possible bias in these estimates is not important at this stage—in fact, qualitative estimates such as high, medium, low may suffice). Observations should be sketched on a map showing major water bodies, land forms, and biotic resources.

Using information obtained from the resource inventory, different approaches can be used to identify species that are potentially exposed to contamination. The first and most direct approach to identifying potential exposure is to ascertain whether anything appears amiss on the contaminated site, based upon ecological knowledge and expectations. Carcasses can be tested directly for evidence of contaminated tissue. Though dead animals are obvious candidates for further study, they are not usually found. Rather, a species may be conspicuous by its absence. For example, an oak wood lot in New York State with no gray squirrels deserves investigation, because it is commonly known that most wood lots support the ubiquitous squirrel, and the squirrel may represent a food source for humans in rural areas. If necessary, this judgment can be quantified through the use of the U.S. Fish and Wildlife Service's Habitat Evaluation Procedures (HEP). These procedures are available by request from the Service, and have been prepared for many species. Caution should be used, however, since they will not necessarily explain the absence of species, as habitat quality is a gradient, representing a (usually) wide range of characteristics that can support a population of varying densities. Finally, even if nothing appears amiss, the terrestrial animals could be incurring long-term chronic effects that are not manifested in apparent population changes.

Because they are usually no carcasses readily available when examining the apparent absence of a species or when looking for long-term effects on present species, the evaluation of potential exposure is necessary. An efficient way to do this is to use transfer pathway analysis.

To attempt this approach, a list of candidate species using the site should first be developed. Attempts should be made to include representatives of the primary guilds on the site. A guild is a group of species within a community that are similar in terms of their space, time, and functional relationships with other species in the community (i.e., large mammalian herbivores, small mammalian herbivores, large water birds, insectivorous birds, etc.) (Whittaker 1975). Other considerations when including species in this list should account for home range requirements and sensitivity to human exposure, since these sites are often restricted in size and in fairly close proximity to man. While a ground squirrel may indeed be found on a 10-acre suburban site, it would be highly unlikely that a badger, for which ground squirrels are a primary prey item, would be present (Lindzey 1982).

An example of an abbreviated data base for a site in the northeast is shown in Table 6-1. Information should be gathered from the literature on home range, habitat, food requirements, reproduction, and other important life history facts relevant to the exposure assessment. This data base serves as the foundation for the identification of possible contaminant intake pathways, and ultimately for the selection of some key species.

Using this data base, one can begin to identify intake pathways that may be of importance in the transfer of contaminants on the site. The process begins with the identification of physiological, behavioral, and ecological aspects that may cause a species to be potentially exposed via one of the intake paths: dermal, inhalation, or ingestion. At this point, knowledge of the contaminant is also important.

The identification of species susceptible to dermal absorption is fairly straightforward. Dermal absorption can occur from most compounds, although Luckey and Venugopal (1973) felt that lipophilic compounds would cross the epidermis more readily than others. Animals that are in frequent contact with the soil or water, and animals that are born naked in burrows, are most susceptible to dermal absorption. Animals that are seldom in contact with the soil or water have little opportunity for dermal contact.

As was the case for dermal absorption, the identification of animals potentially exposed to contamination via inhalation is straightforward. For example, it simply does not make sense to consider the pathway of inhalation if the contaminant does not volatilize under the conditions present at the site (i.e., heavy metals). Furthermore, as previously pointed out, the contaminant must be allowed to become confined in an air space. Generally, the only place that this will occur is in burrows or dens. Therefore,

TABLE 6-1.   Listing of Species Potentially Present on an Hypothetical Stream and Wooded Upland Site in New England.

Mammals	Birds	Reptiles/Amphibians
Red Fox	Eastern Robin	Eastern Garter Snake
White-Tailed Deer	Least Flycatcher	Spotted Salamander
Gray Fox	Wood Duck	Northern Leopard Frog
Coyote	Mallard	Painted Turtle
Mink	Hooded Merganser	Bull Frog
River Otter	Woodcock	
Deer Mouse	Blue Jay	
Eastern Chipmunk	American Crow	
Eastern Gray Squirrel	Great Blue Heron	
Woodchuck	Red-Tailed Hawk	
Beaver	Belted Kingfisher	
Meadow Vole	Great Horned Owl	
Striped Skunk	Ruffed Grouse	
Raccoon		

it is essential to have volatile compounds and burrowing or denning animals before one should consider inhalation as a possible intake pathway at a site.

The identification of species potentially exposed via ingestion is more complex than that for other intake modes, due to the many different avenues, both nontrophic related and trophic related, available to the species. Furthermore, while only a relatively few species may burrow or den, an essential aspect of dermal and inhalation exposure, all species eat. Therefore, there is no immediate screening on that basis. Identification of nontrophic ingestion is probably compound independent, and simply requires the identification of species that may be predisposed to consuming contaminated soil or water.

Trophic-related ingestion of contaminants deserves special attention because it is certainly the most common intake path. It is also the most complex, due to the intricate trophic relationships existing in even small communities. The construction of a simplified trophic web for a contaminated site is a useful aid to evaluating the likelihood of exposure for various species.

The construction of a trophic web requires the identification of the primary routes of energy flow at the site. This is often best accomplished through the use of groups or guilds. Each group or guild will represent species that obtain their food in a similar manner. It is important that this should go beyond simply those animals observed in the reconnaissance. Ecological knowledge and experience should also be applied. Species should be included if the habitat or life requirements for the species (within its natural known range, of course) exist on the site. For example, the fact that no peromyscus (deer mouse genus) were observed in a upland temperature wooded site does not excuse their omission. One should attempt to be robust in the inclusion of species in the guilds, as they could become important players in the future. If that does not come true, then they will drop out of their own accord. Table 6–2 presents important guilds that should be included.

The trophic relationships indentified above, coupled with knowledge of the behavior of the contaminant, will allow identification of those organisms that are potentially exposed. Contaminant behaviors that are useful are persistence, known tendency to biomagnify, and solubility. Persistent is important in the sense that it allows the contaminant time to travel through the food chain. As discussed earlier, short-lived contaminants seldom exhibit extensive food chain effects, other than that on herbivores. When working with compounds that are known to biomagnify, such as PCBs and mercury, the potential exposure for upper-level consumers, top-level carnivores, and scavengers should be closely evaluated, since they will exhibit the highest retention levels (Wren 1987).

Solubility is a factor in a number of ways. Water-soluble contaminants such as selenium can be bioaccumulated by plants, particularly aquatic plants, thereby exposing the herbivores that feed upon them. For example, Hoffman et al. (1990) report that selenium has been known to bioconcentrate to 500 times the background levels in aquatic plants. Animals foraging on such plants would be in danger of receiving significant exposure. Mercury's nearly exclusive link to aquatic systems has resulted in only animals with a trophic link to an aquatic system exhibiting bioaccumulation of the contaminant (NRC 1978). Fat soluable contaminants such as PCB's can be accumulated in the adipose tissue of animals, where the concentration can become high without causing overt adverse effect to the animal. The more fat the animal the more ability the animal has to store the compound. Predators preying on such animals are in danger of receiving significant exposure.

TABLE 6-2. **Representative Guilds for Grouping Species in Exposure Screening Using a Hypothetical Eastern Site.**

Species	Herbivores	Scavengers	Carnivores	Burrowing/ Denning	Aquatic
Mammals					
Red Fox			X*	X	
White-Tailed Deer	X				
Gray Fox	X		X	X	
Coyote	X		X	X	
Mink			X	X	X
River Otter			X	X	X
Striped Skunk	X	X	X	X	
Raccoon	X	X	X	X	X
Deer Mouse	X				
Eastern Chipmunk	X			X	
Eastern Gray Squirrel	X				
Woodchuck	X			X	
Beaver	X			X	X
Meadow Vole	X				
Birds					
Eastern Robin	X		X		
Least Flycatcher			X		
Wood Duck	X				X
Mallard	X				X
Hooded Merganser			X		X
Woodcock			X		
Blue Jay	X	X			
American Crow	X	X			
Great Blue Heron			X		X
Red-Tailed Hawk			X		
Belted Kingfisher			X	X	X
Great Horned Owl			X		
Ruffed Grouse	X				
Reptiles/Amphibians					
Eastern Garter Snake			X	X	
Spotted Salamander			X	X	
Northern Leopard Frog			X	X	X
Painted Turtle	X		X	X	X
Bull Frog			X	X	X

As a result of screening for exposure via the three intake and fate pathways, a number of species may be identified as potentially exposed to contamination. Some animals may only be exposed to one pathway, where others may be exposed via all three pathways. For example, burrowing animals can be exposed through their food, volatilization of contaminants from soils surrounding the burrow, and direct dermal contact with the soils in the burrow. Although intuition may tell you that the nontrophic paths may not be important, they should be considered at this stage, since the path isolation can be more effectively completed at the conclusion of the next stage. The culmination of the process may result in a matrix of pathways, such as shown in Table 6–3.

TABLE 6-3.   **Matrix of Potential Exposures, Based upon Exposure Screening for PCB (P), Selenium (S) and Benzene (B) Found in the Soil and Sediments on a Hypothetical Site in the Northeast.**

Species	Ingestion	Inhalation	Dermal Absorption
Mammals			
Red Fox	P	B	
White-Tailed Deer	S		
Gray Fox	P	B	
Coyote	P	B	
Mink	P	B	P
River Otter	P		P
Striped Skunk	P		
Raccoon	P		
Deer Mouse		B	P
Eastern Chipmunk			
Eastern Gray Squirrel			
Woodchuck		B	P
Beaver	S		
Meadow Vole			
Birds			
Eastern Robin	P		
Least Flycatcher			
Wood Duck	S		
Mallard	S		
Hooded Merganser	P,S		
Woodcock	P		
Blue Jay			
American Crow	P		
Great Blue Heron	P		
Red-Tailed Hawk	P		
Belted Kingfisher	P	B	
Great Horned Owl	P		
Ruffed Grouse			
Reptiles/Amphibians			
Eastern Garter Snake	P		P
Spotted Salamander	P		P
Northern Leopard Frog	P		P
Painted Turtle	P		P
Bull Frog	P		P

Selection of Key Species

Species selected for intensive evaluation of exposure, calculation of effects levels, and (ultimately) the setting of cleanup thresholds are called "key species." The selection of key species allows the analysis to be increasingly refined and managed by being conducted on fewer, identifiable, finite aspects. Therefore, proper selection is critial, because other species will not be evaluated. The key species, as stated by Clark et al. (1988), who called them biomonitors, must have a well-understood natural history and toxicokinetics, in addition to other desirable features such as ease of sampling, a tendency to remain in a locality, and an ability to survive exposure to the levels of contaminant being monitored. Other factors to be considered are commercial importance or charismatic appeal. Finally, when possible, it is preferable that animals are selected for which considerable biological information is available and for which there is toxicity or contamination data available.

By comparing the potentially exposed species, the availability of toxicity information, and the known susceptibility, the identification of some key species is possible. Key species selected should represent critical components in the community, or should represent possible vectors to man or other organisms important in the community.

The primary criteria for the selection of a key species are:

1. history of susceptibility
2. apparent diminished population (based upon first-level study)
3. state or federal species of concern (i.e., endangered or threatened)
4. possible vector to man (food species, food for pets, etc.)
5. logical or apparent life history susceptiblity (top-level carnivore in a system with lipophilic contaminants)
6. representative of a group or guild
7. demonstrated potential exposure

Additional considerations must include an awareness of specific differences, because even closely related species may differ considerably in their sensitivity to a given toxicant (FWS 1980). The generalizations on how the environmental effects of contaminants can differ between reptiles (FWS 1980) can be also applied to other terrestrial animals.

1. Species at higher tropic levels tend to suffer the most from persistent contaminants that cannot be readily detoxified or excreted. For example, FWS (1981) reported that bats encounter more organochlorines in their diets than do herbivores because of food chain effects. The higher metabolic rates of these bats, associated with small size and the activity of flight, demand higher rates of food consumption than found in larger, less active birds and mammals. Greater food intake causes greater amounts of organochlorines to be available for concentration in the fat of bats.
2. The effects of contaminants vary considerably depending on the physiological state of the animal. This is particularly true of fat soluble contaminants whose absorption and mobilization are strongly related to fat body cycling. For example, the FWS (1981) found that bats and small insectivorous birds have a greater risk of adverse effects from organochlorines than other small mammals because of their pronounced fat cycles,with rapid depletion during migration or slow depletion during hibernation.
3. Species with long life cycles are more likely to be affected by persistent contaminants than the short-lived species characteristic of unstable habitats.

At least one species should be selected from each intake transfer pathway represented. Sometimes, it will be necessary to select more than one species from a compartment. However, this should be minimized, since each selected species will add significant additional resources and expenditure as the process continues. It must be remembered that the purpose here is a logical scientifically based *narrowing* of possibilities. The end result should be a conclusion as to the species that are potentially exposed via identified paths. Table 6–4 presents a list of possible key species selected from the exposure pathways represented in Table 6–3.

TABLE 6-4.   Selected Key Species for Intensive Study for a Hypothetical Site in the Northeast Contaminated with PCBs, Benzene, and Selenium.

Species	Contaminant	Rationale
Mink	P,B	High-level aquatic consumer with exposure to all intake pathways. Known susceptibility to PCB. Young are born hairless in dens.
Woodchuck	B	Herbivore that lives in dens. Easy to study and collect samples.
American Robin	P	Eats high levels of worms during some periods. Easy to study and eggs and egg shells can be used for contaminant detection. Documented accumulator of PCBs.
Hooded Merganser	P,S	High-level aquatic consumer. Eggs and egg shells useful for the detection of PCB and selenium.
Mallard	S	Aquatic herbivore. Easy to study and eggs and egg shells useful for the detection of selenium.

Exposure Estimation

Estimates of exposure are necessary to determine potential adverse effects through comparisons to known critical doses. Estimates of exposure are also necessary to quantitatively evaluate the importance of respective intake pathways when providing input to cleanup levels. In practice, exposure evaluation can focus on two basic approaches, estimation of intake dosages or estimation of retention levels. The approach chosen for a particular situation depends upon the ease with which data can be obtained from the key species, and the information available on toxicity. When estimating intake exposure, all intake pathways identified as being present for each key species should be considered, at least qualitatively. If it is obvious that a particular pathway is at least an order of magnitude below other routes of exposure, it can usually be ignored in the quantitative analysis.

The estimation of excretion and detoxification, although of some potential and with some exceptions, is not presently possible because of a lack of toxicological information on wild animals. Furthermore, the use of these would require either extrapolation back to retention levels, intake levels, or environmental levels because of the lack of information on the relationship of detoxification rates and excretion to environmental levels. In this case, one might as well eliminate the uncertainty of an extra extrapolation and work directly with intake or retention. Unless a published reference is available to correlate excretion or detoxification to either retention levels, dosage (intake), or environmental concentration, this is simply not worthwhile. For this reason, the discussion below on exposure estimation will not include consideration of estimation of excretion or detoxification.

Ingestion

As mentioned earlier, contamination can occur from ingesting soil, food, or water. The estimation of each requires different methods. Contaminant exposure from ingesting soil depends upon the contaminant concentration and the amount ingested. The maximum amount of soil potentially ingested depends upon the digestive tract's capacity and passage rate. In most situations, however, the amount of soil ingested will be behaviorally regulated, and will be far below the digestive system capacity. However, one should be aware that, at some points in life cycles, some animals consume considerable amounts of soil. For example, many animals in colder climates go into mineral deficiencies during the winter, when they catabolize tissue for nutrition (Robbins 1983). These animals often attempt to replace minerals, such as calcium, sodium, and phosphorous, by consuming soil at "salt licks." Another example would be animals that burrow or move soil in their mouth. Such animals could, at times, ingest significant amounts of soil.

Maximum intakes can be calculated based on digestive system capacities, but because the actual amount is behaviorally regulated, it will be difficult to obtain realistic estimates. The best solution in this case is to estimate the maximum, then prorate it backwards to some "reasonable judgment on consumption." If it proves to be minimal exposure, even at maximum physiological capacity, there is no need to obtain more precise estimates.

Exposure through the ingestion of contaminated food is one of the better-documented and most frequently analyzed intake pathways. Contamination through food ingestion depends upon the amount of contaminated food eaten and the amount of contaminant in the food. The actual amount of food ingested by key species at the site in question will seldom be available. Rather, an estimation must be completed, using consumption rates or other information available in the literature. The estimation of contaminated food consumption must take into account factors such as food availability, diet composition, and behavior. To estimate exposure through food consumption, it must be determined how much total food is consumed in a day, and then the proportion of the total that the contaminated food represents must also be determined. There are a number of approaches to obtaining an estimate of the total amount of food consumed. They are: digestive system capacity, known consumptions rates, and energy expenditure requirements.

Daily consumption rates may be available for some species. For example, river otters require between 700 g to 1000 g each day (Toweill and Tabor 1982). Daily consumption values simply need to be multiplied by the concentration of the contaminant, to determine the amount of the dose. For example, if an animal living solely on the contaminated site consumes 1kg of food per day and the food is contaminated with 5mg/kg, then the animal's exposure would be 5 mg/day (1kg/day * 5 mg/kg = 5 mg/day). If the animal did not live solely on site, but rather only half of its food was obtained on-site, then its exposure would be 2.5 mg/day. These calculations are expressed in the equation below. Estimates of food intake must be adjusted by the amount of the diet that is comprised of contaminated food.

Exposure = Food consumption * Contaminant concentration * On-site proportion *
Diet proportion

For example, for an animal that consumes 4 kg/day, it may be known that the contaminated prey species contributes only 50 percent of the diet, and it may also be known that the animal only does 50 percent of its foraging on-site. In other words, only half of its food is collected on-site and, of the food collected on-site, only half of that is comprised of the contaminated food source. Therefore, the amount of contaminated food consumed in a day would be calculated as below:

$$4 \text{ kg/day} * .5 * .5 = 1 \text{ kg/day}$$

In some cases, the documented food consumption will be for a food source other than the one of the interest species. In this case, the value should be converted to the prey base of interest on a per weight basis.

For some species, the literature might provide the capacity of the subject species digestive system. Coupled with the passage rate or the number of times that the animal feeds, this can produce the volume of food consumed in a day. This volume must then be converted to the weight of food, by multiplying by the density of the food.

The final method of calculating food intake is to estimate the daily energy requirement for the species and then determining the amount of food required to provide this requirement. For an animal in a steady state (no weight gain or loss), the daily energy requirement equals the daily energy expenditure (DEE). Sometimes, one can find in the literature where a researcher has actually estimated the DEE for the species in question. Since DEE is very dependent on behavior and activity, it is likely that some adjustments are needed to more accurately represent the site-specific case.

If there are no estimates available, there are some generic equations that give rough approximations. There are a number of approaches available for the estimation of DEE, depending upon the species and available information on the species. One approach involves using an estimate of the amount of time spent on each daily activity, and estimates of the energy expended for each activity. More commonly studied species may have the necessary information available. For example, Harrison and Dyer (1984) used estimations of forage intake, home range, and activity to calculate the tolerable lead concentration in mule deer forage. If the necessary information, such as activity times or energy rates, is not available, then some more general equations can be used. For example, DuBowy (1989) multiplied generic basal metabolism values for birds by a factor of three to estimate DEE in marsh birds. Basal metabolism follows general curves, which can be described, for birds, marsupials, and mammals (Robbins 1983). Often, various activities are given as multiples of the basal metabolic rate. Mautz (1978) and Robbins (1983) report energetic rates for various activities for a number of species.

Daily energy expenditure must be converted to weight of food. Various caloric equivalents for forage are available from a multitude of sources. Generally, animal tissue is about 5 to 6 kcal/g, depending upon the fat content, and vegetation is around 4 to 5 kcal/g, depending upon the plant tissue in question. Seeds and other plant parts with greater amounts of fats will have higher values. If forage caloric values are not available, but the forage fat, protein, and carbohydrate composition is known, it is often possible to estimate the caloric equivalents using the following energy coefficients obtained from Robbins (1983).

Cellulose = 4.18 kcal/g dry weight
Protein = 5.65 kcal/g dry weight
Fat = 9.45 kcal/g dry weight

Inefficiencies of the digestive process must also be taken into account, since no forage is digested completely, with 100 percent of its energy available to the consumer. Digestibility trials, aimed at the determination of these inefficiencies, have been conducted on numerous species (Mautz 1978; Robbins 1983). Digestive inefficiencies are food-type specific, but generally vegetation is less digestible than animal tissue. The dry food required to meet the DEE is given by the following equation:

$$\text{Daily dry food} = \text{DEE} / \text{NEC} * \text{DEC} * \text{MEC} * \text{FEC}$$

The digestible energy coefficient (DEC) represents the digestibility of the food (not all food is digested). The metabolizable energy coefficient (MEC) represents after-digestion energy losses due to chemical energy, which is not then available for use (methane, urinary products). The net energy coefficient (NEC) represents the energy lost as heat through the inefficiencies (second law of thermodynamics) of the digestive system and chemical reactions. These energy coefficients, NEC, DEC, and MEC, represent the inefficiencies in the digestive process.

Once the amount of dry food being ingested is determined, it may be necessary to translate this into wet weight, either because the dosage information is in wet weight, or in a unit organism (whole individual). This is simply done by adjusting the weight of the dry food by the moisture content of the food.

$$\text{Wet weight} = \text{Dry weight} / (100 - \% \text{ Moisture})$$

Again (as previously discussed), when applying contaminant information, the amount of food needed to meet the DEE must be corrected for diet proportion and on-site source proportion.

The final way in which animals can ingest contaminants is through their drinking water. As was done for the ingestion of soil and food, the maximum possible water ingestion can be calculated using digestive system capacity and passage rates. This method will, however, probably have less merit, compared to the other intake pathways, because of the considerably more rapid assimilation and excretion of water.

Another method for estimating exposure via the ingestion of water is to calculate the daily drinking water requirement for an animal. The minimum amount of water that an animal must drink (DW) depends upon the animal's total requirement (TR) and the amount of water that it obtains from metabolism (MW) and food (FW), and is expressed by the equation:

$$\text{DW} = \text{TR} - \text{MW} - \text{FW}$$

The total water requirement (TR) has been calculated for numerous species using a variety of methods. If an estimate for the animal in question can not be found in the literature, then one can estimate the TR by using the generic equation provided by Robbins (1983) for mammals:

$$\text{TR} = 0.12 \, (\text{body weight})^{0.84}$$

The calculation of metabolic water requires the estimation of daily energy expenditure (DEE), as described above. Once the DEE is estimated in kcal/day, it should be multiplied by a metabolic conversion factor, such as 0.12 grams water/kcal (Schmidt-Nielsen 1983), to determine the kilograms of metabolic water produced per day. The weight of water can be converted to volume by dividing by the density of water at 38° C, which is approximately 1 kg/liter.

To calculate food water, the amount of food consumed per day must be known or calculated. This can be done in the same manner as described above for food consumption. The amount of water associated with ingested food varies depending upon the moisture content. For example, grasses may range from 50 to 70 percent water wet weight, whereas fish may be as much as 90 percent water wet weight. The following equation provides the weight of the water obtained from forage using the known dry weight of forage eaten and the average wet weight moisture content. Food water is then converted to volume, as above.

$$FW \text{ grams } = \text{ grams dry food } * (\% \text{ moisture})/(100 - \% \text{ moisture})$$

Drinking water requirement (DW) can finally be calculated, by inserting TR, MW, and FW into the above equation. Before determining the amount of contamination receive via drinking water, the proportion of the drinking water requirement obtained at the contaminated water source must be calculated, since animals with larger ranges may obtain water both on- and off-site. This can be approximated by multiplying the DW by the proportion of home range that is represented by the site as opposed to the other nearest water source. If the next nearest water source is greater than the home range, then it can be expected that all of the drinking water is obtained at the site. A ground squirrel will obtain all of its water within a 10-acre hazardous waste site, whereas this may represent only small fraction of the home range for a larger predator. However, if the water on the hazardous waste site is the only water for 5 miles, then it is likely that the predator will also obtain all of its water on-site. Animals with larger home ranges can become habituated to one water source, and use it in greater proportion than indicated by water availability throughout the home range. This can only be reconciled through observation.

Finally, the reader must be aware of seasonal differences that affect animal water budgets. The most obvious is that during warm times of the year most warm-blooded animals use evaporative cooling to thermoregulate. This can increase DW by an order of magnitude or more. For herbivores, changes in the moisture content of forage can also result in multiple differences in water provided by forage. This fact is exacerbated by coinciding with increased water demands for thermoregulation. When the climate is hot and dry, the vegetation provides little water and the animal's DW is at its peak requirement, due to thermoregulation needs.

Dermal Absorption

Dermal exposure requires direct skin contact with the contaminated media. Quantification of such exposure requires knowledge of the amount of time that the animal may be in contact, and the amount of skin in contact, with the soil and the media concentration. In some cases, the critical exposure may be reported in terms that are independent of animal size or the duration of exposure. It is sometimes presented as a concentration of a contaminant applied to the skin of a test animal. (Eisler 1987). In these cases, there is no need to calculate the actual exposure once it is established that the animal is exposed dermally. Burrowing animals will only be potentially exposed for the amount of time that they are in the burrow. Thus, an estimate of that time is necessary. Furthermore, one must estimate the amount of skin in contact with the soil. While many adult mammals are furred, their neonates are often naked. For mammals, one might generally assume that only the underside of the animal is in contact with the soil while lying. The total surface area of the animal can be estimated using the surface area-to-weight relationship provided by Robbins (1983).

$$\text{Surface area } = 10 * \text{Body weight (g)}^{0.667}$$

Following calculation of the surface area, it should be prorated to represent the amount actually in contact with the soil. This value may be difficult to calculate, and may simply require professional guess, such as 0.25. Depending upon the type of toxicity information available for comparison, one may need to convert the exposure into an absolute amount of contaminant. This can be done by assuming an effective contact depth of contaminated soil. This depth is multiplied by the surface area, to yield the contact volume. The contact volume is multiplied by the soil concentration to find the amount of contaminant to which the animal is exposed.

The estimation of dermal exposure to birds through their brood patches could be necessary. This can be done by consulting the literature, where sizes of brood patches might be available. The amount of time the birds are in contact with the contaminated media may have to be estimated, if the critical dose is available in terms of duration or body weight.

Inhalation

The determination of exposure to burrowers from the inhalation of volatile contaminants does not require estimation of animal characteristics, but rather only the concentration of contaminant in the air space in the burrow, because most inhalative chronic toxicities are given in terms of air concentrations and are independent of animal weight. Two assumptions are required. These are that the air space is closed, and that the constituent air gaseous concentrations are at equilibrium. Given these, the air concentration (mg/m^3) of the volatile contaminant then follows the vapor pressure law:

$$\text{Air Conc} = V_{pi} * 1 \text{ Atm} / 760 \text{ mmHg} * MW / (R * K) * 1000 \text{ mg/g}$$

Where V_{pi} = partial pressure of the contaminant, MW = molecular weight, R = the gas constant, K = the temperature in Kelvin.

Retention

The final, and one of the most certain, method for determining the extent of exposure is to sample animal storage or retention of contaminant. Obviously, this can only be effectively accomplished from animals from which samples can be obtained, either dead or alive. Collection permits can often be obtained for the collection of various animals. Animals found dead can also be carefully examined for evidence of contamination. If dead animals are not found or are unavailable samples from live animals can be used. Blood and muscle tissue samples can be collected from many animals with little harm. This approach is not possible for animals that can not be caught or handled, such as bald eagles. The tissue values provide direct proof of exposure, and can be compared directly to documented tissue concentrations.

Critical Thresholds

Two critical thresholds, toxicity levels and effects levels, will be examined in this section. The severity of the exposure or the potential for adverse effects, whether exposure was calculated by estimating ingestion, inhalation, dermal contact, or retention, is evaluated by comparing the exposure to published critical toxicity levels or ecological benchmarks. This marks the last step in the process of exposure evaluation at hazardous waste sites. This final step involves the calculation of contaminant concentrations that are compatible with various ecological endpoints, or effects levels. These effects levels can then provide informed input to decision makers, so they can be considered along with ecomical, engineering, public health, and other factors, to ultimately establish cleanup levels for the site. The process of establishing these levels is frequently a compromise, but if the decision makers do not have confidence in the information provided by the ecologists, ecological concerns may well be on the short end of the compromise. Thus, the thorough analysis, documentation, and clear presentation of the ecological effects level is an important step in integrating ecological concerns into the cleanup process.

Toxicity Levels

The determination of whether the contamination exposure is likely to be causing adverse effects requires a comparison to published critical toxicity levels. This determination first requires a decision on the type of effect that one is concerned about. Toxicity levels will vary considerable for different levels of effect. The preferable values are those that set a standard effect, such as lowest observable effect level (LOEL), no-observable-effect level (NOEL), or some of the lethal dose levels (LD_{50}; LC_{50}). The estimated exposure on the site is then compared to these critical doses. If the animal's exposure exceeds the documented critical exposure, then one can conclude that the animal is at risk of adverse effects. Possible levels of ecological effects, also called endpoints, must also be considered when selecting the toxicity level. These include reproductive inhibition, population effects, individual mortality, or carcinogenesis. A determination of the endpoint should be done before beginning exposure estimation, because it is important that the exposures be estimated in the terms used for the toxicity information available on the selected endpoint. For any one contaminant and pathway, the available toxicity information could be in terms of tissue levels, intake doses, or environmental concentrations. Other factors that influence critical doses could include specific tissues, specific contaminants, and specific species.

Often, available toxicity information will not be directly comparable to the exposure estimated. It is possible that the toxicity information will be presented for different species, contaminant pathways, endpoints, or length of exposure times. The extrapolation of such data, although probably necessary, must be done with the knowledge of possible uncertainty, and with care. When comparing tissue concentrations, for example, the length of exposure must be considered for compounds that may never equilibrate, such as PCBs. Tissue levels lower than that published could be the result of less exposure time, rather than less exposure concentration.

Care must be taken in choosing the tissue on which to test retention levels. As noted earlier, many contaminants selectively locate in particular tissues. For example, wild birds concentrate the highest levels of mercury in the kidney and liver, with much less in the muscle tissue (NRC 1978). Gjerstad and Hanssen (1984) report that lead accumulated more up to 72 ppm in the livers of willow ptarmigans as compared to 0.5 ppm in breast muscle. Information on different tissue retention levels, rather than for critical levels reported in the literature, can lead to difficult comparisons, because of the affinity of specific contaminants for specific organs and tissues.

Different derivatives of contaminants also can exhibit considerably different toxicities. Metabolism of PCBs is correlated with the amount of chlorination (NRC 1978). For example, the EPA (1980) found that mink can ingest up to 25 ppm of Aroclor 1016 in their diet for 18 months without reproductive failure, while Aulerich and Ringer (1977) found that only 2 ppm Aroclor 1254 in the diet of mink from August to June resulted in reproductive failure. Selenium is another example of a contaminant with possible variable effects, depending upon circumstances and compound form. Selenium, as an essential trace mineral, is necessary for proper muscle function. Too little selenium can cause death, while too much selenium can also be toxic. For example, DuBowy (1989) studied the toxic effects of selenium on water birds, while only a few hundred miles away, Flueck (1991) studied selenium deficiencies in deer. Furthermore, Heinz et al. (1989) found that an organic form of selenium was extremely toxic, while inorganic forms were not.

Certain species are more susceptible to adverse effects than other species for the same contaminant. Of course, the susceptibility of mink to PCBs is well documented (Wren 1987). This susceptibility can come from food habits or life cycle phenomenon, such as earlier mentioned with bats and their fat cycles (FWS 1981). Therefore, due to life history difference, and physiological sensitivity, care must be taken when extrapolating toxicity data across species. For example, Schafer et al. (1983) reported that redwinged blackbirds were the most sensitive of 68 bird species tested on a large number of chemicals. They found that the difference in toxicological sensitivity between redwinged blackbirds and quail was 1.4 times and that between redwinged blackbirds and starlings was 2.1 times.

Effects Levels

The general approach to developing and evaluating levels producing various ecological effects is to reconcile a retention (known toxicity level in body) or intake dosage to an environmental concentration. Reconciling retention back to environmental concentration, in the absence of good empirical data, requires more extrapolation (one more step) than does intake dosage, to predict critical environmental concentrations. Usually, some critical toxicity or other standard for acceptable effect will be used and translated into a critical environmental concentration. This environmental concentration would then represent the minimum level of contaminant at which adverse effects would be manifested.

The process of determining critical effects levels, the ultimate goal of the exposure assessment, should only be undertaken if animals are in jeopardy of adverse effects. Jeopardy of adverse effects is determined as stated earlier, by comparing the exposures, in the form of intakes or retention, to known toxicological information. For species where critical toxicity levels are exceeded, soil contamination levels may be calculated that would ensure that critical toxicity thresholds are not exceeded. The effects level should only be calculated for the key species with the greatest potential for adverse effects for each of the three intake pathways, and for retention.

Based on the effects levels, species upon which cleanup levels will be based can be selected on both sociological information and biological information. Selection of a sociologically and ecologically less important species that would require extensive cleanup, over a more sociologically or ecologically important species that would require less cleanup, is not necessarily prudent. Judgments will need to be made, weighing all of the factors, on a situation-by-situation basis. For example, it is possible that a beaver could be chosen for setting cleanup levels on the basis of its known positive impacts on hydrology, ecological importance to the community, and fur value. However, it would not be prudent to set limits specific to beaver if, at a specific location, the animal causes significant property damage by damming drainage areas, resulting in flooding, yet has no monetary value as such, because it is not accessible to trappers.

The first step in determining effects levels is the decision of whether any cleanup at all is necessary. If no animals are being exposed to levels of contaminant that are apparently or known to be harmful, based upon toxicity levels, there is no ecological reason to set cleanup levels. In such incidents, nonecological factors will be the exclusive determinant of cleanup concentrations, and the ecologist must focus on the damage done by the cleanup.

The evaluation of effects levels requires the reconciling of the critical dose, the exposure level, and the level of contamination causing the exposure. The basic objective is to determine the environmental media (soil or water) that will result in animals being exposed to contaminant levels that are below the critical dose.

This needs to be done for only the pathway with the highest risk (exposure exceeding the critical dose by the greatest disparity) for the chosen species. However, it might often be wise, time permitting, to evaluate levels of all paths indicating risk, and more than one species, because different approaches and models might have been used and, therefore, the path with the highest risk may not result in the strictest cleanup levels.

The most straightforward method of calculating effects levels is back-calculation using bioaccumulation factors (BAF):

$$\text{Effects level} = \text{Critical dose} / \text{BAF}$$

or, in nontrophic situations, bioconcentration factors (BCF):

$$\text{Effects level} = \text{Critical dose} / \text{BCF}$$

For example, if the critical dose is 100 ppm, and the BAF is 50, then the effects level required to protect organisms from that critical dose is 100 ppm / 50 = 2 ppm. This means that all areas exhibiting more than 2 ppm of contaminant would have to be cleaned to the level of 2 ppm to prevent the ecological effect associated with the dose.

This is, of course, only valid if the BAF or BCF is known. Often, a factor will be known for a prey species, but not for the animal in question. In these cases, the calculations completed above to determine ingestion dosage must be repeated for the critical dose, to determine acceptable prey concentration. Then the prey concentration is reconciled to an effects level, using known BCF or BAF and the equation above. For example, DuBowy (1989) using a previously published BAF for aquatic insects fed upon by waterfowl, calculated the dietary exposure of selenium, and was subsequently able to determine the water concentration that would result in deleterious effects using the BAF for the insects.

In situations where bioaccumulation or bioconcentration rates are assumed to be constant across a range of contaminant concentrations, the ratio of the critical dose to the corollary animal estimate of exposure can be used when calculating media concentrations required to ensure no effect. This method has the advantage of not requiring direct knowledge of the transfer coefficients. It does, however, require that the environmental concentration of the contaminant causing the exposure be known. If animal exposure and environmental concentration are known, environmental no effect levels can be calculated using the following equation:

$$\text{Environmental no effect level} = \text{EC} * (\text{CD} / \text{AE})$$

where EC is the environmental concentration of the contaminant, CD is the critical dose of the contaminant, and AE is the animal exposure to the contaminant. It is important to recognize that the critical dose and animal exposure are for the same transfer intake or fate. For example, if the critical dose pertains to ingestion, and is given in mg/kg food, then the exposure must also be expressed in mg/kg food. Likewise, if the critical dose pertains to retention tissue levels in mg/kg, then the animal exposure should be in mg/kg for the same tissues.

An alternative to using the transfer coefficient in evaluating effects levels is the back-calculation of toxicity levels through the pathway analysis. This can be a useful method, since transfer coefficients are seldom available for terrestrial systems, and in particular are not likely to be validated for the site-specific case. The level that would protect against a critical dose level is defined in dermal situations as:

$$\text{Effects level} = CD * BW^{0.667} / (CV * SD * RAF)$$

where CD = a critical dose in mg/kg body weight, BW = the body weight in kg, CV = the volume of soil in contact, SD = the soil density, and RAF = a relative absorption factor for that contaminant.

The calculation of soil levels for volatile contaminants can be completed for critical doses (mg/m^3) using the following equation. This assumes that the soil and air are in equilibration and that the concentration of the air will be some fraction of the concentration in the soil.

$$\text{Effects level} = \text{Critical dose} * C * R * K * 760 \text{ atm} / (V_{pi} * MW)$$

Where C = the fractional constant, R = the gas constant, k = the temperature in Kelvin, V_{pi} = the partial pressure of the contaminant, and MW = the molecular weight of the contaminant.

EXAMPLE CALCULATIONS OF EFFECTS LEVELS

In a real situation, effects levels would be calculated for all of the key species selected in the hypothetical site investigation used in this chapter (mink, mallard, woodchuck, American robin, hooded merganser, and mallard). For the purpose of the demonstration here, the mink will be used to demonstrate calculations for PCB ingestion and dermal absorption pathways, and the mallard will be used for calculations of selenium ingestion and retention pathways. The calculation of concentration levels for benzene will not be species-specific, because it is not dependant upon body weight or size. The reader is cautioned that the values used in these calculations are for demonstration purposes only, and many are entirely fictitious. Therefore, the results and inputs into the calculations should not be referenced. The values were selected to be reasonable, but any correlation with published data is coincidental. For any one exposure situation, there are many ways of calculating effects levels, depending on the information available. The calculations below are only a sampling of the many different ways that one could approach the calculation of effects levels.

Ingestion

Assume that one-half of the mink food on the site is contaminated with 5 ppm of PCB. Also assume that the mink eats one half of its daily food requirement on-site. Therefore, its diet dosage equals:

$$\text{Animal exposure} = 5 \text{ ppm} * 0.5 * 0.5 = 1.25 \text{ ppm}$$

Effects would be anticipated, because this is higher than a published critical daily diet concentration of 0.64 ppm. The environmental no effect level is:

$$\text{Environmental no effect level} = \text{Env. conc.} * 0.64 / 1.25$$

If the environmental concentration is 25 ppm, then the environmental cleanup level would have to be 12.8 ppm, to ensure consumption of less than the critical dose. Using this ratio method assumes that the food is contaminated and that its contamination is directly proportional to the media concentrations.

Levels for selenium can be calculated for the aquatic foraging mallard. It must be assumed that the mallard is nesting and obtaining all of its forage from the site. Assume that the aquatic plant BCF equals 500 ppm, and diets of 8 ppm result in embryo mortality in eggs. Assume that the environmental concentration of selenium is 0.1 ppm. The diet exposure then equals:

$$\text{Diet exposure} = \text{EC} * \text{BCF for forage} = 0.1 * 500 = 50 \text{ ppm}$$

Effects levels can be calculated using the ratio method:

$$\text{Effects level} = 0.1 * 8 \text{ ppm}/50 = 0.016 \text{ ppm}$$

Another approach could be to use a critical retention tissue concentration (4 ppm) and a BAF for mallards (1000).

$$\text{Effects level} = 4 \text{ ppm} / 1000 = 0.004 \text{ ppm}$$

The fact that the second approach results in an effects level one fourth that of the first is not uncommon. This will happen due to different sources of data, the quality of the fit from the published information, and its relationship to the site being investigated. Differences between the site-specific characteristics and that of the situation from which the published information was obtained can also result in error. The investigator is left to decide which value represents the best solution for the site in question.

Dermal

The calculation of levels based upon dermal exposure to PCBs requires the following assumptions: mink weight (young in den) = 400 g; proportion of the animal in contact with soil = 0.25; and the relative absorption factor for PCBs to mink = 0.5. If the critical dose is given as a torpical dose in ppm, then there is no need for calculations. If the critical dose is given as 15 ppm, then the soil and water contamination must be below that threshold to prevent the effect. However, the doses are often given on an animal or body basis. If the critical dose equals 0.005 mg/kg, determination of the effects level requires the following calculations:

$$\text{Contact surface} = 10 * 400\text{g}^{667} * 0.25 = 136 \text{ cm}^2$$

The amount of soil that the animal is in contact with can be estimated by converting the surface area to contact volume, by multiplying by 1 cm deep. Therefore, the contact volume = 136 * 1 = 136 cm^3. The weight of this soil is determined by multiplying by a soil density.

$$\text{Soil weight} = 136 \text{ cm}^3 * 2.000 \text{ g/cm}^3 = 272 \text{ g of soil}$$

The dose per unit body weight can then be calculated:

$$\text{Dermal dose} = 272 \text{ g} * 1 \text{ kg} / 1000 \text{ g} * 25 \text{ mg/kg} * = 0.05 / 400 \text{ g}$$
$$= 0.008 \text{ mg/kg}$$

Calculation of the effects level is required because the exposure (0.008 mg/kg) exceeds the critical dose (0.005 mg/kg). The calculation can use the ratio method:

$$\text{Effects level} = 25 \text{ ppm} * 0.005 / 0.008 = 15.6 \text{ ppm}$$

Inhalation

Calculations of the inhalation levels for benzene requires knowledge of the chemical characteristics of the gas, and knowledge of the fraction of soil concentration to air concentration. Assume that the burrow temperature is 280 degrees Kelvin and the published critical dose is 160 mg/m^3. The level for benzene is as follows:

$$\text{Effects level} = 160 * 8.2\text{E-}5 * 280 * 760 / (78 * 125) = 0.286 \text{ mg/m}^3 \text{ of air}$$

The units associated with the values in the equation are: 160 mg/m; 8.2E-5 atm m^3/mole/K; 280K; 760 mmHg/atm; 78 g/mole; and 125 mmHg. If it is assumed that the burrows are in equilibrium and this fraction is 1, then the soil level = 0.286 mg/m^3 of soil. Multiplying this by the density of the soil results in an effects level in mg/kg soil.

$$\text{Effects Level} = 0.286 \text{ mg/m}^3 / 2 \text{ g/cm}^3 / 1{,}000{,}000 \text{ cm}^3/\text{m}^3 * 1000 \text{ g/kg}$$
$$= 1.43\text{E-}4 \text{ mg/kg soil}$$

SUMMARY

Much can be learned about contaminant exposure at a hazardous waste site by using transfer pathway analysis. Pathway transfer analysis provides information from which more certain conclusions can be drawn, because of the sequencing of thorough investigation and modeling techniques. The investigator must exercise flexibility when choosing an approach to estimate exposure and effects levels for use in establishing cleanup concentrations. There are generally any number of approaches for each estimation required. The approach chosen should depend upon the available information and the minimization of uncertain extrapolations. There remains a considerable need for additional research into the modeling of contaminant exposures, effects levels, and cleanup levels. Much more data are needed on toxicological thresholds for wild species. Likewise, the use of transfer coefficients has great promise for the efficient calculation of effects levels. Unfortunately, there are relatively few coefficients available for wild species in natural surroundings. These shortcomings must be overcome through the use of textbook data on different species in different surroundings, which adds to the uncertainty of the analysis. Only with the addition of more specific data on toxicology and contaminant transfer in communities and ecosystems will uncertainty be reduced and the ability to make stronger conclusions created.

REFERENCES

Anderson, L. M., S. D. Fox, D. Dixon, L. E. Beebe, and H. J. Issaq, 1991. Long-term persistence of polychlorinated biphenyl congeners in blood and liver and elevation of liver aminopyrene demethylase activity after a single high dose of Aroclor 1254 to mice. *Environmental Toxicology and Chemistry* 10(5):681–690.

Aulerich, R. J., and R. K. Ringer, 1977. Current status of PCB toxicity in mink, and effect on their reproduction. *Archives Environmental Contamination and Toxicology* 6:279–292.

Borman, F. Herbert, and Gene E. Likens, 1979. *Pattern and Process in a Forested Ecosystem.* New York: Springer-Verlag.

Carson, Rachel, 1962. *Silent Spring.* Boston: Houghton Mifflin.

Clark, T., K. Clark, S. Paterson, D. MacKay, and R. J. Norstrom, 1988. Wildlife monitoring, modeling, and fugacity. *Environmental Science & Technology* 22(2):120–127.

Davidson, Art, 1990. *In the Wake of the Exxon Valdez.* San Francisco: Sierra Club Books.

DuBowy, P. J., 1989. Effects of diet on selenium bioaccumulation in marsh birds. *Journal of Wildlife Management* 53(3):776–781.

Eisler, R., 1987. *Polycyclic Aromatic Hydrocarbon Hazards to Fish, Wildlife, and Invertebrates: A Synoptic Review*. U.S. Fish and Wildlife Service Biological Report 85(1.11).

Flueck, W. T., 1991. Whole blood selenium levels and glutathione peroxidase activity in erythrocytes of black-tailed deer. *Journal of Wildlife Management* 55(1):26–31.

Frenzel, R. W., and R. G. Anthony, 1989. Relationship of diets and environmental contaminants in wintering bald eagles. *Journal of Wildlife Management* 53(3):792–802.

FWS, 1988. *Lead Hazards to Fish, Wildlife, and Invertebrates: A Synoptic Review*. Biological Report 85(1.14), 134 pp.

FWS, 1981. *Bats and Environmental Contaminants: A Review*. Special Scientific Report—Wildlife No. 235.

FWS, 1980. *Effects of Environmental Contaminants on Reptiles: A Review*. Special Scientific Report—Wildlife No. 228.

Gilbertson, M., T. Kubiak, J. Ludwig, and G. Fox, 1991. Great Lakes embryo mortality, edema, and deformities syndrome (GLEMEDS) in colonial fish-eating birds: Similarity to chick-edema disease. *Journal of Toxicology and Environmental Health 33(4):455–520.*

Gjerstad, K. O., and I. Hanssen, 1984. Experimental lead poisoning in willow ptarmigan. *Journal of Wildlife Management* 48(3):1018–1022.

Harris, H. J., H. A. Regier, and G. R. Francis, 1990. Ecotoxicology and ecosystem integrity: The Great Lakes examined. *Environmental Science & Technology* 24(5):598–603.

Harrison, P. D., and M. I. Dyer, 1984. Lead in mule deer forage in Rocky Mountain National Park, Colorado. *Journal of Wildlife Management* 48(2):510–517.

Heinz, G. H., D. J. Hoffman, and L. G. Gold, 1989. Impaired reproduction of mallards fed an organic form of selenium. *Journal of Wildlife Management* 53(2):418–428.

Hirsch, Allan, 1980. Monitoring Cause and Effects—Ecosystem Changes. In *Biological Monitoring for Environmental Effects*. Lexington, MA: D.C. Heath and Company.

Hoffman, D. J., B. A. Rattner, and R. J. Hall, 1990. Wildlife toxicology. *Environmental Science & Technology* 24(3)276–283.

Jones, K. B., 1986. Inventory and monitoring process. In *Inventory and Monitoring of Wildlife Habitat*, A. Y. Cooperrider, R. J. Boyd, and H. R. Stuart (eds.), pp. 1–10. Denver: U.S. Dept. Int., Bur. Land Mgmt. Service Center.

Levine, Adeline, 1982. *Love Canal: Science, Politics, and People*. Lexington, MA: D.C. Heath and Company.

Lindzey, Frederick G., 1982. Badger (*Taxidea taxus*. In *Wild Mammals of North America: Biology, Management, Economics*, Joseph. A. Chapman and George A. Feldhamer (eds.), pp. 653–663. Baltimore: Johns Hopkins University Press.

Luckey, T. D., and B. Venugopal, 1977. *Metal Toxicity in Mammals: Volume 1*. New York: Plenum Press.

Mautz, William, 1978. Nutrition and Carrying Capacity. In *Big Game of North America*, John L. Schmidt and Douglas L. Gilbert (eds.), pp. 321–348. Harrisburg: Stackpole Books.

National Research Council (NRC), 1978. *An Assessment of Mercury in the Environment*. Washington, D.C.: National Academy of Sciences.

Platonow, N. S., and K. H. Karstad, 1973. Dietary effects of PCBs on mink. *Canada Journal Comparative Medicine* 37:391–400.

Robbins, Charles, 1983. *Wildlife Feeding and Nutrition*. New York: Academic Press.

Schafer, E. W. Jr., W. A. Bowles, Jr., and J. Hurlbut, 1983. The Acute oral toxicity, repellency, and hazard potential of 998 chemicals to one or more species of wild and domestic birds. *Archives Environmental Contamination and Toxicology* 12:355–382.

Schmidt-Nielsen, Knut, 1983. *Animal Physiology: Adaptation and Environment*. New York: Cambridge University Press.

Toweill, D. E., and J. E. Tabor, 1982. River Otter (*Lutra canadensis*). In *Wild Mammals of North America: Biology, Management, Economics*, Joseph A. Chapman and George A. Feldhamer (eds.), pp. 688–703. Baltimore, MD: Johns Hopkins University Press.

U.S. EPA, 1991. *Summary Report on Issues in Ecological Risk Assessment*. EPA–625–3–91–018.

U.S. EPA, 1989a. *Summary of Ecological Risks, Assessment Methods, and Risk Management Decisions in Superfund and RCRA*. EPA–230–03–89–046.

U.S. EPA, 1989b. *Risk Assessment Guidance for Superfund: Volume II Environmental Evaluation Manual*. Interim Final. EPA–540–1–89–001.

U.S. EPA, 1989c. *Superfund Exposure Assessment Manual Technical Appendix: Exposure Analysis of Ecological Receptors*. Draft. EPA Environmental Research Laboratory, Athens, GA.

U.S. EPA, 1980. *Toxicity of the Polychlorinated Biphenyl Aroclor 1016 to Mink*. EPA–600–3–80–033.

Weir, David, 1987. *The Bhopal Syndrome*. San Francisco: Sierra Club Books.

Whelan, E., 1985. *Toxic Terror*. Ottawa, IL: Jameson Books, Inc.

Whittaker, Robert H., 1975. *Communities and Ecosystems*. New York: MacMillan.

Wren, C. D., 1987. Toxic Substances in Furbearers. In *Wild Furbearer Management and Conservation in North America*, M. Novak, J. A. Baker, M. E. Obbard, and B. Malloch (eds.), pp. 930–936. Ontario: Canada Ministry of Natural Resources.

7

Evaluation of Contaminants in Sediments

INTRODUCTION

Sediment Definition and Relation to Contaminants

Sediments are the interface of soils or other geologic features and surface waters. Sediments are composed of: native materials that predate inundation by water, such as sand, glacial till, bedrock, and clays; water, from both surface and ground sources; material, both organic and mineral, washed into the water body; and particles generated in the water body, such as plankton frustules, fecal material, or organisms. This complex mixture of materials is susceptible to contamination in streams, rivers, lakes, estuaries, and the open ocean.

The definition of sediments, for this discussion and also reflective of current regulatory thinking, is restricted to areas that are inundated continuously for all but the driest years, and support a self-sustaining truly aquatic community. There are different opinions on the special application of the term sediment, but the importance of considering the hydrologic setting is considered essential (Marcus 1989). This definition excludes the substrate of most wetland types, and also intermittent streams. The reason for restricting the definition is that the methods used in sediment evaluation of the transfer of contaminants between environmental media and exposure to ecological receptors are different from methods used to evaluate contaminates in saturated soils. The limitation of the term to truly aquatic areas does not imply that contamination of saturated soils in wetlands and seasonal water bodies is not a serious environmental concern, only that the methods used to evaluate sediment quality are not universally applicable to those areas. Usually, a selective combination of methods to evaluate soils, such as ecological pathway modeling, exposure assessments, and ecological benchmarks is necessary to assess contamination in wetlands and other semiaquatic habitats.

Except for native material, any of the sediment components listed earlier represent a potential mechanism for the transport of contaminants from a hazardous waste site to or from the sediment. Probably the most common source of sediment contamination associated with hazardous waste is the deposition of eroded particles from sites into downgradient water bodies. Since erosion and surface runoff are ubiquitous phenomena, sediment quality and the potential need for remediation are issues at many hazardous waste sites. In fact, for many contaminants that are relatively insoluble, and thus associated with particles, sediments represent the ultimate fate. When these compounds are environmentally persistent, they can build up in concentration, and thus represent a serious and increasing threat to the environment.

Ecological Concern Over Sediments

Sediments warrant a separate chapter in this book for three reasons. First as discussed earlier, they are frequently the ultimate sink for contaminants from hazardous waste sites. Sediments are also ecologically critical in most if not all aquatic systems affected by hazardous waste sites. Finally, sediment quality is emerging as a new and controversial regulatory concern, and the outcome of the regulatory debate will perhaps have the greatest effect on ecological assessments and remediation of hazardous waste sites.

In most situations related to hazardous waste sites, sediments are the location of the most important secondary production, and thus they are critical in the food chain. There are many critical life functions, such as egg deposition and critical developmental stages, that are in close contact with sediments. There are also processes critical to ecological functions in the overall aquatic environment, such as nutrient storage and regeneration, that occur in the sediments. The organisms and processes associated with the sediments are most often in close physical contact with particles, and because benthic (sediment-dwelling) organisms tend to be sessile and immotile, they are associated with the same specific sediments for long periods, often their entire lives. The sediments are also relatively stable, and thus contamination problems are less likely to self-remediate with time after the primary source is removed, as can often occur with contaminated surface waters. In fact, sediments can frequently constitute a secondary contamination source long after the original source is controlled. For all of these reasons, benthic organisms and ecological functions are sensitive to contaminant effects, and aquatic ecosystems are sensitive to benthic effects. Therefore, the evaluation of sediment contamination and the consideration of sediment remediation are important issues in many RI/FS investigations, and the ecological assessment must provide critical input to the process.

Sediment contamination and the evaluation of sediment quality is an ecological concern for activities not related to hazardous waste sites. Sediment contamination was a major issue in the dredged material disposal program (U.S. EPA and U.S. Army Corps of Engineers 1991). Dredging most often occurred in urban and industrialized areas, and thus contaminants from the runoff and discharges in the tributary drainage basin affected sediment quality. Those charged with the removal and disposal of the sediment for navigation maintenance (predominantly the U.S. Corps of Engineers) have had to address the quality of the material as it affected the environment, during both removal and disposal.

The control of point and nonpoint source pollution originally focused exclusively on surface water. Recently, however, contamination of sediments has also been a serious concern for pollution control (U.S. EPA 1988; Barrick et al. 1989). Because of the extensive concern over sediment contamination and multiple activities affecting sediment, the EPA is currently exercising its authority under Section 304(a) of the

Clean Water Act to develop and implement sediment quality criteria. The intent is to develop national criteria for sediment comparable to the Quality Criteria for Water (U.S. EPA 1986). Some states (e.g., California) are also in the process of developing sediment quality criteria and evaluation methods (Barrick et al. 1989; Zarba 1989).

The development of national sediment criteria has been a controversial and much debated process (Chapman 1991). There are extensive uncertainties in the various methods under consideration for both developing criteria and then measuring compliance with the criteria. Some scientists feel that the conservative assumptions made to compensate for the uncertainties will result in unnecessarily stringent and expensive control of sediment quality and requirements for remediation. There are also those who feel that the controls on discharges to surface waters, such as National Pollution Discharge Elimination System Permits and 404 wetland permits, already in place are adequate to protect sediment quality, and in situ sediment criteria are not necessary (Chapman 1991).

Probably the greatest controversy surrounding Sediment Quality Criteria are their intended uses. The scientists developing the procedures envision true criteria, which can be used as guidance and screening procedures in considering the severity of the problem (Di Toro 1988; Zarba 1989). However, those forced to comply with regulations fear that agency policy personnel will consider the criteria as absolutes or standards, and will mandate unnecessarily stringent restrictions and remediation. The concern is that regulators will not allow consideration of site-specific conditions that mitigate the effects of sediment contamination or render the conservative assumptions used to develop the criteria inappropriate. The U.S. EPA anticipates promulgation of sediment criteria in 1992, and they will hopefully clarify the intended use of the criteria in the codes and guidance that they generate.

Largely spurred by the regulatory mandate for sediment quality criteria, the development of sediment quality evaluation methods has recently proceeded at an increased pace, and there appear to be indications of some consensus on the available technical approaches to evaluating sediment quality. There is less concurrence on the approach for establishing and using sediment quality criteria. The National Research Council (1989), U.S. EPA (1989a), National Oceanic and Atmospheric Administration (Long and Morgan 1989), and the EPA Science Advisory Board (U.S. EPA 1990a) have all commented on the evaluation techniques for sediment, and seem to generally concur on the following items:

- Some form of criteria, or at least screening environmental benchmark value, is desirable for individual contaminants (there tends to be a lack of consensus as to how criteria should be applied once they are developed, either as cleanup levels or as screening techniques).
- Individual site-specific investigations and evaluations will frequently be required to resolve sediment contamination concerns.
- Improved biological and chemical techniques are needed to rapidly and reliably assess the presence and severity of contamination.
- There is a national need to formulate accepted approaches to evaluate the risk from contaminated sediments and to prioritize cleanup.
- Some variation and formalization of the current ad hoc approach, of using a variety of approaches to determine the significance and extent of sediment contamination, may ultimately be the most appropriate practice.

Sediment Evaluation Objectives and Methods

There are two primary objectives of sediment quality evaluation: determining the presence and severity of contamination, and thus the need for cleanup; and establishing how clean is clean, or setting remediation criteria. These objectives should sound familiar, because they parallel two of the primary ecological assessment objectives, as established in Chapter 2. There are methods designed to address each of these objectives.

Approaches to establish the need for remediation are basically measurement techniques to assess the existing conditions. The methods that fall into this category are generally the same as those described by Chapman (1989a), used to evaluate mixtures of chemicals. Such approaches generally deal with specific sediments, and produce yes/no answers concerning ecological effects and thus the potential need for cleanup.

Sediment quality evaluation methods used to establish ecological effects levels are largely predictive tools. They are designed to predict what a certain ecological characteristic will be, given a certain level of a contaminant. The concentration corresponding to the applicable ecological endpoint can then provide input to setting the cleanup criteria. The methods used to address this objective overlap with the approaches that Chapman (1989a) categorizes as individual compound evaluation methods.

The overlap in the classification of methods based on evaluation objectives and Chapman's (1989a) individual versus mixture of compounds is logical. When considering the need for cleanup, the combined impact of a mixture of chemicals is the critical question, whereas remediation criteria must generally be established on a chemical-by-chemical basis, because the range of possible mixtures can be close to infinite. There is considerable overlap in the methods used to address the two objectives, and some are suitable for either. Others, however, although elaborate and intensive, only address one of the objectives, and leave decision makers at a loss when faced with critical and often controversial remediation determinations. Several of the methods build on or compliment other techniques, and can be used in series as part of a phased approach if less intense methods identify, but do not resolve, a sediment quality issue.

The focus of this chapter is on the description, discussion, and assessment of the available sediment quality evaluation methods as they relate to ecological investigations of hazardous waste sites. The methods are organized according to the primary objective that they are intended to address: site-specific evaluation or sediment quality criteria. However, as pointed out earlier and in the following individual discussions, methods can be used for more than one objective, and often some type of site-specific evaluation is a mandatory precursor for a technique to establish criteria.

SITE-SPECIFIC SEDIMENT EVALUATION METHODS

Bulk Sediment Testing

Bulk sediment testing is the most basic, and probably the most widely used, method of evaluating sediment quality. The method involves characterizing the entire sediment sample removed from the area of concern, and is analogous to testing water from a swimming area or drinking water supply source. There are both chemical and biological aspects to the testing. The chemical testing is generally included in the RI/FS as part of the overall site characterization and documentation of the nature and extent of contaminants, and at least some of the data is usually available to determine the need for and specific methods of detailed ecological investigations. A simple approach is to compare the available sediment chemistry data to information from other locations, such as the national inventory of sediment quality (Long and Morgan 1989) as a screening technique to identify sediment toxicity as an ecological area of concern. If the chemical analysis of the bulk sediment reveals values well below known effects levels for similar ecosystems and sediment types, sediment remediation may not be an issue at the site. On the other hand, if the on-site chemical concentration of the sediment is higher than or the same order of magnitude as sediments known to produce ecological effects, additional investigation may be warranted. The bulk sediment chemistry can also be extremely useful in designing the detailed sediment investigation. For example, the location of the additional sampling and analyses can be specified to span the range of concentrations present on-site. The spatial results of bulk sediment measurements can also be an important indication of the contamination source and the ecological resources potentially at risk.

There have been attempts to use bulk chemistry in establishing generally applicable sediment quality criteria. The approach relies on comparing site-specific chemistry to other data. A common tactic is to establish a background concentration of a specific compound, based on measurements unaffected by the hazardous waste site and in an otherwise relatively pristine area. When sediments significantly exceed these background levels, they are considered above the criteria. Use of this approach gives no consideration to the ecological effects of sediments, only an indication of increased levels of contaminants.

In 1977, Region V of the U.S. EPA established guidelines for the evaluation of sediments in the Great Lakes using bulk sediment analysis. The guidelines were directed at the disposal of dredged materials, and the impacts in the area of dredging were not a primary concern. The approach was to classify sediments as nonpolluted, moderately polluted, or heavily polluted for each individual compound of concern, and the classification was not always based on a full suite of scientific evidence (Fitchko 1989).

The Wisconsin Department of Natural Resources derived a sediment criteria approach based on bulk sediment analysis in 1985 (Fitchko 1989). The approach was to define background contaminant levels based on samples in unpolluted areas and preindustrialization geologic sediments. The acceptability of in-water disposal of sediments was based on the relationship of the concentration in the dredged material to background levels. As with the EPA Region V guidelines, these criteria were only applied to the disposal of dredged material and did not consider other types of sediment contamination issues.

The National Oceanic and Atmospheric Administration also monitors sediment quality in marine waters, and periodically reports on findings (Long and Morgan 1989). They have developed descriptors for the low (ER-L) and median (ER-M) concentrations found nationwide for 12 trace metals, 18 petroleum hydrocarbons, and 11 synthetic organic compounds. This has been useful in identifying "hot spots" and areas on which to focus additional investigation, but they emphatically point out that these levels are not criteria, are not based on any effects level, and should not be used to establish cleanup levels. They can, however, be a very useful tool to screen concentrations found at a specific site to determine how it ranks in a national context.

Biological testing of bulk sediments, or sediment bioassay, is the foundation of sediment quality evaluation as part of an ecological assessment. This method was the

TABLE 7-1.   Summary of Available Bulk Sediment Bioassay Methods.

Test Organism	Duration (Days)	Endpoints	Additional References
Chironomus tentans (12 ± days old) or *C. riparius* (<3 days old)	10–14 \ 30±	growth survival emergence	Adams et al. 1985 Nebeker et al. 1984 Giesy et al. 1988 Ingersoll and Nelson 1989
Hyalla azteca (<7 days old)	10–30	survival behavior growth reproduce	Nebeker et al. 1984 Nebeker and Miller 1988 Ingersoll and Nelson 1989
Rhepoxynius abronius (acclimated from field)	>10	behavior mortality	Swartz et al. 1979 and 1985 DeWitt et al.1988 Robinson et al. 1988

Source: Summarized and updated from U.S. EPA 1989a.

primary tool used in the early assessment of sediments, including the initial ocean dumping regulations promulgated by the EPA in 1977, which included bulk toxicity testing of sediments to evaluate the suitability of dredged material for disposal. All other methods of evaluating sediments must be based on (or otherwise related to) direct sediment bioassays, to establish legitimacy. The method has weaknesses and limitations, as discussed later, but it measures a first principle because it evaluates the direct and simple link of sediment quality to observed biological effect.

The basic bioassay method is the addition of organisms to test chambers containing bulk sediment and overlying water. The tests are generally static (i.e., no replacement of water or sediment) and continue for one to four weeks. At the end of the tests, the sediment is examined and the condition of the added organisms observed. As summarized in Table 7–1, there are a number of organisms and endpoints that have been standardized for sediment testing. Other test organisms under consideration for standard tests are the marine amphipods *Eohoaustorius estuarius, Ampelisca abdita,* and *Grandidieriella japoinica* (U.S. EPA 1989a and Swartz 1989). Development of bivalve larvae and bioluminescence are other sediment bioassay endpoints commonly used (Swartz 1989). While these endpoints are expedient and extremely useful to evaluate relative toxicity, they are measurement endpoints and do not always automatically directly relate to ecological assessment endpoints of concern. Therefore, the link between these measurements and the ecology of the site must be understood and related to the decision makers. There does not necessarily have to be a direct link, if the case for surrogates can be established.

Sediment bioassays have also been conducted using a variety of organizational endpoints, including:

Biochemical	Cellular
Physiological	Individual
Pathological	Community
Reproductive	Population
Developmental	Behavioral

Swartz (1989) summarizes these endpoints and cites examples of their use. The American Society of Testing and Materials (ASTM 1988) is continually updating and attempting to standardize sediment bioassay methods, and should be consulted in the design of any sediment bioassay program.

The greatest advantage of sediment bioassay as part of an ecological assessment is the direct and straightforward measure of the ecological suitability of the sediment. The site-specific characteristics, including sediment type and synergistic effects, are incorporated, thus reducing the need for data manipulation and therefore sources of variance or error. The bioassay can also be designed for the specific organism and endpoints important to the site, and further reduce the need for correction factors. If the test results are unambiguous, and the agreed-upon endpoints are adequately represented, the bioassay program can clearly define and justify the need for and location of sediment remediation. Because of the direct and simple relationship between sediment bioassay results and sediment quality, this is a widely accepted and relatively standardized method of sediment quality evaluation. The scientific, regulatory, and legal communities have accepted and used the method (U.S. EPA 1989a), and regulatory entities (i.e., the EPA and U.S. Army Corps of Engineers) as well as professional societies (i.e., ASTM) are constantly developing and improving standardized methods. Sediment bioassays also represent a common base for evaluating or supplementing other more speculative sediment quality evaluation methods.

Toxicity testing of bulk sediment does have certain technical and policy limitations. The collection, storage, or handling of sediments prior to testing can alter the biological, chemical, or physical characteristics, which can, in turn, contribute to or reduce the toxicity. For example, if a set number of organisms are to be added to the sediment, and then survival measured after exposure, the indigenous sediment population must be eliminated to measure percent survival. This is generally accomplished by freezing, dehydration, or removal, all of which can alter the sediment. As in all biological testing, there is always the uncertainty of laboratory artifacts and in situ factors (such as competition and natural environmental variables) not replicated in the laboratory test. The advantage of simplicity in the tests can also be a limitation, because the mode of toxicity is not identified and the effects of individual chemicals can not be discriminated.

From a policy standpoint, the method is generally limited to the evaluation of specific sediments and a determination of need/no need for remediation. The transfer of the information to other sediments, sometimes even at the same site, or identification of the extent of required remediation is limited. Also, the effects of individual chemicals is often unclear, so if assigning the source of effects or controlling specific compounds is an issue, sediment bioassay may provide only limited information. Under most circumstances, the results are only applicable to specific biological conditions and, due to the length of the test, steady state bioaccumulation is not achieved. Consequently, the results of the tests are not always easily used to evaluate food chain effects or public health concerns. As an added limitation to using the tests, the species used are generally not used for human consumption, so food chain effects must be assumed in order to evaluate public health effects.

Pore Water Testing

In all saturated sediments, there exists pore, or interstitial, water among the inorganic and organic particles. The physical (i.e., grain size, compaction, and density) and chemical characteristics dictate the relative amount of pore water within any specific sediment, or the water content of the sediment. In most sediments, 20–50 of the volume consists of pore water (U.S. EPA 1989a). Contaminants, as well as nutrients and oxygen, are dissolved in the sediment pore water, and are generally considered to be the primary mode of availability, and thus exposure, for organisms associated with the sediments. A close correlation between pore water concentrations and sediment toxicity has been reported for a variety of situations, and thus supports the importance of pore water as an exposure pathway (Adams et al. 1985, Swartz et al. 1985, Connell et al. 1988, DiToro 1988, Knezovich and Harrison 1988, Swartz et al. 1988). Therefore, testing of the pore water could provide a direct indication of the potential toxicity of specific sediments.

The simplest application of pore water testing as a sediment quality evaluation method would be the collection and chemical analysis of pore water from the sediment of concern. The results could then be compared to water quality criteria or other ecological benchmarks for each contaminant of concern. This approach assumes a similar sensitivity of the indigenous benthic organisms to those used to develop the criteria or benchmarks. If such an assumption is unwarranted, the benchmarks can be adjusted for the species of concern. Such an approach is probably most suited to a screening-level analysis, to identify sediment contamination as an issue and, if warranted, to provide input to the design of a detailed sediment testing and evaluation program.

Chemical testing of pore water and comparisons to benchmarks is a more appropriate method for screening than using bulk sediment chemistry. Since pore water is the acknowledged primary mode of contaminant transfer, the assumptions relating bulk sediment to pore water chemistry are not necessary. Also, the variability in toxicity due to the physical and noncontaminant chemistry of the individual sediments is eliminated. Therefore, if pore water data is available, it would normally be used in preference to bulk sediment chemistry. Pore water testing might be necessary, as a screening technique, if there were no ecological benchmarks for bulk sediment but accepted water toxicity data existed for the contaminants of concern.

If the evaluation of sediment quality is warranted beyond the screening phase, biological testing of pore water can be a useful tool. The approach parallels bioassay for aqueous solutions, as described in Chapter 8, with the added complication of collecting the test solution from the sediment instead of the water column. The test organisms, bioassay endpoints, and other characteristics of the test must be selected to reflect the benthic environment and site-specific ecological endpoints.

The biological testing of the sediment pore water identifies the ecological suitability of specific sediments. However, because it involves added steps, and the associated potential for error in the extracting, transporting, and storing of pore water, it is not as realistic a measure as bulk sediment toxicity testing. Therefore, if sediment-specific testing is deemed necessary, and the suitability of the specific sediment is the only issue, the added complexity and difficulty of pore water testing is generally not warranted. However, pore water testing offers options not available from bulk sediment testing. The additional advantages include the identification of individual contaminants producing effects, and input to remediation criteria.

If the sediment of concern is shown to be toxic, and remediation is warranted, pore water bioassay techniques can provide significant input to the selection of the appropriate remediation method. The process follows the toxicity identification evaluation (TIE) procedure developed by the U.S. EPA for the evaluation of wastewater effluent

(Mount 1988; Mount and Anderson-Carnahan 1988a and b). The procedure is a phased approach, to characterize, identify, and then confirm the toxic components of a complex aqueous solution. The solution is subjected to a series of treatments to remove or render inactive various categories of compounds. If the treatment does not alter the toxicity of the pore water compared to the untreated pore water, then the category of compounds vulnerable to the specific treatment can be eliminated as contaminants of concern.

A variety of treatment or inactivating methods are available to isolate the contaminants responsible for the toxicity. Adjusting the pH can be used to alter ammonia toxicity, and can also influence the effects of various metals. Aeration is a useful technique to eliminate volatile or oxidizable contaminants. Chelating agents can also be used to produce nontoxic complexes of many metals.

TIE is an interactive process used to focus on the specific compounds responsible for the toxicity observed in the pore water and, by extension, in sediments. The procedure can also be used to assign the degree of observed toxicity attributable to each compound when a number of contaminants are present in the sediment. Once the causative compounds have been identified, the method can be useful in developing remediation criteria. By testing various dilutions of pore water, and thus concentrations of the contaminants of concern, an acceptable concentration in pore water can be identified. The pore water concentrations can then be related to a sediment concentration, and sediment treatment methods compared to the level of toxicity in the sediment.

There are two major advantages of pore water testing as a sediment quality evaluation approach: flexibility and acceptance. The method is adaptable to identifying and evaluating individual compounds causing toxicity and remediation criteria. The well-established and accepted methods for the bioassay of aqueous solutions and the TIE procedures are significant benefits available when pore water testing is used for the evaluation of sediment quality at hazardous waste sites.

Complexity, and thus cost, can be a significant disadvantage of the method. Obtaining and processing the pore water requires significant effort, and when added to the analytical and bioassay costs, the total expenditure can easily exceed that of other sediment evaluation methods. If a TIE is performed, the effort escalates rapidly as a function of the complexity of the chemical composition and the degree of precision required. Thus, pore water testing is generally not appropriate if a simple yes/no determination of toxicity is the issue.

The major drawbacks of pore water testing as a sediment evaluation method are the difficulty in obtaining pore water and the potential for altering the toxicity characterics in the process. Pore water can be extracted from the sediments by a variety of methods, including compression techniques, displacement via inert gases, centrifuging bulk sediment, direct sampling, and microsyringe sampling (Knezovich et al. 1987; Knezovich and Harrison 1988; Sly 1988). None of these methods have been extensively used and verified, and thus are not widely accepted. Since a minimum of 1.5 liters of pore water (which represents at least 8 liters of sediment) are generally required for testing, a large volume of sediment and increased expenditure are usually necessary for pore water testing. If TIE is required, the volume of pore water can be an order of magnitude higher.

Depending on the extraction and processing method used, the chemical characteristics, and thus the toxicity of the solution collected, can be altered compared to the in situ pore water. The association with particles and other factors effecting bioavailability, such as pH and oxygen, can be significantly altered during extraction. Thus, the results of laboratory testing of the solution can vary from in situ effects. The approach incorporates the assumption that exposure and intake of pore water is the sole method of contaminant transfer. Although the correlation between pore water concentration of contaminants and toxicity is well documented, there is some debate as to the importance of other contamination pathways such as dermal contact and ingestion of contaminated particles. The testing of pore water as a sediment evaluation method ignores these pathways.

Even with these limitations, the analysis of pore water can be an important tool in evaluating sediment contamination. It has the advantage of accepted water analytical procedures with the results interpreted from the extensive body of accepted aquatic toxicity data. It also provides the unparalleled advantage of manipulating the toxic characteristics of the media in order to determine the causative agent (and thus the potentially responsible party) and to evaluate alternative remediation criteria and treatment methods. The method has the ability to address mixtures of compounds as well as individual contaminants, not influenced by the toxic characteristics of other chemicals in the sediments, which is unique to the pore water approach. The major limitation is the collection of the pore water and many feel that this is a technological problem that can be overcome through minor modifications and validation of existing methods (Tishler/Kocuek Consultants 1990).

Benthic Community Structure

The benthic, or bottom-dwelling, community is intimately associated with sediment particles and pore water, relying on the association for habitat, food, exchange of gasses, and nutrient uptake. The benthic species have adapted to maximize the association, especially for the exchange of materials and behavioral characteristics, such as feeding and escape from predation. The close physical contact and highly developed mechanisms for exchange of material between organisms and sediment results in parallel exposure pathways for contaminants within the sediment complex. The relatively static nature of the sediment, compared to the water column or a terrestrial system, is reflected in the sedentary nature of most benthic organisms. Thus, the categorically close association of sediment and the benthos tends to extend to specific sediments and particular populations. Consequently, the characteristics of the benthic community are strongly affected by, and in turn reflect, the contamination of the particular sediment where the organisms are found.

The analysis of the benthic community structure is therefore the biological equivalent of bulk chemical testing of sediment. The approach measures the in situ biological character of the sediment, and uses the information to evaluate sediment quality. Several states have recognized the appropriateness of benthic communities as an indication of environmental health, and have incorporated benthic community evaluation into their overall approach to environmental protection and regulation (Davis and Lathrop 1989). At individual hazardous waste sites, the examination and evaluation of the benthic community structure can similarly be a useful tool in the assessment of sediment quality.

The approach to a benthic evaluation is quite straightforward, and not unique to the evaluation of hazardous waste sites. The technique has been used for some time to assess the effects of wastewater discharges, by examining the community up- and downstream of a pollution source (Hynes 1969). The basic steps are to collect samples of sediment and associated organisms from the area of concern. The organisms are then identified and enumerated, and the resulting information used to characterize the community using a number of standard qualitative and quantitative techniques. These techniques include population density, number of species present, and a combination of these two community characteristics as an estimate of species diversity. The characteristics of the dominant species, particularly their tolerance to specific contaminants, can also be useful in evaluating sediment quality. The community characteristics are then compared to the "expected" parameters determined from available benthic community information or a synoptic evaluation of a reference site. The comparison is then viewed in relation to the factors that could account for the variations (e.g., physical differences in substrate, nutrient availability, and water depth) and the influence of sediment quality, particularly of the concentration of contaminants of concern.

Benthic community analysis, as a sediment quality assessment tool, is well suited to a screening analysis as part of a phased approach. The U.S. EPA's Rapid Bioassessment Protocol (U.S. EPA 1989b) was developed specifically to quickly and efficiently detect the presence and areal extent of sediment contamination. As summarized in Chapter 4 of this book, the method is best suited to lotic (i.e., flowing water) systems, but the same approach and philosophy can be adapted to lentic (i.e., standing water) and estuarine environments. The results of the largely qualitative first-tier investigations can be used to define the absence of a sediment contamination concern or, alternatively, to establish the need and study methods for additional, more extensive sediment investigations.

The classic benthic investigation methods for detailed studies involve collecting a volume of sediment, using a grab or dredge. The sediment is then returned to the laboratory for the removal, enumeration, and identification of organisms. The resulting data can be standardized by the area sampled and subsamples can be used for chemical analyses. In lotic systems, or littoral zones of estuaries and lakes, nets can be used to collect organisms without large volumes of associated sediments. The results from net samples can be standardized by sampling a specific area, as in the Surber Sampler (Surber 1937), or a specific time interval.

In situations where the source of the toxicity or other stress to the benthic community is an issue, artificial substrates can be employed, alone or in combination with sediment samples, to evaluate sediment quality. Artificial substrates, as the name implies, are non-native or standardized native devices deployed in the aquatic system. Common devices used are stacked wood-product plates (Hester-Dendy sampler), baskets filled with rocks, or uniform-area concrete pieces. The devices are suspended just above the natural substrate for about a month, and allowed to be colonized with indigenous fauna. The degree and characteristics of colonization in the area of concern, compared to a reference area, gives information as to environmental stress. Sediment or indigenous benthos can be collected from the artificial substrate site at the time of the substrate retrieval. The relationship of the artificial substrate to the in situ benthic community can then be considered to segregate effects due to in-place sediment, as opposed to water quality of other nonsediment factors. Specific field sampling methods and protocols for benthic investigations are well documented in a number of agency and society publications (U.S. EPA 1973; U.S. EPA 1978; U.S. EPA 1989b; U.S. EPA 1990b; Holme and McIntyre 1984; and ASTM 1988).

Except in the rare event where the sediment is found to be abiotic, the success of benthic community structure analysis, as a sediment evaluation technique, is dependent on the detection of often subtle biological alterations. This requires establishment of the normal or unaffected community. A reference area is the most common method for defining the bounds of the unaffected community. If a stream or river is the area of concern, it is often relatively easy to locate an immediately upstream reference area with gradient and substrate similar to the stream reach of concern. For small streams, a similar system in the same drainage basin is generally available as a reference site. Lakes, estuaries, and rivers with multiple pollution sources pose more of a challenge in finding a site comparable to the area of concern without the influence of a hazardous waste site.

Once the unaffected benthic community has been characterized, it must be compared to the community associated with the sediment of concern. Since natural variability prevents any two systems, even with no or identical environmental stress, from supporting identical communities, a procedure for detecting greater than normal differences is required. Characteristics of benthic communities amenable to comparison between communities include species composition, density, and functional information. A number of numerical indices, many of which have been reviewed by Washington (1984) and Pielou (1977), have been developed to represent these characteristics in a uniform fashion suitable to comparisons between locations or over time. Diversity indices, which numerically integrate relative species abundance, are frequently useful to compare community structure at both the reference and hazardous waste sites. Biotic indices assign a tolerance value to each species, and then sum the values for the sample. Thus, a comparison of these indices between sites yields additional numerical information related to the type of species occurring in the two areas. Biotic indices have historically been used for the assessment of organic enrichment, but the EPA and others have evaluated species tolerance to other common contaminants (Hart and Fuller 1974; U.S. EPA 1990b). Thus, appropriate values could be developed for biotic indices at individual hazardous waste sites. There are also techniques for combining a variety of numerical parameters into a composite index, and thus attempting to incorporate maximum information in comparing benthic communities (Ohio EPA 1987, also cited in Davis and Lathrop 1989).

Once the evaluation of the benthic community structure is complete, at best the result is a delineation of the area of sediment supporting an altered benthos. Sometimes, the degree of effect can be established in relative terms. The community structure analysis alone produces little or no information as to cleanup levels, contaminants causing effects, or the mechanisms causing the environmental stress. However, as described later for the Sediment Triad method and Apparent Effects Threshold technique, benthic structure analysis can be used in combination with other methods to address larger issues.

The major advantage of the technique is its direct measure of effects, closely related to ecological endpoints. Because of the limited anticipated conclusion (i.e., sediment does/does not cause benthic effects), few extrapolations or assumptions are involved. The evaluation is totally of an in situ situation, so laboratory artifacts do not produce limitations. As a screening tool, the evaluation of benthic community structure can be an extremely efficient technique, sometimes requiring a one-day field visit to identify the extent of sediment effects. Another advantage of the approach is the wide acceptance of the necessary field and laboratory methods. There is also a wealth of statistical and other data-evaluation techniques available to characterize and compare communities, and often good historical or regional data to supplement the comparison to the reference area.

The limited objectives of benthic community analysis are a disadvantage of the technique for comprehensive sediment evaluation. Although this limitation can be rectified by combination with other techniques, the total effort for both detailed benthic studies plus the other approaches can be prohibitive, for all but the largest hazardous waste site investigations. The natural variability inherent in benthic communities can also interfere with comparing site and reference area characteristics, and thus limit the confidence in the results or require substantial sample replication, with the associated costs.

A thorough benthic investigation can be extremely effort and time intensive, and thus outside the scope of a normal RI/FS investigation. A complete program must include sufficient replicates for statistical analysis, and full coverage of the area of concern plus a reference area, so that a single sampling event could easily require in excess of 75 individual samples. Without an extensive multiseasonal data collection program, the approach can yield a false negative as to benthic community structure effects. If a healthy community is observed in the spring, following maximum reproduction and high stream flow, the sediment might be judged uncontaminated. Later in the year, however, chronic toxicity and lower stream flows could produce an extremely stressed benthos. Thus, at least two seasonal samplings are generally required. Also, if the effects are subtle, extensive sample replication, detailed taxonomy, and sophisticated statistics may be necessary to elucidate the effects. The U.S. EPA (1989b) estimates that processing and taxonomic identification averages approximately 11 hours per sample. Collection, QA/QC, data management and analysis, and report writing could add at least another five to seven hours. A complete study, with adequate replicates and seasonal evaluations, could thus easily approach 3,000 labor hours. Benthic studies of this nature are more common as academic or research endeavors, and outside the scope and resources of a hazardous waste site investigation.

Tissue Residue

The tissue residue approach is similar to the benthic community structure method for evaluating sediment quality. However, the residue technique differs in the measurement endpoint: tissue concentration of contaminants of concern, rather than community characteristics. Organisms are collected from the area of concern and chemically analysed. The concentration of the contaminants associated with the hazardous waste site are determined for the whole organisms or specific organs. If human consumption is a major concern, concentration in edible tissue (usually muscle tissue) is an appropriate endpoint. When ecological food chain implications are an issue, whole-organism concentration is the usual measurement, because most aquatic predators do not differentiate specific organs. The sensitivity of the approach can be increased by analyzing organs, such as brain or liver, where contaminants concentrate due to the biochemistry or physiological function of the organ.

The tissue residue approach can be used for a variety of objectives over a range of methodological complexities. The range of objectives spans from delineation of affected area, to quantification of ecological implications, to the development and evaluation of ecological effects levels. For the first of these, defining the affected area, the process parallels establishing a "normal" community by examining a reference area, as used in the benthic community structure approach. A similar approach can be employed to assess tissue residue by determining levels of tissue concentration at unaffected but similar sites. Such an approach is useful in delineating the extent of contamination, but alone yields little or no information as to ecological effects, implications, or the need for remediation.

The next level of objective and intensity of investigation requires a benchmark based on the correlation of tissue concentration to ecological effects, rather than just a concentration above the background level. A common and straightforward example of this approach is a benchmark based on the acceptable dose of contaminant to the predator likely to consume the organism measured. The acceptable level can be related to survival of the individual predator, or to more subtle effects on the predator, such as reproduction, but must be directly related or transferable to the predetermined assessment endpoint. This is the approach used for human health assessments, and thus for many compounds there are acceptable tissue concentrations established for human consumption. Since the concentrations for human health protection are largely based on animal experiments, it is frequently possible to review the sources used to establish the levels for human consumption and develop ecologically appropriate benchmarks.

Determining suitable tissue levels for other ecological effects can be more difficult, because there is less research on the relation of tissue concentration to individual or population effects. However, such effects are often more appropriate to set benchmarks for, because the individual or population exposure to contaminated sediments is more direct and continuous than the episodic events of predators consuming benthic animals associated with an often-limited portion of their range. Therefore, attempts should be made to include benchmarks for the effects of various tissue levels on individuals and populations.

In situations with a large, accurate, and precise data base, the utility of the method can be expanded to the most complex level of investigation, and an attempt at establishing effects levels. With sufficient data, including bulk sediment chemistry, correlation of tissue concentration and various ecological effects, and appropriate test organisms (i.e., those in equilibrium with the sediment at the time and location of collection), it may be possible to correlate sediment and tissue concentration for contaminants of concern. The correlation can then be used to establish the sediment remediation concentration that will be protective of the ecological endpoints. In such cases, the delineation of the remediation area can be extrapolated far beyond the area of tissue collection. This applies not only to areal extent, but also to the depth of excavation and removal. The acceptability of returning treated sediments can also be evaluated, using the correlation of tissue and sediment concentration.

The tissue residue approach has been successfully used to establish criteria for specific sediments and situations where there is an adequate data base. Acceptable TCDD releases from the Hyde Park superfund site were established based on acceptable sediment concentrations in Lake Ontario protective of lake trout (Cook et al. 1989; Carey et al. 1989, as cited in U.S. EPA 1989a). Tissue concentration following extended exposure is also currently used by the U.S. EPA and Army Corps of Engineers to evaluate the suitability of dredged material disposal (U.S. EPA and U.S. Army Corps of Engineers 1991).

Regardless of the objective of the tissue residue approach, the species to be analyzed is critical to the successful implementation of the approach for sediment quality evaluation. The species must, of course, be present at the site, capable of being collected in sufficient mass for chemical analysis, and directly associated with the sediment of concern. If the organisms are only migrants, seasonal inhabitants, or random frequenters of the specific area, their tissue concentrations reflect their life history and not solely the sediment affected by the specific hazardous waste site under investigation. Species that fit this criterion are often small, sessile infauna, and thus it is at best difficult to collect sufficient biomass for analysis. There must also be sufficient existing literature on the selected or a similar species, so that defensible benchmarks can be established. As with all ecological endpoints, the selected species must be important to, or indicative of, an ecosystem function, in order to be meaningful in the overall RI/FS process. It is often difficult to identify a species that meets all of these requirements. If contamination effects are present, species sensitive to the contaminant of concern, which are common endpoint species, may not be present due to the toxicity of the sediments, and consequently not available for tissue analysis. This limitation can sometimes be overcome by deploying caged organisms within the sediment of concern. Mollusks and demersal fish are the most frequently used organisms in the tissue residue approach, because they do meet many of the criteria.

As with the benthic community structure approach, a large advantage of the tissue residue evaluation is its direct measurement of in situ conditions. It is therefore less subject to laboratory artifacts or alterations during collection or processing than some of the other sediment evaluation methods. The endpoint for the measurement is widely understood and accepted as important, thus the method can be useful in reaching concurrence on remediation. The individual tissue analyses follow well-established analytical procedures and are generally highly accurate and precise, compared to measures of community structure. Thus, it is often possible to statistically demonstrate contaminant effects. The approach is also compound-specific, which can be useful in determining source, cause, and treatment. The data requirements for the method are so similar to those for human health assessment that the approach can often be used with no additional data collection.

The tissue residue approach is limited by difficulties in setting benchmarks. As discussed earlier, with the exception of values for the protection of consumers, there are significant complications in establishing what are often the most appropriate measurement endpoints. The mobility of species is also a drawback when the objective is the characterization of a specific area.

The approach is also susceptible to failure after a large expenditure of effort. It is not unheard for the collection effort to yield no specimens for analysis. Such a finding is significant if community structure is the endpoint, but it is just a waste of time if tissue analysis is the objective. It is also possible to totally fail to define a significant correlation between tissue and sediment concentration, due to natural variability and species mobility. Due to the effort required, it is often difficult to precisely delineate the area for remediation based on tissue concentration alone, so that if a correlation can not be established, the benefits of the approach are reduced.

Tissue analysis only detects the effects of compounds that bioconcentrate. This is usually restricted to metals and persistent organic contaminants. Sometimes, the approach can be applied to other contaminants, if metabolites accumulate in measurable concentrations.

Sediment Quality Triad

From the above descriptions of sediment quality evaluation approaches, it becomes apparent that aspects of the various methods overlap and can be complimentary. The sediment quality triad method takes advantage of the overlap by drawing from the advantages of each technique and employing additional approaches to compensate for the disadvantages. The basic method employs three separate procedures for sediment characterization: benthic community structure, sediment bioassay, and chemical analysis (Long and Chapman 1985). The mixture of contaminants, and thus possible causes and sources of biological effects, are identified by the chemical analysis. The bioassay expands the data base by assessing bioavailability, and thus toxicity, of the contaminants contained within the bulk sediment. The toxicity tests are conducted under controlled situations where effects from biotic and abiotic factors not related to contaminants can be greatly limited. The community analysis can corroborate results from the other measures under in situ conditions, and can greatly expand available endpoints to community and ecosystem characteristics (Long 1989). The triad approach can easily be expanded to include other techniques of sediment quality evaluation, such as tissue analysis or any other measures critical to a specific situation, contaminant, or site.

The combined evaluation of results generally provides a broader and more detailed picture than the sum of information derived individually from each method. For example, chemical analysis could strongly indicate highly toxic sediments. Yet, due to in situ factors limiting bioavailability, the area might support an unaffected benthic community. For such a sediment, the bioassay may produce an intermediate result, because only some of the factors limiting bioavailability are present in the bioassay test chambers. There could also be a situation where the benthic community appeared severely affected, yet chemical analysis and bioassay indicated clean sediment. Further reflection on the situation could identify heavy sedimentation of uncontaminated material as the culprit. The cause could be contaminated groundwater denuding wetland vegetation adjacent to the affected sediment, which resulted in increased erosion and sedimentation. The actual situation would call for a much different remediation to restore sediment quality then the sediment removal solution indicated by community structure analysis alone. Chapman et al. (1987) and Long (1989) report actual cases where the results from the three methods have not always indicated the same source or degree of sediment impacts, and thus the combined evaluation provides an expanded understanding of sediment quality alteration cause and effect.

The sediment quality triad approach does not involve any field or laboratory investigations not included in the above descriptions of evaluation methods. The approach does, however, introduce a greatly expanded opportunity for data evaluation. The basic data evaluation tool, summarized by Long (1989), calls for combining the measures of sediment quality for all three methods. Since each method involves different types and units of measurement, they can not be directly combined. However, the data can be normalized to the reference site or to values for the reference sediment by developing a ratio-to-reference value. The value is expressed as a percent of the comparable reference measurement (i.e., sediment of concern/reference value) such that 1.0 indicates an identical quality to the reference. Once the data is normalized, the results of the three methods can be combined or compared on a similar scale. If a particular area has a ratio-to-reference value well above the 1.0 for community structure and bioassay, there is corroborating evidence of significant sediment effects. If only one, or a small number, of contaminants have values above 1.0, the cause of the impacts can be determined. Conversely, if community structure values are above 1.0, the bioassay results are variable, and the chemical analyses result in ratio-to-reference values of 1.0 for the contaminants of concern then the cause may not be related to sediment contamination from the hazardous waste site. Although such information alone does not define remediation criteria, it is extremely useful information for evaluating removal and treatment options.

Findings from one component of the triad can also be used to interpret results from other sediment evaluation techniques. For example, where biological effects are observed, the chemical analysis can provide evidence of specific contaminant and possibly physiological causes, and also possibly the source of the contaminant, so that appropriate treatment or cost recovery can be implemented. In cases where remediation may be extremely costly, the burden or preponderance of evidence provided by all three methods may aid in making a tough decision. Conversely, where concentrations are high compared to marginally accepted benchmarks, a lack of in situ effects may provide justification for no action, if the economic and environmental (e.g., construction impacts on existing unaffected habitat) costs of remediation are high.

The sediment quality triad approach is useful in defining the nature and extent of sediment contamination. By addressing both chemical and biological attributes, a more comprehensive delineation of the affected area, and understanding of the magnitude of effects, is possible. The combination of in situ and laboratory tests also lends credence to the characterization of a specific area as "clean" or "dirty." The evaluation of one set of data in light of findings from other procedures adds to the depth of understanding regarding the behavior, source, migration, and toxicity of contaminants. The approach can not easily be directly used to develop remediation criteria, either for a specific site or in general. However, the apparent effects threshold approach, which is a close relative of the sediment quality triad method, was developed for, and is well suited to criteria development. The effects threshold approach is discussed later.

The main advantage of the method is the detailed documentation and understanding of the problem, or lack of problem, that can be developed. The approach generally represents a wish list for the ecologist. The extensive documentation produced by the method can also be useful if the source, cause, and extent of remediation is extremely controversial, and must be justified to skeptics during cost recovery. Because in-depth knowledge of the sediment quality is developed by employing the approach, it represents an excellent tool to identify "hot spots" in any particular system, and the ability to prioritize remediation with a great deal of confidence.

The disadvantages of the approach are the extensive effort required and the lack of direct application to ecological effects levels or remediation criteria or a definition of the area to be remediated. The lack of direct input to criteria and definition of area can be rectified by using the method as part of the apparent effects threshold approach. The elaborate and detailed effort required for the approach can sometimes also serve to reduce overall investigation costs. The sediment quality triad approach could achieve efficiency in a situation where chemical data has already been collected and indicates a need for remediation, but detailed and corroborative evidence is needed to justify remediation. In such a situation, an extensive benthic community structure sampling and analysis program, if used alone, would be required to substantiate the chemical findings and the need for remediation. However, the benthic program could be much reduced by applying the sediment quality triad approach and using the existing chemical data. Examples of a reduced benthic program might be fewer stations, fewer replicates, smaller samples (and thus less processing and taxonomic identification effort), or a benthic profiling camera rather than sediment samples.

EVALUATION METHODS FOR SEDIMENT QUALITY CRITERIA

Apparent Effects Threshold

The apparent effects threshold (AET) method for evaluating sediment contamination was developed during investigations of Puget Sound, has been used for a number of other applications in Washington State, and is in trial use in other areas (U.S. EPA 1989a). The method was used as part of the Commencement Bay Superfund site RI/FS, and has since been used as part of the Puget Sound Dredged Disposal Analysis program, to determine the need for additional testing and evaluation of alternative disposal practices. The method is also being used as part of the Puget Sound Urban Bay Toxics Action Program, to define and prioritize sediment contamination problems. Practical application of the method has expanded into policy development, as the Washington State Department of Ecology sediment management standards are partially based on the AET approach and specific values. Outside of Washington State, chemical and biological data from San Francisco Bay, the Southern California Bight, and San Diego Bay are being considered by the California State Water Resource Control Board in the development of area-specific AETs (U.S. EPA 1989a).

In essence, the AET method is a correlation of effects observed in the field with the measured sediment concentration of a contaminant at issue (Barrick et al. 1989). To evaluate the correlation, a series of sediment collections must be made from the area of concern and a comparable reference area. The sediment must then be analyzed for chemical concentration and biological effects. The validity of the method is highly dependent on the similarity of the material analyzed for contaminant concentration and ecological effects. The optimum situation is analysis of one subsample from each individual sediment sample for chemical concentration and another subsample for the biological parameter of concern. Unfortunately, this is not always possible if collection methods for the chemical and biological samples are not compatible, or field conditions prevent subsampling. For example, the same methods can not generally be used to collect fish or large mollusks and homogenous undisturbed sediments for chemical analysis. Also, the potential for contamination sometimes prevents unrestricted sub-sampling for chemical analysis. A typical application of subsampling would be when infauna (i.e., smaller animals living within the sediment) are the biological parameter of concern, and a large undisturbed sediment sample can be collected. In such a case, the complete sample can be retrieved and, before it is disturbed, a small amount of sediment extracted from the interior without touching the sampling device, and thus contaminating the sediment for chemical analysis. The remainder of the sample can then be processed and analyzed for infaunal characteristics. If subsampling is not feasible, the sediment collected for chemical analysis should cover the area analyzed for biological effects, and the two collections should be made as close to simultaneously as possible. Once the matched chemical and biological samples have been analyzed, the highest concentration in the sample with no biological effect establishes the AET for each contaminant of concern.

There are a number of procedural considerations critical to the AET approach. First, the measurement endpoints signifying an effect must be established. Endpoint designation must relate to overall project objectives and other selection criteria, as discussed in Chapter 4. In selecting the measurement endpoints for the AET, particular attention must be given to the amenability of rigorous statistical testing. The endpoints must be tested to determine significant effects in comparison to the biological conditions in the reference area, or to some other predetermined definition of an unaffected biological characteristic. Pair-wise t tests have been the statistical method widely used in Puget Sound applications of the AET approach (Barrick 1989 et al.). Multiple endpoints and associated ecological effects levels can always be used to provide decision makers with alternative remediation criteria. For example, one endpoint might correspond to suitable conditions for egg deposition and hatching, while a second less stringent concentration might not protect egg hatching, but is suitable as unaffected feeding habitat. Typical endpoints used are alterations of infaunal community structure, tissue concentration of contaminant, physiological indicators, the histopathological condition of individuals, and abnormalities in individuals. The sediment can also be used in bioassays to determine the effects level. Multiple endpoints could also be used to address different types of effects resulting from various contaminants of concern. For example, an organic pollutant might reduce species diversity, while a heavy metal would produce food chain effects via bioconcentration. In such a situation, the determination of the AET would be more complicated, but would follow the same approach.

In determining the AET, it is important that *all samples* above the AET show an effect. A sample can be below the AET and still have an effect, but it could be caused by another contaminant in the sediment. As shown in the hypothetical example in Figure 7–1, the AET can be determined by rank-ordering the sediment samples by concentration for each contaminant of concern. The presence of biological effect is then indicated, as by shading in the example, for each sample in the sequence. If multiple definitions of effects are employed for different endpoints, then a different type of shading would indicate effects for the different endpoints. In the example shown in Figure 7–1, the endpoints of reduced egg hatching and unacceptable food supply are both illustrated. After the concentration and effects data is plotted, it is a simple matter of identifying the point in the sediment concentration gradient above which *all* samples are shaded to show an adverse effect. In the example, the AET for acceptable egg hatching would be 2.9 parts per million (ppm), because all sediments with higher concentrations are associated with egg hatching effects. Although effects are observed at lower concentrations of 1.1 ppm (Station QB2), 1.6 ppm (Station HB3), and 2.1 (Station IH3), the effects are assumed not to be due to the contaminant of concern, because no effects were observed at a concentration of 2.9 ppm (Station QB3). Similarly, the AET for an endpoint of acceptable feeding habitat would be 5.6 ppm. This exercise would be repeated for each compound of concern, and if multiple endpoints to represent a range of ecological effects levels are warranted, an AET for each alternative endpoint recommended.

The screening level concentration approach is a special case of the AET technique (Chapman 1989b, Neff et al. 1988). This technique uses the benthic species present as the measurement endpoint. The presence of selected benthic species, indicative of unstressed conditions, is correlated with the concentration of the contaminant of concern in the bulk sediment on a sample-by-sample or station-by-station basis. The sediment contaminant level at the station with the highest sediment concentration that also supports 95% of the indicator species is established as the screening level concentration.

The AET approach is well suited for a detailed evaluation of a specific, relatively homogeneous area. If there is full agreement on the endpoints to be used in establishing

OBSERVED EFFECT UNACCEPTABLE FOOD					■		■		■	■
POOR EGG HATCHING		▨		▨	▨		▨	▨	▨	▨
STATION	IH1	QB2	IH2	HB3	IH3	QB3	HB1	QB1	IH4	IH5
CONCENTRATION ppm	1	1.1	1.3	1.6	2.1	2.9	4.5	5.6	9.6	9.8

FIGURE 7-1.   Display of hypothetical data used to determine Apparent Effects Threshold (AET) concentrations.

the need for remediation, it is an excellent method for identifying areas for cleanup. It can also be used to directly establish criteria, not only to determine the extent of remediation, but also for evaluating the level of treatment, if the treated sediment is to be redistributed on-site. At a specific site, the AET (because it is essentially determined in situ) incorporates the additive and synergistic effects of all contaminants in the sediment, and thus addresses a significant concern of individual contaminant criteria. When a large data base has been established for a particular area, such as Puget Sound, or possible a particular ecosystem (e.g., a temperate estuary with silt/clay sediments), regional or ecosystem-wide AETs could be established. Sediment investigation of a specific site could then be limited to analysis to confirm compatibility with the general AETs, rather then establishing unique values.

The AET approach is limited because of the assumption that a given concentration seen to have no effect in one area would have no effect anywhere on the site. This approach ignores noncontaminant parameters that could influence the effects level, such as bioavailability. If sediment characteristics at the point of measurement bind much of the contaminant, toxicity could be low. Yet at another location, most of the contaminant is available and toxicity could be significantly higher. In Puget Sound, this concern has been considered by normalizing the concentrations to organic carbon and fine-grain particles, but little advantage over simply using dry weight was identified (U.S. EPA 1989a). There are other factors independent of individual contaminant concentration, such as biological interactions and concentrations of other compounds, that can alter the application of a no-effects level from one specific location to a larger area. Because the AET method does not assess cause and effect, the individual contribution of different contaminants or other sediment characteristics in not quantified. These limitations of the AET method can be controlled by restricting AETs to similar areas. Also, as the data base is enlarged, a better understanding of the AETs can be developed, and corrections made to account for other factors.

An extensive sample collection and analysis program is required to fully implement the AET approach. Sufficient pairs of matched samples must be considered to establish the statistical significance of effects, and to cover a wide range of concentrations for each contaminant of concern. This is usually greater than 10–15 stations, and at least three replicates are generally required at each station for the biological analysis. Depending on the endpoints, the area of the site, and the number of contaminants of concern, such an effort can be well beyond the scope of a typical RI/FS. The magnitude of the investigation can be reduced by employing a modified AET as a screening technique. Based on limited site-specific field data or information from the literature, ranges of AETs can be determined. For areas or contaminants orders of magnitude above or below the ranges identified, detailed investigations may not be necessary, and the investigation can then focus on specific areas, endpoints, and contaminants where effects may exist.

The conservative nature of an AET can be a significant limitation. By design, the AET is the highest concentration of no observed effects. If there is a wide gap in concentration between the sample with the highest concentration showing no effect and the sample with the lowest concentration showing an effect, it is never clear where in this gap the true effects threshold occurs. Thus, an apparent threshold is established to provide a margin of safety. Similarly, because the approach does not truly establish cause and effect, the additive effects of other compound can be so significant in certain situations that little or no significant toxicity is produced by the compound of concern, yet a very low AET is established.

The AET approach is most useful for evaluating sediment quality as part of a hazardous waste site evaluation if there is a large existing data base for similar conditions. For sites in Puget Sound, the method should definitely be given strong consideration. AETs already have a strong regulatory and scientific acceptance factor in Washington State, where AET values have been established for over 60 compounds (U.S. EPA 1989a). When AET were established and then compared to a different data base, the correct prediction ranged from 85% (benthic infauna and amphipod bioassay endpoints) to 95% (oyster larvae and microtox bioassay endpoints) (Barrick et al. 1989). Also, The U.S. EPA Region X has developed SEDQUAL, which is a menu-driven sediment evaluation program and data base for use on personal computers (Nielson 1988). The program stores the matched biological and chemical sediment data, calculates the AET, and compares the AET to the larger data base for Puget Sound to calculate the rate of correct predictions.

Equilibrium Partitioning

The equilibrium partitioning approach differs from the other sediment quality evaluation methods in that it has been specifically developed to generate universally applicable sediment quality criteria. In contrast, the other methods are outgrowths of procedures to evaluate certain impacts and ecological characteristics of specific sediments. As a result of this difference, many of the applications strengths and weaknesses are a mirror image of the other sediment evaluation approaches.

The basic theory of equilibrium partitioning, as originally presented by Pavlou (1987) and much refined and explored by others (U.S. EPA 1989c; DiToro 1988; Zarba 1989), consists of three assumptions. The first is that pore water is the toxic component within the sediment, which impacts benthic organisms. The second is that benthic organisms respond to pore water concentrations in a manner similar to the response of water column biota to water column concentrations of the same compound. The final assumption is that the concentration of the contaminant is determined by a quantifiable relationship between the bulk sediment concentration of the contaminant and a measurable sediment characteristic that binds or complexes the contaminant. This complexing agent is responsible for partitioning and producing an equilibrium between the dissolved bioavailable and particulate-bound forms of the contaminant, hence the term *equilibrium partitioning approach*.

If these three assumptions hold, the evaluation of sediment quality becomes a relatively simple task based on bulk chemical data alone, and establishing ecological effects levels becomes a mathematical exercise. The concentration of contaminant in the pore water is calculated from the total mass of contaminant in the sediment, using laboratory-derived coefficients partitioning the contaminant between the water and the complexing substance. The pore water concentration is then compared to well-established water quality criteria, such as the U.S. EPA Quality Criteria for Water (U.S. EPA 1986). Sediment concentrations with a pore water concentration higher than the criteria are considered effected, and those with lower concentrations are deemed acceptable. Sediment ecological effects levels could be determined without any site-specific data, by simply reversing the calculation and expressing the criteria normalized to the sediment characteristic responsible for complexing or binding the contaminant. The calculation is as follows (Zarba 1989; U.S. EPA 1989c):

$$rSQC = Kp * cWQC$$

where:

$cWQC$ = Water Quality Criterion

$rSQC$ = Sediment quality criterion (μ/kg sediment)

Kp = Partitioning coefficient for the chemical
(l/kg sediment) between the sediment and water

Each of the three assumptions required for the equilibrium partitioning approach have been questioned, and proponents of the approach have developed arguments for accepting the assumptions. The correlation of organism response to pore water concentration, as opposed to bulk sediment content of contaminant, is summarized by the U.S. EPA (1989a) and Zarba (1989). For the compounds examined, the toxicity correlation with pore water appears to be much better than the correlation with total sediment concentration. Although log transformation of the data is sometimes required, the prediction of toxicity can be quite good. The relationship seems to apply to a number of ecological endpoints including survival and tissue concentration. There seems to be a general consensus that pore water concentration does, in fact, relate closely to benthic organism toxicity and other responses. However, there is concern that pore water is not the only exposure pathway. Dermal contact, and (more importantly) ingestion of contaminated sediment particles, are cited as additional methods of exposure and uptake. The proponents make the argument that the organism is in something approaching equilibrium with the pore water, and concentration in the pore water reflects the sum of all exposure regardless of pathway. Therefore, if an animal accumulates contaminant through feeding at a higher concentration than the equilibrium with pore water, the organism would reestablish the equilibrium by losing contaminant to the pore water. Thus for this argument ecological effects are a simple function of pore water concentration.

If pore water is the cause, or at least a good surrogate for the cause, of sediment toxic effects, then why not just measure pore water and apply established pore water criteria? The answer lies in practicality. As discussed earlier under the pore water bioassay approach, collection of pore water is very difficult, and there are no generally accepted standard methods. In all likelihood, there are sites and situations where it is impossible to collect sufficient pore water unaltered from in situ conditions for analysis.

Therefore, a practicable method for estimating pore water concentration is necessary as part of the sediment quality evaluation approach. The assumption that the pore water is in equilibrium with the bulk sediment concentration is generally accepted, but the factors effecting relative concentrations in each fraction, and thus the bioavailability, are not as clear. It does appear that the complexing agents differ by the class of contaminants. There is extensive documentation of complexing of nonpolar organic contaminant by total organic carbon in the sediment (U.S. EPA 1989c; DiToro 1988; Zarba 1989). For a sediment with a total organic content greater than about 0.5%, the partitioning in the sediment can be estimated by the octanol water coefficient for the nonpolar compound. The extent of sediment toxicity from a nonpolar organic contaminant can then be expressed as the mass of contaminant per mass of organic carbon. The U.S. EPA has employed this technique to develop interim sediment quality criteria for the following compounds:

PAHs	Pesticides
Acenaphthene	Chlorodane
Aniline	Chlorpyrifos
Phenanthrene	DDT
	Dieldrin
PCBs	Endrin
Total	Ethyl
	Parathion
	Heptachlor
	Heptachlor epoxide
	Gamma-hexachlorocyclohexane

The complexing agents and mechanisms for polar organics are only poorly understood. However, such compounds are not seen as important in sediment quality, and thus equilibrium partitioning has not been attempted (Zorba 1989).

Metal toxicity in sediments appears to be affected by a number of binding agents and sediment characteristics (DiToro 1988; Zarba 1989). The particulate organic carbon in the sediment seems to affect the bioavailability of metals and thus should be considered when determining sediment toxicity. In aerobic sediments oxides of iron and manganese influence pore water concentration, and thus toxicity. Sulfides in marine sediments, and possibly some freshwater systems appear to be an important variable in the partitioning of metal contaminants. The exact relationship between these factors and metal toxicity in sediments is not as well understood or quantified as the parallel case with nonpolar organic contaminants and organic carbon (Luoma 1989). The relationship is currently under development, and the EPA is considering developing sediment quality criteria for metals based on some combination of these complexing agents (U.S. EPA 1989c). DiToro et al. (1992) have shown acid volatile sulfides to be important to the toxicity of cadmium and nickel in sediments.

The final assumption of the equilibrium partitioning method, the similar sensitivity of benthic and water column organisms, has been explored by the U.S. EPA (1989a). They found that when all benthic organisms are compared to pelagic species, the benthos is somewhat less sensitive. However, when epibenthic fauna are excluded, and the comparison is limited to infauna, which are in more direct contact with sediments, the pelagic species may still be somewhat more sensitive, but the comparison is very close.

For the purposes of evaluating the need or criteria for remediation at a specific site, a comparison of pelagic and benthic organism sensitivity is not always critical. A sediment benchmark can be developed for the species of concern and the ecological endpoint at the particular site, employing the equilibrium partitioning approach. Thus, the acceptable sediment concentration would be based on the endpoint and species for the site, rater then the universal Quality Criteria for Water. The complexing agent for class of contaminant, and the partitioning coefficient and calculation, would be the same as for the general approach for equilibrium partitioning.

There are numerous advantages to the equilibrium partitioning approach, and these have resulted in at least interim acceptance by both the U.S. EPA and the state of California as the method for determining sediment quality criteria. A major advantage of the method is that it makes use of the extensive data base used to develop the

Quality Criteria for Water (U.S. EPA 1986). As additional aquatic toxicity data become available, they can easily be used to update the sediment criteria. This advantage also applies to individual hazardous waste sites, as the data from a few aquatic bioassays could be applied to sediments of different contaminant concentrations but otherwise similar characteristics within the site. The advantage in establishing nationwide criteria is the application to a wide variety of sediment types, using the same coefficients and toxicity information. Thus, only bulk sediment contaminant data and the measurement of a limited number of other sediment characteristics is required to apply the criteria. This advantage also applies to hazardous waste sites with variable sediment types. Perhaps the greatest advantage of the method, as part of an ecological assessment at a hazardous waste site, is the application of the method as a screening tool. Usually, the contamination of sediments can be examined at the screening level with available data. Areas or contaminants that fall well below the criteria calculated by the equilibrium partitioning method do not pose an ecological risk, and ecological investigations can focus on other issues. If the results of screening using the approach are variable, the site-specific testing program can be designed to address narrow issues, rather than starting with broad questions concerning the extent and cause of sediment effects, which requires very extensive and expensive toxicity testing programs.

The greatest technical limitations of the method are estimating the partitioning of contaminants into the pore water, and bioavailability once in the pore water. There is a wide range of reported partitioning factors, and not all complexing agents or factors are always considered. The development of a larger data base, and field checking of the coefficients and assumptions, as recommended by the U.S. EPA Science Advisory Board (U.S. EPA 1990a), can minimize this limitation.

The method is also restricted to the evaluation of single compounds. Synergistic effects are not directly considered, but this limitation also applies to water quality criteria, and has not limited their widespread use. The formal equilibrium partitioning approach is also limited to only those compounds for which there are water quality criteria. However, the theory can be applied to a broader range of compounds, as site-specific or general research develops additional toxicity data.

The eventual use of criteria developed by the approach is another potential limitation of the method. The U.S. EPA technical personnel anticipate that equilibrium partitioning-based numbers will rarely be used as mandatory cleanup levels, and that they should be used to: show the nature and extent of contamination; identify the need for additional site-specific investigation; and facilitate regulatory decisions (Zarba 1989). However, there is the real possibility that the numbers will be misused in hazardous waste investigations. Once decision makers see sediment concentrations above the published levels, there will be a tendency to require remediation without reflecting on site-specific ecological endpoints, the assumptions and uncertainties in deriving the criteria, or site-specific goals and conditions.

Spiked Sediments

The testing of spiked samples is the sediment equivalent of the toxicity testing used to develop water quality criteria. A gradient of sediment concentrations is established by adding, or spiking, the contaminant of concern to the sediment. Alternatively, the gradient could be established be adding clean sediment to contaminated sediment from the site, to produce a range of concentrations. The series is then tested using bioassay techniques, and the highest concentration supporting the desired ecological endpoint can be identified as the sediment criterion. The process is repeated for each contaminant of concern. Gradients of mixtures of contaminants can also be simulated, if multiple contaminants are an issue. The procedure most often used for the test is addition of a known number of animals to the sediment, and then measuring those surviving after a fixed period. Tests have also been conducted by passing ambient water with a homogeneous density of pelagic larvae of benthic species over the sediment, and then measuring the rate of colonization. Specific and detailed methods of sediment spiking tests are given in:

Birge et al.	1987	Francis et al.	1984
Hansen and Tagatz	1980	Kemp and Swartz	1988
Oliver	1984	Schuytema et al.	1984
Swartz et al.	1986	Swartz et al.	1988
U.S. EPA	1989d	Giesy et al.	1990

The method produces a dose (or concentration) -response curve, and is thus well-suited to correlating concentrations to effects levels. Based on the ecological endpoint, and thus the effects level, remediation criteria can easily be developed from the response curve. The method can be used as a supplement to other methods to verify toxicity predictions, or to develop numerical criteria once the contaminants of concern and the need for remediation have been identified. The method also provides an excellent approach to verifying or adjusting criteria developed using the equilibrium partitioning method.

The greatest drawbacks of the sediment spiking approach relate to potential difficulties in simulating in situ conditions. The selection of sediment to be spiked is often the first difficulty encountered. The sediment must be free of the contaminant to be added, or at a consistent and well-documented concentration so that the gradient of concentration can be constructed with confidence. The sediment must also be free of other factors that could effect bioassay endpoints, such as predators which might consume the test organisms, or other contaminants. Also, the sediments must be sterilized prior to testing, so that the indigenous fauna is not confused with the added test organisms at the end of the bioassay. The sterilization method used can alter the toxicity of the sediments, and thus significantly effect the results.

The physical, geological, and noncontaminant chemistry of the sediment to be used can also be limitation to the sediment spiking approach. As described earlier for the equilibrium partitioning method, the toxicity of sediment is often as much a function of bioavailability as bulk sediment concentration. Zarba (1989) reports that sediment toxicity, as a function of total chemical concentration of the contaminant, can differ by greater than an order of magnitude due to varying sediment characteristics. Thus, since availability is determined by sediment characteristics such as organic carbon content or acid volatile sulfides, the results of the sediment spiking experiment can be largely dependent on the specific characteristics of the sediment used. If the factors affection bioavailability for the particular contaminant of concern are well known and quantifiable, the sediment used for spiking can duplicate the critical characteristic, or the results can be mathematically adjusted by normalizing the contaminant concentrations (as in equilibrium partitioning). At hazardous waste sites, where the sediment is heterogenous with respect to characteristics governing bioavailability, the sediment spiking approach can become extremely complicated. If multiple contaminants are at issue, the complexity increases accordingly, plus mixtures may have to be considered in addition to single compounds. If different factors govern availability for different contaminants in the mixture, the complexity and possibility for error increases drastically. If there is an unknown and heterogenous characteristic of the sediment which limits bioavailability, the results of the tests could be meaningless.

For areas of similar sediments, the limitations associated with reproducing in situ conditions are reduced. For systems with similar sediments, such as depositional areas in a single estuary or lake embayment, development of a large sediment spiking data base could be effective. The data base could be used to develop criteria for individual compounds, and perhaps to refine analysis methods for partitioning or normalizing toxicity data for the development of additional criteria, or for applying the data base to other systems.

SUMMARY OF SEDIMENT EVALUATION TECHNIQUES

It is clear that sediment contamination, and therefore remediation, is a critical issue at many, if not most, hazardous waste sites with ecological issues (Fitchko 1989). There is no single universally accepted and standardized "cook book" approach to addressing the sediment quality issues of the need for remediation or cleanup criteria. There are methods that are intended to be universally applicable, such as equilibrium partitioning, at one extreme, and other approaches at the other extreme, such as sediment bioassay, which only addresses a specific sediment, sometimes limited to the sample tested. The more universal approaches rely on many (and sometimes weak) assumptions which are critical to the sediment quality evaluation method. At best, these methods produce results with a wide confidence interval. At the other extreme, the sediment-specific methods often produce very accurate and reproducible results. However, when the results are transferred to other sediments, or manipulated to define cleanup criteria, the accuracy of the method is not always retained.

Both the universal and site-specific approaches can play important roles in the ecological assessment of a hazardous waste site. The more theoretical, universal methods can be applied to evaluate the existence of a potential sediment quality problem and the delineation of areas of concern. If sediment quality does prove to be an issue, these general methods can also be essential in establishing complete, precise, and efficient detailed site sediment investigations. In most cases, before economically and ecologically expensive sediment remediation is established as a requirement, some method of site-specific sediment testing is appropriate. Such site-specific methods are useful in confirming predictions of theoretical methods and in establishing more precise criteria. It is also generally necessary to perform location-specific sediment evaluation to establish the precise limits of remediation. Hopefully, as the national sediment quality and toxicity data base expands, the general methods will become more precise, and the site-specific investigations needed to refine and implement results of the general methods will become increasing less complex.

REFERENCES

Adams, W. J., R. A. Kimerle, and R. G. Mosher, 1985. Aquatic safety assessment of chemicals sorbed to sediments. In Aquatic Toxicology and Hazard Assessment, Seventh Symposium. R. D. Cardewll, R. Purdy, and R. C. Bahner (eds.), pp. 429–453. ASTM STP 854. Philadelphia, PA: American Society for Testing and Materials.

ASTM, 1988. Annual book of ASTM standards: water and environmental technology. Vol. 11.04. Philadelphia, PA: American Society for Testing and Materials.

Barrick, Robert, Harry Beller, Scott Becker, and Thomas Ginn, 1989. Use of the Sediment Quality Triad in Classification of Sediment Contamination. In *Contaminated Marine Sediments: Assessment and Remediation*. National Research Council. Washington, D.C.: National Academy Press.

Birge, W. J., J. Black, S. Westerman, and P. Francis, 1987. Toxicity of Sediment-Associated Metals to Freshwater Organisms: Biomonitoring Procedures. In *Fate and Effects of Sediment Bound Chemicals in Aquatic Systems*. K. L. Dickson, A.W. Maki, and W. A. Brungs (eds.) pp. 199–218. New York: Pergamon Press.

Carey A. E., N. S. Shifrin, and A. C. Roche, 1989. Lake Ontario TCDD bioaccumulation study final report. Cambridge, MA: Gradient Corporation.

Chapman, P. M., 1991. Criteria: What type should we be developing? *Environ. Sci. Technol.* 25:1353–1359.

Chapman, P. M., 1989a. Current approaches to developing sediment quality criteria. *Environ. Toxicol. Chem.* 8:589–599.

Chapman, P. M., 1989b. Sediment quality triad approach. In *Classification Methods Compendium*. U. S. EPA Draft Final Report, June 1989.

Chapman, P. M., R. C. Barrick, Ferry M. Neff, and Richard C. Swartz, 1987. Four independent approaches to developing sediment quality criteria yield similar values for model contaminants. *Environ. Toxicol. Chem.* 6:723–725.

Connell, D. W., M. Bowman, and D. W. Hawker, 1988. Bioconcentration of chlorinated hydrocarbons from sediment by oligochaetes. *Ecotoxicol. Environ. Safety* 16:293–302.

Cook, P. M., A. R. Batterman, B. C. Butterworth, K. B. Lodge, and S. W. Kohlbry, 1989. Laboratory study of TCDD bioaccumulation by lake trout from Lake Ontario sediments, food chain and water. U.S. Environmental Protection Agency, Environmental Research Laboratory: Duluth, MN.

Davis, W. S., and J. E. Lathrop, 1989. Freshwater benthic macroinvertebrate community structure and function. In *Sediment Classification Methods Compendium*. U. S. EPA. Draft Final Report, June 1989.

DeWitt, T. H., G. R. Ditsworth, and R. C. Swartz, 1988. Effects of natural sediment features on the phoxocephalid amphipod *Rhepoxynius abronius;* Implications for sediment toxicity bioassays. *Mar. Environ. Res.* 25:99–124.

DiToro, D. M., 1988. Equilibrium Partitioning Approach to Generating Sediment Quality Criteria. Report to the U.S. Environmental Protection Agency Science Advisory Board, December 1988. Washington D.C.

DiToro, D. M., J. D. Mahony, D. J. Hansen, K. J. Scott, A. R. Carlson, and G. T. Ankley, 1992. Acid volatile sulfide predicts the acute toxicity of cadmium and nickel in sediments. *Environ. Sci. Technol.* 26:96–101.

Fitchko, J., 1989. *Criteria for Contaminated Soil/Sediment Cleanup*. Northbrook, IL: Pudvan Publishing Co. Inc.

Francis, P. C., W. J. Birge, and J. A. Black, 1984. Effects of cadmium-enriched sediment on fish and amphibian embryo-larval stages. *Ecotoxicol. Environ. Safety* 8:378–387.

Giesy, J. P., C. J. Rosiu, R. L. Franey, and M. G. Henry, 1990. Benthic Bioassays with toxic sediment and pore water. *Environ. Toxicol. Chem.* 9:233–248.

Giesy, J. P., R. L. Graney, J. L. Newsted, C. J. Rosiu, A. Benda, R. G. Kreis, and F. J. Horvath, 1988. Comparison of three sediment bioassay methods using Detroit River sediments. *Environ. Toxicol. Chem.* 7:483–498.

Hansen, D. J., and M. E. Tagatz, 1980. A laboratory test for assessing impacts of substances on developing communitites of benthic estuarine organisms. In *Aquatic Toxicology*. J. G. Eaton, P. R. Parrish, and A. C. Hendricks (eds.), pp. 40–47. ASTM STP 707. Philadelphia, PA: American Society for Testing and Materials.

Hart, C. W., and S. H. Fuller, 1974. *Pollution Ecology of Freshwater Invertebrates*. New York: Academic Press, Inc.

Holme, N. A., and A. D. McIntyre, 1984 *Methods for the Study of Marine Benthos*. Boston, MA: Blackwell Scientific Publications.

Hynes, H. B. N., 1969 *The Biology of Polluted Waters*. Liverpool, U. K.: Liverpool, University Press.

Ingersoll, C. G., and M. K. Nelson, 1989. Solid-phase sediment toxicity testing with freshwater invertebrates: *Hyalella azteca* (Amphipoda) and *Chironomus riparius* (Diptera). In *Aquatic Toxicology Risk Assessment: Proceedings of the Thirteenth Annual Symposium*. ASTM STP. Philadelphia, PA: American Society for Testing and Materials.

Kemp, P. F., and R. C. Swartz, 1988. Acute toxicity of interstitial and particle-bound cadmium to a marine infaunal amphipod. *Mar. Environ. Res.* 26:135–153.

Knezovich, J. P., and J. L. Harrison, 1988. The bioavailability of sediment-sorbed chlorobenzenes to larvae of the midge *Chironomus decorus*. *Ecotoxicol. Environ. Safety* 15: 226–241.

Knezovich, J. P., F. L. Henderson, and R. G. Wilhelm, 1987. The bioavailability of sediment-sorbed organic chemicals: A review. *Water Air Soil Pollution* 32:233–245.

Long, E. R. 1989. The use of the sediment quality triad in classification of sediment contamination. In *Contaminated Marine Sediments: Assessment and Remediation*. National Research Council. Washington, D. C.: National Academy Press.

Long, E. R., and P. M. Chapman, 1985. A sediment quality triad: measures of sediment contamination, toxicity and infaunal community composition in Puget Sound. *Mar. Pol. Bull.* 16:405–415.

Long, E. R., and L. G. Morgan, 1989. The Potential for Biological Effects of Sediment-Sorbed Contaminants Tested in the National Status and trends Program. National Oceanic and Atmospheric Administration Technical Memorandum. NOS OMA 52.

Luoma, S. N., 1989. Can we determine the biological availability of sediment-bound trace elements. *Hydrobiologia* 1176:379–396.

Marcus, W. A., 1989. Regulating contaminated sediments in aquatic environments: A hydrologic perspective. *Environmental Management* 13:703–731

Mount, D. I., 1988. Methods for aquatic toxicity identification and evaluations: Phase III toxicity confirmation procedures. EPA/600–3–88/036. U. S. Environmental Protection Agency, Duluth, MN.

Mount, D. I., and L. Anderson-Carnahan, 1988a. Methods for aquatic toxicity identification and evaluations: Phase I toxicity characterization procedures. EPA/600–3–88/034. U.S. Environmental Protection Agency, Duluth, MN.

Mount, D.I., and L. Anderson-Carnahan, 1988b. Methods for aquatic toxicity identification and evaluations: Phase II toxicity identification procedures. EPA/600–3–88/035. U. S. Environmental Protection Agency, Duluth, MN.

National Research Council, 1989.*Contaminated Marine Sediments: Assessment and Remediation*. Washington, D. C.: National Academy Press.

Nebeker. A. V., and C. E. Miller, 1988. Use of amphipod crustacean *Hyalella azteca* in freshwater and estuarine sediment toxicity tests. *Environ. Toxicol. Chem.* 7:1027–1034.

Nebeker, A. V., M. A. Cairns, J. G, Gakstatter, K. W. Maluet, G. S. Schuytema, and D. F. Krawczyk, 1984. Biological methods for determining toxicity of contaminated freshwater sediments to invertebrates. *Environ. Toxicol, Chem.* 3:617–630.

Neff, J. M., B. W. Conagy, J. M. Vaga, T. C. Gulbransen, and J. A Scanlon, 1988. Evaluation of the screening level concentration approach for validation of sediment quality criteria for freshwater and saltwater ecosystems. In *Aquatic Toxicology and Hazard Assessment,* 10th Volume. Philadelphia, PA: American Society for Testing and Materials.

Nielson, D., 1988 SEDQUAL Users Manual. Prepared for Tetra Tec, Inc. and U.S. Environmental Protection Agency Region X, Office of Puget Sound. Bellevue, WA: PEI Environmental Services.

Ohio Environmental Protection Agency, 1987. Biological Criteria for the Protection of Aquatic Lives. Columbus, OH: Ohio EPA, Division of Water Quality Monitoring and Assessment, Surface Water Section.

Oliver, B. G., 1984. Bio-uptake of cholorinated hydrocarbons from laboratory-spiked and field sediment by oligochaete worms. *Environ. Sci. Technol.* 21:785–790.

Pavlou, S. P., 1987, The use of the equilibrium partitioning approach in determining safe levels of contaminants in marine sediments In *Fate and Effects of Sediment-Bound Chemicals in Aquatic Systems*. K. L. Dickson, A. W. Maki, and W. A. Brungs, (eds.). Toronto, Ontario: Pergamon Press.

Pielou, E. C., 1977. *Mathematical Ecology*. New York: John Wiley & Sons. York.

Robinson, A. M., J. O. Lamberson, F. A. Cole, and R. C. Swartz, 1988. Effects of culture conditions on the sensitivity of phoxocephalid amphipod *Rhepoxynius abronius* in two cadmium sediments. *Environ. Toxicol. Chem.* 7:953–959.

Schuytema, G. S., P. O. Nelson, K. W. Malueg, A. V. Nebeker, D. F. Krawczyk, A. K. Ratcliff, and J. G. Gakstatter, 1984. Toxicity of cadmium in water and sediment slurries to *Daphnia magna*. *Environ. Toxicol. Chem.* 3:293–308.

Sly, P. G., 1988. Interstitial water quality of lake trout spawning habitat. *J. Great Lakes Res.* 14:301–315.

Surber, E. W. 1937. Rainbow trout and bottom fauna production in one mile of stream. *Trans. Amer. Fish. Soc.* 66:193–202.

Swartz, R. C. 1989. Marine sediment toxicity tests. In *Contaminated Marine Sediments: Assessment and Remediation*. National Research Council Washington, D.C.: National Academy Press.

Swartz, R. C., W. A. DeBen, and F. A. Cole, 1979. A bioassay for the toxicity of sediment to marine macrobenthos. *J. Water Pollution Control Fed.* 51:944–950.

Swartz, R. C., W. A. DeBen, J. K. P. Jones, J. O. Lamberson, and F. A. Cole, 1985. Phoxocephalid amphipod bioassay for marine sediment toxicity. In *Aquatic Toxicology and Hazard Assessment: Proceedings of the Seventh Annual Symposium*. R. D. Cardwell, R. Purdy, and R. C. Bahner (eds.). ASTM STP 854 Philadelphia. PA: American Society for Testing and Materials.

Swartz, R. C., G. R. Ditsworth, D. W. Schults, and J. O. Lamberson, 1986. Sediment toxicity to a marine infaunal amphipod: cadmium and its interaction with sewage sludge. *Mar. Environ. Res.* 18:133–153.

Swartz, R. C., P. F. Kemp, D. W. Schults, and J. O. Lamberson, 1988. Effects of mixtures of sediment contaminants on the marine infaunal amphipod *Rhepoxynius abronius*. *Environ. Toxicol. Chem.* 7:1013–1020.

Tishler/Kocuek Consultants, 1990. Sediment Quality Criteria: Evaluation of Development Approaches. Prepared for the Chemical Manufacturers Association, 2501 M Street NW, Washington, D.C. 20037.

U.S. EPA, 1990a Report of the Sediment Criteria Subcommittee of the Ecological Processes and Effects Committee (Science Advisory Board): Evaluation of the Equililbrium Partitioning Approach for Assessing Sediment Quality. EPA-SAB- EPEC-9--006.

U.S. EPA, 1990b. Macroinvertebrate field and laboratory methods for evaluating the biological integrity of surface waters. EPA 600/4–90/030.

U.S. EPA, 1989a. *Sediment Classification Methods Compendium*. Draft Final Report, June 1989.

U.S. EPA,1989b. Rapid Bioassessment Protocols for Use in Streams and Rivers: Benthic Macroinvertebrates and Fish. EPA/444/4–89–001. Edited by James A. Plafkin.

U.S. EPA, 1989c. Briefing report to the EPA Science Advisory Board on the Equilibrium Partitioning Approach to Generating Sediment Quality Criteria. EPA 440/5–89–002.

U.S. EPA 1989d. Guidance Manual: Bedded Sediment Bioaccumulation Tests. EPA/600/x–9/302.

U.S EPA, 1988. Draft and Final Supplemental Environmental Impact Statement for Boston Harbor Wastewater Conveyence System. U.S. EPA Region I. Boston, MA.

U.S. EPA, 1986. Quality Criteria for Water. Office of Water Regulation and Standards. EPA/440/5–86/001.

U.S. EPA. 1978. Techniques for Sampling and Analyzing the Marine Macrobenthos. EPA/600/3–78–030.

U.S. EPA. 1973. Biological Field and Laboratory Methods for Measuring Quality of Surface Waters and Effluents. EPA/670/4–73–001.

U.S. EPA and U.S. Army Corps of Engineers, 1991. Evaluation of Dredged Material Proposed for Ocean Disposal: Testing Manual. EPA/503/8–91/001.

Washington, H. G., 1984 Diversity, biotic and similarity indices: A review with special relevance to aquatic ecosystems *Water Resources* 18:653–694.

Zarba, C. S., 1989. Equilibrium Partitioning Approach. In *Sediment Classification Methods Compendium*. U.S. EPA Draft Final Report, June 1989.

8

Ecotoxicology and Ecological Assessment at Hazardous Waste Sites

Anthony F. Maciorowski

U.S. Fish and Wildlife Service

INTRODUCTION

The past three decades have witnessed the increasing use of toxicity tests and other toxicological and ecological methods to identify, characterize, and assess the ecological hazards of chemicals and chemical wastes. This growth is directly related to the passage of comprehensive environmental legislation. In the 1960s and early 1970s, the Clean Water Act (CWA) and the National Environmental Policy Act (NEPA) set the stage for sweeping toxic substances legislation throughout the late 1970s and 1980s. Laws requiring an evaluation of the effects of chemicals and chemical wastes on ecosystems or their inhabitants include: the Marine Research, Protection and Sanctuaries Act (MRPSA); the Federal Insecticide, Fungicide and Rodenticide Act (FIFRA); the Toxic Substances Control Act (TSCA); the Resource Conservation and Recovery Act (RCRA); the Comprehensive Environmental Response Compensation and Liability Act (CER-CLA); and the Superfund Amendment and Reauthorization Act (SARA).

A significant outgrowth of such legislation has been a continuing synthesis of pollution biology, toxicology, and ecology into the discipline of ecological toxicology or ecotoxicology (MacIntyre and Mills 1974; Neuhold 1986; Dickson et al. 1979; Levin et al. 1988; Cairns 1992). Ecotoxicology is perhaps unique in that it combines aspects of toxicology and ecology [both of which are, in and of themselves, synthetic multidisciplinary sciences (Casarett and Doull 1975, Odum 1977)]. As a basic science, ecotoxicology is a field in rapid transition and is only beginning to synthesize its diffuse components into a unified discipline (Maciorowski 1988). From an applied science perspective, hazard evaluation and risk assessment are major subactivities within eco-toxicology, and are currently the primary focus for future regulatory actions (Science Advisory Board 1990a,b,c,d,e). Finally, ecological assessment at hazardous waste sites should be recognized as one portion of the broader-based hazard evaluation and risk assessment process.

The preceding paragraphs may seem a roundabout way of introducing the ecological assessment of hazardous waste sites. However, such investigations are only one aspect of a dynamic, rapidly changing field. As such, ecological assessments at hazardous waste sites are subject to scientific debate, as well as rapidly changing terminology, methods, and regulations common to the field as a whole. A practical understanding of the ecological assessment of hazardous waste sites, therefore, requires comprehension of these broader issues.

The present chapter provides a brief overview of ecotoxicology, followed by a discussion of the hazard evaluation and risk assessment process. These sections set the scientific context in which ecological assessments at hazardous waste sites are performed. Subsequent sections then discuss the approaches, and an overview of the methods, necessary to implement site-specific ecological investigations at hazardous waste sites.

BASIC AND APPLIED ECOTOXICOLOGY

Ecotoxicology is a complex array of concepts, methodologies, procedures, effects determinations, and remediation and management decisions. As such, ecotoxicology is at once a basic science directed toward the description and elucidation of complex biological and chemical mechanisms and interactions, and an applied science focused on environmental management and regulatory compliance and enforcement actions. The field has advanced so rapidly that the differences between basic and applied ecotoxicology are often blurred. Further, ecotoxicologists must often understand, differentiate, and reconcile intricate scientific and regulatory policy issues in a public forum.

As a basic science, ecotoxicology is concerned with the description, explanation, and elucidation of phenomena that advance knowledge in areas of technical specialization or the field as a whole. The complexity of ecotoxicology is apparent by its subject matter. Chemicals may affect every level of biological organization (molecules, cells, tissues, organs, organ systems, organisms, populations, communities) contained in ecosystems. Any one level is a potential unit of study for the field, as are the interdependent structures and relationships within and between levels. Superimposed on this biological hierarchy are equally complex abiotic chemical and physical phenomena that may affect the environmental fate and toxic responses of a chemical at one, several, or all hierarchical levels. Therefore, basic ecotoxicology provides the fundamental concepts, unifying principles, and theories that advance the overall science.

In contrast, applied ecotoxicology is often directed toward identifying and resolving discrete site-specific environmental problems caused by existing point and nonpoint pollutant sources. In many situations, this occurs through legislatively mandated regulatory compliance and enforcement actions. Once the emphasis shifts to the regulatory arena, data is rarely collected solely for its intrinsic scientific value. Rather, ecotoxicological data is used to identify ecological damage, verify cause and effect, and to determine the effectiveness of remediation options. Under these circumstances, data may become evidence for use in litigation or other regulatory actions. In the regulatory domain, science remains important, but so too are social, economic, political, and legal concerns. As such, conservative methods promulgated through a public rule-making process have been preferred over open-ended research approaches. In practice, this has resulted in a strong reliance on single-species toxicity tests and routine ecological surveys that are amenable to standardization and well-defined quality assurance and quality practices.

The pragmatic "testing" approach commonly used in applied toxicology often seems contrary to the development of basic ecotoxicology. Various authors have eloquently stated the need for the mechanistic, explanatory, and theoretical considerations necessary to realistically predict long-term effects on chemicals on whole ecosystems (Krenkel 1979; Grey 1980; Cairns 1980, 1992). The Committee to Review Methods for Ecotoxicology (Environmental Studies Board, National Research Council, 1981) concluded that the prediction of population and ecosystem effects of chemicals requires a detailed multilevel assessment approach, containing the following elements:

1. Characterization of the test substance, to include chemical and physical properties, estimates of fate within and among ecosystems, and estimates of dose and exposure times for biotic components of ecosystems.
2. Physiological responses of representative species indicating morphological, biochemical, genetic, and pathological changes related to exposure.
3. Multispecies responses, to include information on alterations of organismic interactions, population or system structure, and patterns involving interspecies interactions.
4. Ecosystem responses, to include changes in functional processes that affect the resistance and resilience of ecosystems.

Scientific debate over ecological assessment approaches often revolves around "proper" objectives, approach, scale, and resolving power. For ecotoxicology to mature and prosper, there is a continued need for explanatory concepts and unifying theory to predict the effects of chemicals on whole ecosystems. Yet, there is also a societal and legislative mandate to identify and resolve problems at existing contaminated sites. Scientifically defensible predictive capabilities are needed in both basic and applied ecotoxicology. However, one must also recognize that few regulatory compliance and enforcement actions are currently designed to be truly *predictive* at the whole ecosystem level. More commonly, they are remedial investigations designed to identify and reduce source contamination at existing sites. As such, applied toxicologists generally employ *descriptive* and *comparative* studies of ecosystem components to judge the relative condition of the target site. The point to underscore is that descriptive, comparative, and predictive approaches are all valid, depending upon the objectives of a particular study. More importantly, ecotoxicologists must begin to recognize and differentiate between the scientific and regulatory policy issues and objectives that often serve as points of contention.

Scientific debate over predictive ecological assessment strategies will undoubtedly continue until experimentation, validation, interpretation, and synthesis improve convergence between method and theory. The latter process represents the essence of scientific development, and accounts for the inherent lag time between the development and routine application of science. Advances in applied research are necessarily dependent on advances in basic research. However, unlike basic research, practical remedial solutions for contaminated environments often require timely management decisions. As such, ecotoxicologists would do well to heed the words of Hull (1970), who distinguished between the development of theory and the application of methods in science by viewing them within the context of the words *theoretical significance* and *usefulness*.

"An extremely accurate scientific theory of great scope will certainly be useful, but there are many things which are useful though of little theoretical significance."

Toxicity tests and ecological field surveys are profoundly useful in evaluating the impact of chemicals and chemical wastes on the environment. Further, their selection, performance, and interpretation are rarely a theoretically neutral exercise when viewed in an appropriate assessment framework.

ECOLOGICAL ASSESSMENT, HAZARD ASSESSMENT AND RISK ASSESSMENT

Assessment is commonly used throughout the biological and environmental science literature. Yet, it exhibits a gradation of meanings, which are often confounded by qualifying adjectives such as *biological, ecological, impact, hazard,* or *risk* (Table 8-1). Much of the terminology problem is semantic. *Biological, ecological,* and *impact assessment* generally imply the measurement of adverse effects (Cairns and Dickson

TABLE 8-1.   Commonly Encountered Assessment Terminology.

Effects Measurement

> *Biological Assessment* is a measurement of the biological condition of a specified habitat or site. Biological assessment has two major components: toxicity tests, which may be performed *in situ* or in the laboratory; and field surveys, which measure the structure and function of biota inhabiting a specified ecological habitat.

> *Ecological Assessment* is the quantification of ecological effects occurring at a site or other landscape scale, principally the population and community level effects on terrestrial and aquatic biota and biological processes.

> *Impact Assessment* is the measurement of an environmental stress caused by physical perturbation or chemical contamination of the environment.

The Two-Phased Hazard Assessment Process

> *Hazard evaluation* represents the testing or data collection phase to characterize chemical properties, processes, and adverse toxicological and ecological effects of a particular chemical substance or contaminated environment.

> *Hazard assessment* represents the data synthesis and interpretation of the hazard evaluation data to determine the potential hazard or safety of a chemical.

The Integrated Risk Assessment Process

> *Risk management* is a decisionmaking process that involves all the human-health and ecological assessment results, considered with political, legal, economic, and ethical values, to develop and enforce environmental standards, criteria, and regulations.

> *Risk assessment* is a scientific process that includes hazard identification, receptor characterization and endpoint selection, stress-response assessment, and risk characterization.

> *Risk reduction* is an engineering process to prevent or minimize contaminant release and exposure.

1972; U.S. EPA 1989a). The latter three phrases remain in widespread use and generally predate the development of hazard evaluation and hazard assessment as a formal process.

By the late 1970s, toxicity test methods and ecotoxicological data were proliferating (e.g., Maciorowski et al. 1980; Mayer et al. 1980; Spehar et al. 1980). Increasing development and sophistication of ecotoxicology led to increased criticism of toxicity tests as the primary approach to ecological assessment. Various authors questioned the ecological relevance of toxicity test approaches (Sprague 1976; Mount 1977a,b; Doudoroff 1977; Krenkel 1979; Grey 1980), with a common conclusion that test methods were far ahead of an ability to interpret results. At issue was the need for an equitable process to integrate toxicological, ecological, chemical, and physical data with environmental exposure data in a decisionmaking framework. Two workshops on ecological hazard evaluation and assessment were ultimately convened to meet this need (Cairns et al. 1978; Dickson et al. 1979; Cairns and Maki 1979; Cairns 1980).

Ecological hazard evaluation and assessment (Table 8-1) emerged as a two-phased process consisting of data collection (*hazard evaluation*) and final synthesis and interpretation of the collected data (*hazard assessment*). The process focused on estimating the potential environmental exposure of a chemical based on knowledge of its environmental fate, persistence, and use and production patterns. The estimated environmental concentration could then be compared to an experimentally derived concentration at which no unacceptable adverse environmental impact was expected to occur. Although largely presented as a scientific exercise, the need for a process that could be followed to an identifiable end for regulatory decisionmaking was recognized (Maki 1979).

Recent emphasis on risk-based approaches for environmental regulations (Thomas 1987; U.S. EPA 1989c, Science Advisory Board 1990a,b,c,d) is leading to increased integration of science, engineering, and societal values for environmental decisionmaking. The integrated risk assessment process (Table 8-1) involves three interactive phases, each of which has discrete process steps and associated activities (Table 8-2). *Risk management* is a policy-based activity that defines assessment questions and endpoints to protect human health and ecosystems. Of necessity, risk management incorporates societal norms by considering socioeconomic and legal values in establishing assessment endpoints. Risk management, therefore, defines the broad-based objectives for subsequent scientific and engineering activities (Table 8-2). *Risk assessment* is the scientific phase of the process, and *risk reduction* involves engineering solutions to mitigate source contamination.

The interactive nature of the integrated risk assessment approach is depicted in Figure 8-1, with risk management centered in the box containing human health and ecosystem assessment endpoints. Risk assessment is the scientific exercise that: identifies the hazard and resources at risk; defines measurement questions and endpoints to determine the condition of the resources at risk; links cause and effect through stress-response studies; and determines environmental exposure through monitoring or modeling. Integrating the collected data and information then allows a weight-of-the-evidence approach for characterizing risk. If the scientific and societal risk is unacceptably high, the source of the chemicals can be mitigated by engineering activities to minimize the exposure of resources at risk to the hazard.

The preceding overview demonstrates that *assessment* has multiple meanings, which may imply either measurement (more aptly termed hazard evaluation) or a broader-based synthesis and interpretation process for environmental decisionmaking (hazard

TABLE 8-2.   **Process Steps and Associated Activities for Integrated Ecological Risk Management, Ecological Risk Reduction, and Ecological Risk Assessment.**

ECOLOGICAL RISK MANAGEMENT

Process Step	Activity
Assessment questions and endpoints	Risk interpretation through development of criteria, standards, or data quality objectives and enforcement

ECOLOGICAL RISK ASSESSMENT

Process Step	Activity
Hazard identification	Experiential or analytical
Receptor characterization	Determine resources at risk
Measurement questions and endpoint selection	Select appropriate measures that indicate the condition of the receptor at risk
Stress-response assessment	Toxicity tests and/or field measures
Exposure assessment	Monitoring or modeling
Risk characterization	Information integration

ECOLOGICAL RISK REDUCTION

Process Step	Activity
Release minimization	Process engineering
Exposure minimization	Remediation
Contaminant prevention	Environmental compatibility

assessment or risk assessment). Currently, the integrated risk assessment process is the predominant scientific and regulatory context for ecological evaluations at hazardous waste sites. The process requires a clear understanding of the roles and responsibilities of regulatory decisionmakers (risk management), scientists (risk assessment), and engineers (risk reduction). Without such an understanding, there is a tendency to become embroiled in highly technical debate over specific measurement questions and endpoints, often to the detriment of the broaderbased assessment questions and endpoints that drive the process.

ENDPOINTS AND IMPLICATIONS

In the preceding section, emphasis was placed on the integrated risk assessment process, and the differences between assessment and measurement endpoints were introduced. The focus on process was deliberate. Without an appropriate context, the selection of assessment and measurement endpoints can be difficult. Assessment endpoints represent formal expressions of the environmental or ecological attributes to be protected

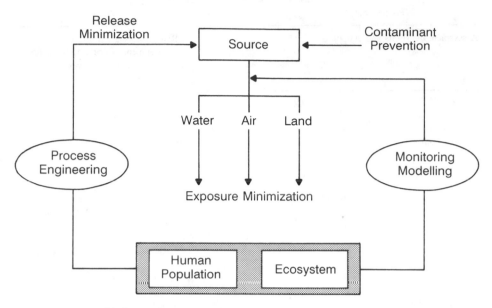

FIGURE 8-1.   The integrated risk assessment and risk reduction framework.

(U.S. EPA 1989a). Once an assessment endpoint is established, determining whether the assessment endpoint is affected at the hazardous waste site is the next order of business. This is accomplished by selecting appropriate measurement endpoints. Measurement endpoints are actual effects caused by the hazard that can be measured by field or laboratory methods and procedures. Desirable attributes of assessment and measurement endpoints (Table 8-3) are detailed in U.S. EPA (1989a,b). If the measurement endpoints indicate that the assessment endpoint is significantly affected (risk assessment), remediation is warranted (risk reduction).

The relationship between assessment and measurement endpoints cannot be overstated. Without this relationship, investigators are faced with sorting through a vast array of approaches and methods to measure ecological damage that may have varying degrees of relevance to science and society. Clear delineation of assessment and measurement endpoints is necessary because chemicals may affect one or several levels of biological organization, and different methods are needed to determine effects at different levels (Figure 8-2). More importantly, not every measurable effect of potential interest to science is of equal interest to society at large. The integrated risk assessment process provides a mechanism for reconciling scientific, regulatory, and societal values.

Differences of style in science must be recognized before investigators will respect the influence of others working across disciplinary barriers (Meier 1972). This is particularly true of ecotoxicology, which has basic and applied practitioners trained in classical toxicology, as well as in classical ecology. Toxicologists have traditionally conducted reductionist studies with organisms or their component parts, as the experimental units of concern. Such studies have provided a wealth of information on

TABLE 8-3.   Characteristics of Assessment and Measurement Endpoints.

ASSESSMENT ENDPOINTS
        Social relevance
        Biological relevance
        Unambiguous operational definition
        Measurable or predictable
        Susceptible to the hazard
        Logically related to management decision
MEASUREMENT ENDPOINTS
        Corresponds to or is predictive of an assessment endpoint
        Readily measured
        Appropriate to the scale of the site
        Appropriate to the exposure pathway
        Appropriate temporal dynamics
        Low natural variability
        Diagnostic
        Broadly applicable
        Standard
        Existing data series

Source: U.S. EPA 1989a.

the biochemical, physiological, genetic, and pathological effects of chemicals to a variety of species. Nevertheless, ecologists have been reluctant to extrapolate laboratory-derived single-species effects to free-ranging populations, arguing that laboratory data lack the holistic biotic and abiotic interactions that characterize communities and ecosystems.

Odum (1984) summarized the difficulty in dealing simultaneously with the whole and its parts, and the fact that reductionism and holism are often pitted against each other. Indeed, some factions almost suggest that only one or the other is correct. Proponents of reductionism may be primarily interested in ascertaining what occurs in a system to the greatest possible degree. Advocates of holism may be more interested in overall inputs and outputs to a particular system, and often care little about homeostatic mechanisms if they do not affect net output (Crovello 1970). Ecologists generally favor a holistic approach, arguing that toxicant impact at higher levels of biological organization cannot be explained through a study of its component parts, because new properties unique to that level emerge as levels of biological hierarchy are ascended. A literal interpretation of reductionism assumes that phenomena at ecological levels of organization are governed by molecular-level events. However, commonly available molecular procedures are rarely capable of predictably specifying an ecological event (Galluci 1973). A key point is that both holism and reductionism are integral to science, and both are necessary for measuring and predicting chemical hazard in ecotoxicology.

The hierarchical continuum of biological responses to chemical exposure (Figure 8-2) adapted from Adams et al. (1989), can help to explain the relationships of different kinds of data in ecotoxicological investigations. Effects at organismic and suborganismic levels are easily measurable, have high toxicological relevance, and are manifested much more rapidly than long-term population and community-level responses. Population and community-level effects can also be measured, and have greater ecological relevance than organismic responses. However, differentiating direct chemical responses from stochastic population and community adjustments to seasonal and

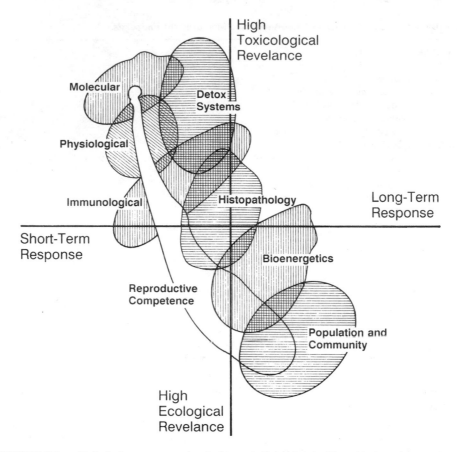

FIGURE 8-2.   Biological responses to chemical hazard, illustrating the hierarchical continuum relative to response time and toxicological and ecological relevance.

spatial variability can be difficult. Ecosystem processes such as resiliency and stability are under increasing investigation (Webster et al. 1975; Holling 1986; Westman 1978; Kelly and Harwell 1990), and have been proposed as promising models to predict large-scale ecosystem perturbation. Advances in basic research may one day allow the use of higher dimensional phenomena in landscape scale assessment scenarios (Rapport 1989). However, such responses are currently well beyond the scope of site-specific ecological assessments at hazardous waste sites.

ECOLOGICAL RISK ASSESSMENT APPROACHES AT HAZARDOUS WASTE SITES

In the years to come, science and regulatory policy will continue to advance and interact. In all likelihood, future developments will move toward predictive estimation methods for landscape levels larger than sites. Nevertheless, descriptive and comparative ecotoxicological methods currently remain commonplace for most hazardous waste site investigations. More importantly, hazardous waste sites are regulated by the Comprehensive Environmental Response Compensation and Liability Act (CER-CLA), as amended by the Superfund Amendment and Reauthorization Act of 1986 (SARA). As such, the primary objective of such studies is the scientific diagnosis of unacceptable ecological hazards within a regulatory framework. Generally, regulatory objectives focus on the identification of chemical hazards, the status of the biological community in the affected area, the estimated magnitude of toxic effects, and verification that observed effects can be linked to the chemical hazard. The regulations and associated technical guidance for ecological assessment at hazardous waste sites (U.S. EPA 1989a,b) are basic references for anyone involved in ecological assessment studies at hazardous waste sites.

Different hazardous waste sites will exhibit different topographical, geological and ecological characteristics. Different sites will also be subject to widely divergent chemical hazards. Nevertheless, they usually share common assessment questions and basic methodological approaches (Table 8-4). Chemical analyses, toxicity tests, and field surveys have been the traditional methodologies for ecological assessment. More recently, biomarkers have become prominent in CERCLA and SARA technical guidance (U.S. EPA 1989a), and will likely see increased use in the future.

Each methodology provides a specific type of information, and no single approach is superior. Rather, each provides data important to the overall ecological risk assessment process. Chemical analyses identify a hazard by providing a catalog of chemicals and their relative concentrations at the site. If a site contains one, or at most a few well-known chemicals, chemical analyses alone may be appropriate. However, complex chemical mixtures and difficult environmental matrices (e.g., soil, sediment) often confound precise identification and quantification of chemical constituents and their environmental interactions. Further, attempts to estimate the toxicity of complex mixtures by chemical measurements alone have proven unsuccessful.

The toxicity-based approach directly measures toxicity of a contaminated environmental sample or waste material, and therefore serves as an integrated measure of toxicity, regardless of the chemicals present. Acute toxicity tests are short-term tests usually designed to measure lethality. Chronic toxicity tests utilize long-term exposure to measure more subtle effects such as effects on growth, reproductive impairment, or other sublethal responses. Toxicity tests link the occurrence of chemical contamination to biological effects. However, toxicity tests are usually performed with laboratory animals in laboratory settings, and therefore only indicate the potential for toxic effects to free-living populations and communities. Biomarkers are biochemical or physiological responses of organisms that indicate chemical exposure or sublethal stress. Biomarkers represent a powerful bridge between laboratory toxicity tests and field surveys, because they can be used in both modes of investigation. Demonstration and quantification of natural population and community responses to chemicals requires field surveys in the area of the site to verify structural or functional changes caused by contaminants.

The advantages and limitations of toxicity tests, field surveys, and biomarkers are detailed in U.S. EPA (1989a). Suffice it to say that ecological risk assessment requires the formulation of several different basic questions (Table 8-4), and that the different questions require different methodological alternatives. By combining methodological approaches, data can be combined into a weight-of-the-evidence strategy to characterize overall ecological conditions, which cannot be determined from a single methodological approach.

TABLE 8-4.   Questions and Approaches for Ecological Assessment at a Hazardous Waste Site (HWS).

Questions	Approach
Are soils, water, or sediments at the HWS contaminated?	Chemical analyses Toxicity tests Field surveys
Are the contaminated soils, water, and sediments at the HWS toxic or hazardous to living organisms?	Acute and chronic toxicity tests Biomarkers of sublethal stress Chemical analyses
Are organisms at the HWS exposed to these hazardous contaminants?	Biomarkers of exposure
Have biological communities or populations, on-site or off-site, been measurably impacted at the HWS?	Field surveys Chemical analyses
Are the effects on populations or communities at the HWS caused by the presence of hazardous wastes?	Weight of the evidence using all of the above data

ECOLOGICAL RISK ASSESSMENT STUDY DESIGN AND METHODS SELECTION

Although only a few basic biological methodologies are routinely used in ecological risk assessment (Table 8-4), investigators are faced with numerous methods, procedures, and techniques within each methodology. Procedures have been developed for a variety of site-specific conditions that vary dramatically with respect to dominant organisms, ecosystem type, and the contaminant of concern. Toxicity tests are available for freshwater, estuarine, marine, and terrestrial organisms. Further, test methods for a particular species or environmental medium (e.g., water, sediment, or soil) often require substantially different technical considerations and expertise. Test species may include bacteria, algae, plants, invertebrates, fish, or other vertebrates. Toxicity test endpoints may include death, growth, reproductive impairment, gross or histological pathology; as well as physiological, genetic, or behavioral responses. For chemicals that bioconcentrate, bioaccumulate, or biomagnify, tissue residue concentrations may be measured or experiments performed to model chemical uptake and depuration. Finally, simulated ecosystem experiments or ecological field studies may be needed to determine chemical effects on populations, communities, or higher dimensional ecosystem phenomena.

The sheer number of potentially useful methods, techniques, and procedures in ecological risk assessment can be overwhelming. Fortunately, not every method is germane to every ecological risk assessment situation. Crovello (1970) offered two essential considerations concerning methods selection in ecological studies which are equally appropriate for hazardous waste sites. Foremost is whether a particular analysis will fulfill its stated objective. That is, will an analysis be responsive to its purpose, or will it only provide uncertain information about the system under study? The second is whether an analysis program will have sufficient power to discriminate between alternative hypotheses formulated specifically for the study. The foregoing questions cannot be resolved by the methods themselves. Rather, they must be addressed by understanding the characteristics of the site in question; deliberate conceptual review of the relationship between objectives and methods; careful study design; planning and implementation of sampling logistics and data collection; and appropriate data analysis and interpretation.

Preliminary evaluation of the hazardous waste sites under investigation (Table 8-5) is necessary to identify critical site-specific issues, information gaps that must be considered in the final study design, and the environmental compartments and ecological resources that are affected or at risk. The resultant information provides im-

---

TABLE 8-5.   Preliminary Site Reconnaissance Evaluation Steps for Hazardous Waste Site (HWS) Studies.

Review Existing Site Information

> *Site History*—Review available information on industrial activities to provide insights into the nature, sources, and event of contamination.

> *Chemistry Data*—Review available chemistry data for site soils, sediments, surface water and ground water to assist in the identification of sample sites for chemical and biological samples in the definitive study, hot spots, or areas of chemical gradients.

> *Fate and transport model results*—Review available data concerning movement and transformation of contaminants to provide information on sources and routes of travel to identify locations and ecosystem components that may be impacted.

> *Existing ecological data*—Review available historical data and define natural background conditions expected at the site, and determine data needs for the definitive study.

Site Reconnaissance

> *Major landscape features*—Map site topography and the distribution of major habitat types (e.g., grasslands, forests, lakes, streams, wetlands).

> *General physical and chemical characteristics of the terrestrial environment*—Describe soil type(s) and local geology.

> *General physical and chemical characteristics of the aquatic environment*—As appropriate, describe lake area and depth, stream size and flow, types of bottom substrate, temperature, water clarity, and general water quality characteristics such as conductivity, salinity, hardness, pH, temperature, alkalinity, and dissolved oxygen levels.

> *Vegetation types*—Identity of dominant species and classification of the major vegetation community types.

> *Important terrestrial and aquatic animals*—Qualitative observations of birds, mammals, fish, stream benthos, and other animals inhabiting the HWS, or the apparent absence of organisms considered typical of the HWS habitats.

> *Areas of contamination and ecological effects*—Locations of obvious zones of chemical contamination and ecological effects, ranked by apparent severity.

portant information necessary for methods selection. For example, if preliminary chemical data and environmental fate and transport information indicate that only soils are impacted, the terrestrial components of the ecosystem become the primary focus of the assessment. If contamination affects soils, water, and sediments, aquatic and terrestrial ecosystem compartments may both require assessment. Additionally, the preliminary evaluation provide basic insights necessary to determine whether whole communities, a particular community segment, other defined species assemblages, or a single population should be the primary target for the integrated ecological assessment.

Following the preliminary assessment, choices must be made about site-specific measurement objectives, and the type and number of measurement endpoints. A guiding principle to keep in mind during study design and methods selection for hazardous waste sites is that there is no universally applicable method or set of methods that can be applied to all ecological risk assessment problems. It is equally important to remember that the goal of hazardous waste site assessments is to determine unacceptable ecological damage that can be linked directly to the contaminant source. Under the varying circumstances and situations, a weight-of-the-evidence approach, which combines chemical, ecological, and toxicological data selected with *a priori* knowledge of the site will provide the soundest scientific assessment. Only then can methods selection be finalized.

Comprehensive review of the various toxicity test, field survey, and biomarker methods cannot be accommodated in a single book chapter. Fortunately, detailed descriptions of methods, procedures, and techniques are available in existing methods manuals, guidance documents, and other technical literature. Key documents for hazardous waste sites include *Ecological Assessment of Hazardous Waste Sites: A Field and Laboratory Reference* (U.S. EPA 1989a) and *Risk Assessment Guidance for Superfund Volume II: Environmental Evaluation Manual* (U.S. EPA 1989b). These two documents provide an invaluable overview of basic terminology, study design, methods, quality assurance and quality control considerations, and reporting requirements. Readers with limited experience in ecological assessment at hazardous waste sites should use the aforementioned U.S. EPA documents as a starting point for information on methods. Field survey methods are addressed separately in Chapter 4, and are not considered further in the present chapter. Suffice it to say that, in the overall ecological assessment process, field surveys generally provide the data to address higher-dimensional ecological effects. In contrast, single-species toxicity tests are used to identify contaminant sources and the relative toxicity of contaminated environmental media.

Toxicity test methods used at hazardous waste sites are directed toward determining the toxicity of contaminated water, soil, and sediments found at the site. Toxicity tests can be conducted *in situ* or in the laboratory. *In situ* tests involve exposing caged animals or vegetation directly to the medium of concern. *In situ* tests are relatively easy to perform and provide a realistic estimation of ambient environmental toxicity. However, *in situ* methods generally lack formal standardization and the investigative control typical of laboratory studies. As such, laboratory toxicity tests are more common, and involve shipping water, soil, and sediment samples directly to a testing facility. Sample collection and shipping procedures are critical components of laboratory toxicity tests, and are subject to stringent quality assurance practices for chemical cleanliness, temperature, and holding times prior to testing to ensure maintenance of sample integrity and experimental results. The importance of Quality Assurance/Quality Control (QA/QC) for hazardous waste studies cannot be overstated. General QA/QC guidelines (U.S. EPA 1984a,b; 1986), as well as guidelines specific to hazardous waste sites (U.S. EPA 1989a,b) and toxicity tests (U.S. EPA 1985a,b; 1988) are available.

The specific test species, test endpoint, and type and length of exposure used in toxicity tests are dependent on the specific test objective and the environmental media of concern. Water and sediment toxicity is usually measured by the death, growth, or reproductive impairment of aquatic organisms (e.g., fish, daphnids, crustaceans, insects, algae, duckweed), whereas soil toxicity is determined using similar endpoints on terrestrial species (e.g., birds, mammals, insects, earthworms, insects, plants). In addition to whole samples, toxicity tests may also be performed on eluates prepared from soils and sediment samples. The eluate tests are used to determine the mobility of chemical constituents from hazardous wastes and may employ aquatic (e.g., fish, daphnids, algae) or terrestrial species (earthworms, plant seed germination, and root elongation). In addition to higher animal and plant test methods, microbial toxicity tests may be performed on water or sediment and soil eluates using growth, enzymatic activity, bioluminescence, and other physiological (e.g., ATP) or nutrient cycling endpoints (e.g., carbon, nitrogen, sulfur, phosphorus transformations). As indicated earlier, toxicity tests for hazardous waste site studies have been extensively discussed (U.S. EPA 1989a) and are only briefly summarized here. Suffice it to say that toxicity test methods for aqueous samples are better developed (U.S. EPA 1985a,b; 1988) than sediment (Swartz 1987) and terrestrial toxicity tests (U.S. EPA 1982a,b,c; 1989a). The latter tests have largely been developed in response to pesticides and toxic substances registration. Although they have not been widely applied to hazardous waste sites, they can be modified for use on environmental samples.

Biomarker methods (U.S. EPA 1989) have not been extensively used in ecological assessments at hazardous waste sites, but the biomarker approach is an extremely active area of research and is destined to be increasingly prominent in the future. Biomarkers can be used independently, or incorporated into both laboratory toxicity tests and field studies. Two major biomarker categories include exposure indicators and sublethal stress indicators. Exposure biomarkers consist of direct indices of exposure (e.g., tissue residues of bioaccumulative trace metals and organic compounds), as well as indirect biochemical responses to exposure (e.g., enzyme induction in response to chemical exposure). Biomarkers of sublethal stress are generally adapted from the biomedical fields and may include generalized organismic responses to toxicants (e.g., histopathology, skeletal abnormalities), as well as contaminant-specific biochemical markers (e.g., delta-aminolevulinic acid dehydrase for lead exposure, cholinesterases in response to organophosphate and carbamate insecticides).

REFERENCES

Adams, S. M., K. L. Shepard, M. S. Greeley, Jr., B. D. Jimenez, M G. Ryon, L. R.-Shugart, J. F. McCarthy, and D. E. Hinton, 1989. The use of bioindicators for assessing the effects of pollutant stress on fish. *Marine Environmental Research* 28:459–64.

Cairns, J., Jr., 1980. Estimating hazard. *Bioscience* 30(2):101–107.

Cairns, J., Jr., 1992. Paradigms flossed: The coming of age of environmental toxicology. *Environmental Toxicology and Chemistry* 11:285–87.

Cairns, J., Jr. and K. L. Dickson (eds.), 1972. Biological Methods of the Assessment of Water Quality. ASTM STP 528. Philadelphia, PA: American Society for Testing and Materials, ASTM Special Technical Publication.

Cairns, J., Jr., K. L. Dickson, and A. W. Maki, 1978. Estimating the Hazard of Chemical Substances to Aquatic Life. ASTM STP 657, Philadelphia, PA: American Society for Testing and Materials.

Cairns, J., Jr. and A. W. Maki, 1979. Hazard evaluation in toxic materials evaluation. *Journal of the Water Pollution Control Federation* 51:666.

Casarett, L. J. and J. Doull (eds.), 1975. Toxicology: The Basic Science of Poisons. New York: Macmillan Publishing Co.

Crovello, T. J., 1970. Analysis of character variation in ecology and systematics. *Annual Review of Ecology and Systematics* 1:55–98.

Dickson, K. D., A. W. Maki, and J. Cairns, Jr. (eds.), 1979. Analyzing the Hazard Evaluation Process. Washington D.C.: Water Quality Section, American Fisheries Society.

Doudoroff, P., 1977. Reflections on pickle-jar ecology. *In* Biological Monitoring of Water and Effluent Quality (J. Cairns, K. Dickson, and G. Westlake, eds.). ASTM STP 607, Philadelphia, PA: American Society for Testing and Materials.

Environmental Study Board, National Research Council, 1981. Testing for Effects of Chemicals on Ecosystems. Washington, D.C.: National Academy Press.

Galluci, V. F., 1973. On the principles of thermodynamics in ecology. *Annual Review Ecology Systematics* 4:329–57.

Greene, J. C., C. L. Bartels, W. J. Warren-Hicks, B. R. Parkhurst, G. L. Linder, S. A. Peterson, and W. E. Miller, 1988. Protocols for short-term toxicity screening of hazardous waste. Corvallis, OR: U.S. EPA.

Grey, J. S., 1980. Why do ecological monitoring? *Marine Pollution Bulletin* 11:62.

Holling, C. S., 1986. The resilience of terrestrial ecosystems: Local surprise and global change. *In* Sustainable Development of the Biosphere (W. C. Clark and R. E. Munn, eds.). Cambridge, U.K.: Cambridge University Press.

Hull, D. L., 1970. Contemporary systematic philosophies. *Annual Review of Ecology and Systematics* 1:19–54.

Kelly, J. R. and M. A. Harwell, 1990. Indicators of ecosystem recovery. *Environmental Management* 14:527–45.

Krenkel, P. A. 1979. Problems in the establishment of water quality criteria. *Journal of the Water Pollution Control Federation* 51:2168–88.

Levin, S. A., M. A. Harwell, J. R. Kelly, and K. D. Kimball (eds.), 1988. Ecotoxicology: Problems and Approaches. New York: Springer-Verlag.

MacIntyre, E. D. and C. F. Mills (eds.), 1974. Ecological Toxicology Research. New York: Plenum Press.

Maciorowski A. F., L. W. Little, and J. Sims, 1980. Bioassays—Procedures and results. *Journal of the Water Pollution Control Federation* 52:1630–56.

Maciorowski A. F., 1988. Populations and communities: Linking toxicology and ecology in a new synthesis. *Environmental Toxicology and Chemistry* 7:667–68.

Maki, A. W., 1979. An analysis of decision criteria in environmental hazard evaluation programs. *In* Analyzing the Hazard Evaluation Process (K. L. Dickson, J. Cairns, Jr., and A. W. Maki, eds.). Washington, D.C.: Water Quality Section, American Fisheries Society.

Mayer, F. L., Jr., P. M. Mehrle, Jr., and R. A. Schoettger, 1980. Trends in aquatic toxicology in the United States: A perspective. *In* Proceedings of 3rd USA–USSR Symposium on the Effects of Pollutants upon Aquatic Ecosystems: Theoretical Aspects of Aquatic Toxicology. Duluth, MN: U.S. EPA.

Meier, R. L., 1972. Communications stress. *Annual Review of Ecology and Systematics* 1:55–98.

Mount, D. I., 1977a. Biotic monitoring. *In* Biological Monitoring of Water and Effluent Quality (J. Cairns, Jr., K. L. Dickson and G. F. Westlake, eds.). ASTM STP 607. Philadelphia, PA: American Society for Testing and Materials.

Mount, D. I., 1977b. Present approaches to toxicity testing, a perspective. *In* Aquatic Toxicology and Hazard Evaluation (F. L. Mayer and J. L. Hamelink, eds.). ASTM STP 634. Philadelphia, PA: American Society for Testing and Materials.

Neuhold, J. M., 1986. Toward a meaningful interaction between ecology and aquatic toxicology. *In* Aquatic Toxicology and Environmental Fate (T. M Poston and R. Purdy, eds.). ASTM STP 921. Philadelphia, PA: American Society for Testing and Materials.

Odum, E. P., 1977. The emergence of ecology as a new integrative discipline. *Science* 195:1289–93.

Odum, E. P., 1984. The mesocosm. *Bioscience* 35:165–71.

Rapport, D. J., 1989. What constitutes ecosystem health? *Perspectives in Biology and Medicine* 33:120–32.

Science Advisory Board, U.S. Environmental Protection Agency, 1990a. Reducing risk: Setting priorities and strategies for environmental protection. SAB-EC-90-021. Washington, D.C.: U.S. EPA.

Science Advisory Board, U.S. Environmental Protection Agency, 1990b. The report of the Ecology and Welfare Subcommittee: Relative risk reduction project, reducing risk, Appendix A. EPA SAB EC-90-021A. Washington, D.C.: U.S. EPA.

Science Advisory Board, U.S. Environmental Protection Agency, 1990c. The report of Human Health Subcommittee: Relative risk reduction project, reducing risk, Appendix B. EPA SAB-EC-90-021B. Washington, D.C.: U.S. EPA.

Science Advisory Board, U.S. Environmental Protection Agency, 1990d. The report of the Strategic Options Subcommittee: Relative risk reduction project, reducing risk, Appendix C. EPA SAB-EC-90-021C. Washington, D.C.: U.S. EPA.

Science Advisory Board, U.S. Environmental Protection Agency, 1990e. The report of the Ecological Monitoring Subcommittee: Evaluation of the ecological indicators report for EMAP. EPA SAB-EC-91-001. Washington, D.C.: U.S. EPA.

Spehar, R. L., R. W. Carlson, A. E. Lemke, I. I. Mount, Q. H. Pickering, and V. M. Suarski, 1980. Effects of pollution on freshwater fish. *Journal of the Water Pollution Control Federation* 52:1703–68.

Sprague, J. B., 1976. Current status of sublethal tests of pollutants on aquatic organisms. *Journal of the Fisheries Research Board of Canada* 33:1988–92.

Swartz, R. C., 1987. Toxicological methods for determining the effects of contaminated sediments on marine organisms. *In* Fate and Effects of Sediment Bound Chemicals in Aquatic Systems. (Dickson, K. D., A. L. Maki, and W. A. Brungs, eds.) New York: Pergamon Press.

Thomas, L. M., 1987. Environmental decision-making today. *Environmental Protection Agency Journal* 13:2–5.

U.S. Environmental Protection Agency, 1982a. Environmental effects tests guidelines. EPA/560/6-82/002. Washington, D.C.: U.S. EPA.

U.S. Environmental Protection Agency, 1982b, Pesticide assessment guidelines. EPA/540/9 82/018 through 028. Washington, D.C.: U.S. EPA.

U.S. Environmental Protection Agency, 1982c. Toxic substances test guidelines. EPA/6-82-001 through 003. Washington, D.C.: U.S. EPA.

U.S. Environmental Protection Agency, 1984a. Guidance for preparation of combined work/quality assurance project plans for environmental monitoring. Report No. OWRS QA-1. Washington, D.C.: U.S. EPA.

U.S. Environmental Protection Agency, 1984b. The development of data quality objectives. EPA Quality Assurance Management Staff and the DQO Workgroup. Washington, D.C.: U.S. EPA.

U.S. Environmental Protection Agency, 1985. Short-term methods for estimating the chronic toxicity of effluents and receiving wastes to freshwater organisms. EPA/600/4-85/014. Cincinnati, OH: U.S. EPA.

U.S. Environmental Protection Agency, 1986. Development of data quality objectives. EPA Quality Assurance Management Staff and the DQO Workgroup. Washington, D.C.: U.S. EPA.

U.S. Environmental Protection Agency, 1989a. Ecological assessment of hazardous waste sites. EPA 600/3-89/013. Corvallis, OR: U.S. EPA.

U.S. Environmental Protection Agency, 1989b. Risk assessment guidance for Superfund. Volume II. Environmental evaluation manual. Interim final. EPA/540/1-89/001. Washington, D.C.: U.S. EPA.

U.S. Environmental Protection Agency, 1989c. Water quality standards for the 21st century: Proceedings of a national conference. Washington, D.C.: U.S. EPA.

U.S. Environmental Protection Agency, 1989d. Water quality criteria to protect wildlife resources. EPA/600/3-89/067. Corvallis, OR: U.S. EPA.

U.S. Environmental Protection Agency, 1989e. Rapid bioassessment protocols for use in streams and rivers: Benthic macroinvertebrates and fish. EPA/444/4-89-0001. Washington, D.C.: U.S. EPA.

Webster, J. R., J. B. Waide, and B. C. Patten, 1975. Nutrient recycling and the stability of ecosystems. *In* Mineral Cycling in Southeastern Ecosystems (F. G. Howell, J. B. Gentry, and M. H. Smith, eds.). ERDA CONF-740513. Springfield, VA: National Technical Information Service.

Westman, W. E., 1978. Measuring the inertia and resilience of ecosystems. *Bioscience* 28:705–711.

9

Pine Street Canal Ecological Assessment: A Case Study

William B. Kappleman

Metcalf & Eddy, Inc.

INTRODUCTION

The Pine Street Canal (PSC) Superfund site is an uncontrolled hazardous waste site located on the eastern shore of Lake Champlain in Burlington, Vermont (Figure 9-1). The original study area occupied approximately 32 ha, while the current site boundary, delineated using contaminant concentrations from on-site sampling efforts, encompasses approximately 19 ha (M&E 1992). The PSC site is situated in a highly industrialized area located less than 1 km south of the center of Burlington. The site lies in a topographically low area and includes an abandoned barge canal, a turning basin, adjacent filled-in boat slips, and vegetated wetlands south, east, and west of the canal. The canal is hydrologically connected to Lake Champlain through a restricted inlet/outlet under an active railway and has a small input stream entering from the south. Several culverts connected to the Burlington storm water system also discharge into the surface waters of the canal. The canal and associated wetlands on the PSC site form a distinct ecological community which is unique within the surrounding urbanized setting. While the site has been dramatically altered by human activity, the cessation of industrial operations on portions of the site within the last two decades has allowed these disturbed areas to revert back to a more natural state, characterized by early successional vegetation and wildlife not common to an urban setting. Access to the site is largely unrestricted by fences or other obstructions and portions of the site have recent evidence of trash dumping, campfires, and fishing, indicating that the site is often used by local residents.

The primary environmental concern on the site is contamination resulting from the past operation (early 1900s to 1966) of a coal gasification plant near the southern end of the canal. Process wastes, including coal tar, from this plant were reportedly disposed of, or leached into, the canal and adjacent wetlands. These wastes have been detected in groundwater, sediments, and soils throughout the site. The PSC site was placed on the National Priority List by the U.S. Environmental Protection Agency (U.S. EPA)

223

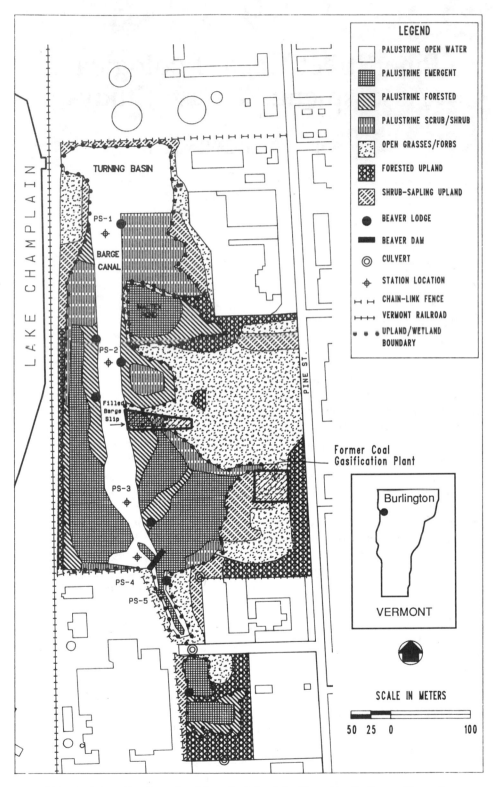

LEGEND

- PALUSTRINE OPEN WATER
- PALUSTRINE EMERGENT
- PALUSTRINE FORESTED
- PALUSTRINE SCRUB/SHRUB
- OPEN GRASSES/FORBS
- FORESTED UPLAND
- SHRUB-SAPLING UPLAND
- ● BEAVER LODGE
- ▬ BEAVER DAM
- ◎ CULVERT
- ✛ STATION LOCATION
- CHAIN-LINK FENCE
- VERMONT RAILROAD
- UPLAND/WETLAND BOUNDARY

LAKE CHAMPLAIN

TURNING BASIN

PS-1

BARGE CANAL

PS-2

Filled Barge Slip →

PS-3

PS-4

PS-5

PINE ST.

Former Coal Gasification Plant

Burlington

VERMONT

SCALE IN METERS

50 25 0   100

FIGURE 9-1. Pine Street Canal Superfund site, Lake Champlain, Burlington, Vermont.

224

in November 1981. In 1985, U.S. EPA Region I conducted an emergency removal of approximately 400 cubic meters of contaminated soils and sediments from Maltex Pond. Remedial Investigation/Feasibility Study activities commenced in 1987.

This chapter describes the ecological assessment conducted for the PSC site to elucidate the existing and potential future effects of site contaminants on ecological resources. This case study will outline the process utilized during the ecological assessment, pointing out the various decisions that had to be made as the study progressed. Many of these decisions, and the methodology employed, were driven by the scarcity of toxicity data relating the effects of the primary site contaminants (polynuclear aromatic hydrocarbons [PAHs] and the volatile organic compounds [VOCs] benzene, toluene, and xylene [BTX]) to wildlife resources.

This case study was conducted, and the results in this chapter are organized, consistent with a number of U.S. EPA guidance documents on ecological assessments and risk assessments (U.S. EPA 1989a; 1989b; 1989c; 1989d), and is divided into three main parts: (1) nature and extent, (2) ecological conditions, and (3) ecological risk assessment. Nature and extent involved a review of data on the nature and extent of site contamination and the ecologically important fate and transport mechanisms for the various site contaminants. This information was used to determine the contaminants (PAHs and BTX) and media (upland soils, wetland soils, and canal sediments) of concern, and was instrumental in guiding the development of a biological field sampling plan to assess the ecological resources present on the site, as well as to assess the potential effects of the contamination on these resources.

The second part of this chapter, ecological conditions, describes the methodology used and the results of the biological field sampling program. Based on the nature and extent of contamination, aquatic sampling was emphasized, and included quantitative studies of benthos, fish, and zooplankton at the PSC site and at an uncontaminated reference site. Biological sampling in wetland and terrestrial habitats was more qualitative and consisted of wetland delineation, wetland functional assessment, vegetation (habitat) mapping, and wildlife surveys, including small mammal trapping. Based upon the results of the biological field studies, the site appears to currently have significant ecological value. The biological field studies did identify effects of site contamination, involving the benthic infauna and possibly muskrats. The scientific names of plants and animals mentioned in the text are listed in Appendix 1.

Information described in the first two parts of this chapter was then used in a baseline ecological risk assessment to quantify the existing risks to ecological resources and to determine ecological effect levels (concentrations of the contaminants of concern in the media of concern, below which no demonstrable reproductive effects to ecological receptors are likely to occur). The baseline ecological risk assessment consisted of determining exposure pathways, species or species groups potentially at risk of exposure, and the toxicity of the contaminants of concern to selected ecological receptors. Ecological effect levels were developed based on established numerical criteria (e.g., interim sediment quality criteria) or on exposure pathway modeling using site-specific data, as well as information obtained from the literature.

The decision to remediate, and the development of potential ecological endpoints for the PSC site following remediation, represents the classic ecological conflict at hazardous waste sites (U.S. EPA 1989c). The site currently has significant ecological value, and only a few readily demonstrable effects of site contamination were apparent. There is the real possibility that the removal of contaminated media could destroy some of the site's existing ecological value. Some of the features that contribute to this value, such as mature forested wetlands, would be difficult to recreate within a reasonable time-frame. However, there is also the possibility that remediation will ultimately enhance the site's overall ecological value. Preserving the site's unique ecological attributes, while at the same time enhancing the site's overall ecological value through remediation, is a difficult task, and will usually involve various trade-offs to achieve the proper balance.

NATURE AND EXTENT

Review of Contaminant Data

Volatile organics (benzene, toluene, ethylbenzene, xylenes [BTEX], and styrene), semivolatiles (mostly two-six ring PAHs), and inorganics (metals and cyanide) associated with wastes from the former coal gasification plant have been found at various concentrations in different media throughout the PSC site. This section will briefly summarize, by media, the nature and extent of this contamination. A full evaluation of the contamination at the site can be found in M&E (1992).

Groundwater

Groundwater underlying the wetlands west of the former coal gasification plant was found to be the most contaminated, having relatively high concentrations of PAHs, BTEX, and cyanide. Other organic compounds were also detected at lower concentrations relative to the BTEX compounds. Numerous metals were detected in groundwater throughout the site. Pesticides and polychlorinated biphenyls (PCBs) were either not detected or detected at very low levels.

Surface Water

In comparison to groundwater, PAHs were detected less frequently and in lower concentrations in canal surface waters. PAHs detected included two of the lower molecular weight, noncarcinogenic PAHs (naphthalene and acenaphthylene). Most individual PAH compounds were not found at levels exceeding sample detection limits.

Volatile organics detected in on-site surface waters included BTEX compounds, ketones, chlorinated volatiles, and carbon disulfide. Except for elevated concentrations of benzene and toluene in one (non-canal) sample, concentrations of volatile organics in on-site surface waters were near or below sample detection limits and much lower than concentrations detected in groundwater. Detections of BTEX were much more common in samples from the vegetated wetlands on the southern portion of the site than in the waters of the canal and turning basin.

Pesticides were detected in only one surface water sample, and PCBs were not detected in any samples. Metals were detected in both canal and non-canal surface water samples throughout the site. Concentrations of individual metals were generally less than concentrations detected in on-site groundwater. Cyanide was detected in wetland surface water samples west of the former coal gasification plant. Cyanide was not detected in samples taken directly from the canal.

Contamination of canal surface water appeared to be minimal compared to other media. Metal concentrations were generally less than those found in groundwater, and relatively low levels (at or near detection limits) of volatile organics and semivolatiles were found in on-site surface waters. These results indicated that surface waters did not appear to retain volatile organics, and that the semivolatiles present were relatively insoluble in water. Adjacent Lake Champlain waters did not appear to contain elevated levels of toxic compounds.

Canal Surface Sediments

Surficial sediment samples from the canal, taken within 0.7 m of the surface in permanently submerged areas, showed extensive levels of PAH contamination. The PAH and inorganic contaminants found at the PSC site are likely to accumulate in sediments because these compounds have a strong tendency to adsorb onto sediment particles. The canal sediments, especially at the canal's southern end, contain visible coal tar product at or just below the surface. High BTEX and PAH concentrations were apparent in the southern portion of the canal; the levels of these contaminants in canal sediments generally declined going northward up the canal and into the turning basin. The primary volatile organic compounds detected in canal sediments were BTEX compounds. Concentrations of noncarcinogenic PAH compounds were generally higher than carcinogenic PAH compounds.

PCBs were not detected in surface sediments, and pesticides were detected only infrequently at relatively low concentrations. Metals were generally elevated compared to background levels; levels were elevated in bordering wetlands as well as in the canal itself. Cyanide was detected in most of the canal surficial sediment samples. Overall, cyanide concentrations in sediments tended to be lower in comparison to soils from the wetlands west of the former coal gasification plant.

Surface Soils

Surface soils were defined as soils collected from depths of 0 to 15 cm in upland areas and wetlands not permanently submerged, although data from surface borings from depths of 0 to 1.5 m were also used where surface soil samples were unavailable. PAHs were relatively widespread throughout the surface soils of the site, with concentrations generally highest between the canal and the former coal gasification plant, especially in the wetlands located west of the former coal gasification plant. Lower molecular weight PAHs were detected less frequently than higher molecular weight, carcinogenic PAHs.

Surface soils were contaminated with low levels of volatile organics, primarily BTEX, over much of the site. BTEX compounds are less persistent than most PAHs in surface soils because of physical and chemical processes (e.g., volatilization) that naturally occur. As indicated both by the association of BTEX with coal gasification wastes and the frequent detection of BTEX compounds in subsurface soils and groundwater, BTEX compounds may have been present at higher concentrations in the past. Leaching of BTEX to groundwater and partitioning to the atmosphere have probably occurred over time, resulting in reduced BTEX concentrations in surface soils.

Pesticides were detected in only one surface soil sample and PCBs were not detected in any on-site surface soil samples. Concentrations of some metals, primarily chromium, antimony, barium, lead, and zinc, were slightly elevated relative to background surface soil concentrations. Elevated levels of cyanide were detected in surface soils in the wetlands west of the former coal gasification plant.

Subsurface Soils

High levels of coal tar, PAHs, BTEX, and cyanide were detected in wetland subsurface (more than 1.5 m below the surface) soils west of the former coal gasification plant. The highest levels of contamination existed to depths of 12 m. Lower levels of these contaminants, plus metals, were detected in subsurface soils throughout much of the site. PCBs were detected in only one subsurface sample, and pesticides were detected infrequently at low concentrations.

Ambient Air

Limited monitoring of contaminant levels in ambient air was conducted on the PSC site. These studies did not detect PAHs at levels exceeding detection limits, although volatile organics were detected.

Contaminant Summary

In summary, groundwater and subsurface soils were among the most contaminated on-site media, containing elevated levels of PAHs, BTEX, and cyanide; metal levels were also elevated in these media. PAH, BTEX, and inorganic (metals and cyanide) compound concentrations were elevated in surface sediments, primarily in the southern end of the canal, and lower levels of these compounds were generally present in surface soils. Elevated concentrations of PAHs were found in upland surface soils at the location of the former coal gasification plant, in the wetland soils west of this location, in the canal sediments, in the wetland soils of the northern barge slip (off the northern edge of the turning basin), and in the upland soils of the petroleum storage area north of the turning basin. Contamination of surface water was minimal compared to other media. Pesticides and PCBs do not appear to be present in significant levels in any on-site media.

Hazard Assessment

The purpose of a hazard assessment is to identify the potential contaminant(s) and media of concern to ecological receptors. To accomplish this, information on the site's history was reviewed to determine potential sources and types of contaminants present on the site, and data from sampling of on-site media were reviewed to determine the nature and extent of contamination of various media. In combination with information on the relative toxicity and bioavailability of the detected compounds, and the types of organisms and habitats present on-site, this allowed the various contaminant(s) and media to be preliminarily screened, thereby identifying the contaminant(s) and media of potential concern. The history of the PSC site and the nature and extent of contamination were discussed earlier. Ecological conditions are described in the following section.

Based upon: (1) the available history of the site, (2) the extensive sampling of ground and surface water, soils, and sediments, (3) the observed contaminant levels in these media, and (4) the relative toxicity and bioavailability of the detected compounds, the media of potential concern were identified as: (1) the surface (depths of 0–1.5 m) sediments in the canal and turning basin, (2) the surface (depths of 0–1.5 m) soils in the adjacent vegetated wetlands, and (3) upland surface (depths of 0–1.5 m) soils, especially in the vicinity of the former coal gasification plant. The contaminants of concern were identified as polynuclear aromatic hydrocarbons (PAHs) and the volatile organics benzene, toluene, and xylene(s) (BTX).

Groundwater and subsurface (depths greater than 1.5 m) soils were eliminated as media of concern since ecological receptors have extremely limited direct contact with these media. Although no Ambient Water Quality Criteria (U.S. EPA 1986) have been developed for the contaminants of concern, surface water was also eliminated due to low observed contaminant concentrations; PAHs in particular were very rarely detected in on-site surface waters. Air is not formally considered a media of concern, since the source of contaminants in air is likely to be from volatilization of contaminants off of surface soils and exposed sediments; the soils and sediments are the source of the contaminants. Thus, exposure pathways (inhalation) involving air will be discussed as a function of contaminant concentrations in soils and sediments.

Ecologically Important Fate and Transport Mechanisms

This subsection summarizes the pertinent information concerning the fate and transport of the contaminants of concern as it applies to ecological receptors. The fate and transport of contaminants in the environment depends on the properties of both the contaminant and the environmental medium in which it occurs. For each compound, the physical and biological pathways are identified, as are the storage or degradation mechanisms present in the environment. This latter category includes the rate at which these compounds are metabolized within biological organisms.

Polynuclear Aromatic Hydrocarbons

Polynuclear aromatic hydrocarbons (PAHs), also known as polycyclic aromatic hydrocarbons or polycyclic organic matter (POM), consist of hydrogen and carbon atoms in the form of two or more fused benzene rings. The majority of PAHs present at the PSC site are likely the result of past coal tar disposal practices; PAHs are a major component of coal tars.

Environmental concern has focused on PAHs, which range in molecular size from two- to seven-ring structures. The number of rings on the molecule strongly affects its biochemical interactions in the environment. Consequently, the fate, transport, and toxicity of PAHs correlate strongly with the specific size of the PAH molecule. For example, unsubstituted PAHs containing two or three rings can cause acute toxicity and other adverse effects in some organisms, but are noncarcinogenic. In contrast, four- to seven-ring compounds are much less acutely toxic, but many are demonstrably carcinogenic or mutagenic to a wide variety of organisms (Eisler 1987). Relatively little is known about the fate and transport of specific PAH compounds. Information on PAHs as a group is largely inferred from information on benzo(a)pyrene and mixtures of PAHs (Clement Associates 1985).

In water, PAHs may volatilize, disperse into the water column, become incorporated into bottom sediments, concentrate in aquatic biota, or experience chemical oxidation and biodegradation (Eisler 1987). The chemical properties of PAHs suggest that the most likely fate is adsorption onto suspended particulate matter, especially particulates high in organic content. Approximately two-thirds of PAHs in water are typically associated with particulates, and one-third are dissolved (Eisler 1987). The ultimate fate of PAHs on aqueous particulates is sedimentation followed by photooxidation, chemical oxidation, biotransformation, or biodegradation by bacteria and other benthic organisms (Eisler 1987). Under these circumstances, the specific fate and transport of PAHs would depend largely on the hydrologic condition of the environment (U.S. EPA 1979). PAHs in aquatic sediments generally degrade more slowly than PAHs in the atmosphere. Furthermore, in the absence of penetrating radiation or oxygen, PAHs in aquatic sediments degrade extremely slowly and may persist indefinitely (Eisler 1987). In sediments with high levels of organic carbon, however, PAHs tend to be strongly adsorbed and thus have limited bioavailability. Under appropriate hydrologic conditions (e.g., turbulent water), volatilization could be as important an aquatic transport process as adsorption (U.S. EPA 1979), particularly for the lower molecular weight PAHs. The remaining third of PAHs in water, existing in dissolved solution, may be degraded by rapid photolysis and, to a lesser extent, by oxidation (U.S. EPA 1979). Oxidation by chlorine and ozone may be the most important fate process for aqueous PAHs when these oxidants are available in sufficient quantities (Clement Associates 1985).

At PSC, the aquatic system is a highly depositional environment, due to low surface water velocities. Thus, contaminants associated with deposited particles would tend to concentrate in the sediments. In addition, the canal waters have a low light transmission rate, due to high levels of tannins, organic matter, particulates, and plankton present in the water column. Thus, photooxidation is probably minimal. Therefore, it is expected that the vast majority of PAHs present in the PSC aquatic system would be associated with particulates and would be deposited in the canal sediments, where they would accumulate and persist. This is supported by on-site sampling results, which show significant PAH contamination of the sediments, but little or no PAH contamination of the water column.

The fate of PAHs in the atmosphere depends on the size of the specific compound. Many lower molecular weight PAHs are volatile. Adsorption onto airborne particles is the likely fate of many larger molecular weight PAHs; these airborne particles may be inhaled by organisms, resulting in exposure to the PAHs. These adsorbed PAHs photodecompose readily in the atmosphere by reaction with ozone and various oxidants. However, photooxidation can produce reaction products that are known to be carcinogenic or mutagenic (Eisler 1987). Degradation times range from less than one day to several weeks, depending on the size of the compound and the size of the particle it is adsorbed to. Airborne PAHs that do not photodecompose are eventually returned to aquatic and terrestrial systems by precipitation (Clement Associates 1985).

PAHs in surface soils may be volatilized into the atmosphere. PAHs in subsurface soils may be assimilated by plants (see later), degraded by soil microorganisms, or accumulated to relatively high levels in the soil. High PAH concentrations in the soil can lead to high microorganism populations capable of degrading the compounds (Eisler 1987).

Biodegradation and biotransformation by benthic organisms, including microbes and invertebrates, are the most important biological fate processes for PAHs in sediments (U.S. EPA 1979). Most animals and microorganisms can metabolize and transform PAHs to intermediate compounds that may ultimately experience complete degradation (Eisler 1987). PAHs with high molecular weights are degraded slowly (half-lives of up to a few years) by microbes and readily by multicellular organisms (U.S. EPA 1979). Biodegradation probably occurs more slowly in aquatic systems (especially anaerobic systems) than in soil (Clement Associates 1985).

Some PAHs rapidly bioaccumulate in animals because of the high lipid solubility of PAHs (Eisler 1987). The rate of PAH bioaccumulation is inversely related to the rate of PAH metabolism, and is also influenced by the concentration of PAH to which the organism is exposed. Both rates are dependent on the size of the specific PAH molecule; PAHs with less than four rings are readily metabolized and not bioaccumulated, while PAHs with more than four rings are more slowly metabolized, and tend to bioaccumulate on a short-term basis (U.S. EPA 1979; Clement Associates 1985; Eisler 1987). PAHs of all sizes show little tendency for long-term bioaccumulation despite their high lipid solubility, probably because most PAHs are rapidly and extensively metabolized (U.S. EPA 1979; Clement Associates 1985; Eisler 1987). Bioaccumulation is thus not considered an important fate in most multicellular organisms because it is usually a temporary process (Eisler 1987).

Large differences in the ability to absorb and assimilate PAHs from food have been reported among species of aquatic organisms. In all cases where assimilation of ingested PAHs has been demonstrated, the metabolism and excretion of PAHs was rapid. Thus, little potential exists for food chain biomagnification of PAHs, at least in aquatic systems (Eisler 1987).

In some plants growing in highly contaminated areas, assimilation may exceed metabolism and degradation, resulting in accumulation in plant tissues (Eisler 1987). Laboratory experiments have demonstrated that plants can bioaccumulate PAHs to levels above those found in the environment, although this has not been conclusively demonstrated in field-grown plants (Eisler 1987). Uptake can be by both leaves (from atmospheric deposition) and roots (from soils or sediments) with subsequent translocation to other plant parts (Edwards 1983; Eisler 1987). Although much of the data are from domestic crops, leaves tend to have the highest concentrations of PAHs, followed by roots, and fruits and seeds (Sims and Overcash 1983). Uptake is also variable by plant species and soil conditions (e.g., pH). Little data are available on bioaccumulation by vegetation and trophic transfer to higher-level consumers in terrestrial and aquatic food chains (Eisler 1987).

Benzene

Benzene is a single-ring hydrocarbon that, in a pure state, exists as a clear, colorless liquid. The predominant process of transport and removal of benzene in the environment is volatilization to the atmosphere, followed by photooxidation involving a reaction with hydroxyl radicals (U.S. EPA 1979; Clement Associates 1985; U.S. PHS 1987a). The atmospheric residence time of benzene ranges from a few hours to a few days, depending on the concentration of available hydroxyl radicals (U.S. PHS 1987a). Undoubtedly, some atmospheric benzene is returned to soil and water in the form of precipitation (U.S. PHS 1987a).

Most benzene in surface water volatilizes readily and is easily transported through the air. The half-life of benzene in surface water has been estimated to be between four and five hours (U.S. PHS 1987a). Sorption processes are potential removal mechanisms in both surface water and groundwater (Clement Associates 1985). Both the vapor pressure and solubility of benzene in water are fairly high, however, and the persistence of some benzene in the water column would be expected in most cases. While oxidation is the primary mode of benzene degradation in the atmosphere, oxidation in water is unlikely (U.S. EPA 1979; U.S. PHS 1987a). A more probable fate of aqueous benzene is aerobic biodegradation; there is evidence of gradual biodegradation of benzene at low concentrations by aquatic microorganisms (Clement Associates 1985; U.S. PHS 1987a). Anaerobic decomposition may also occur, but this process is likely to be much slower than aerobic decomposition (U.S. PHS 1987a). Apparently, the rate of benzene degradation is enhanced when other hydrocarbons are present (U.S. EPA 1979; Clement Associates 1985). Nonetheless, benzene degradation in water and soil is slower in most cases than degradation in air.

Benzene released to the soil can be biodegraded by microorganisms or can be transported to the air through volatilization, to surface water through runoff, and to groundwater as a result of leaching. In soils and sediments, the physical characteristics of benzene indicate that adsorption onto organic material would be significant only under conditions of constant exposure (Clement Associates 1985). Biodegradation is probably the ultimate fate process in soils. However, organisms that biodegrade benzene appear to be present in soil only in low numbers (U.S. PHS 1987a). Volatilization and runoff would only occur if benzene were near the surface; benzene deeper in soils would likely be biodegraded or leached to groundwater (U.S. PHS 1987a). Factors that might affect the rate of leaching include the soil type (e.g., sand or clay), amount of rainfall, and depth to groundwater (U.S. PHS 1987a).

At PSC, most of the benzene was detected in groundwater, probably due to leaching from subsurface soils; subsurface soils also had relative high levels of benzene. Concentrations in surface waters and surface soils were relatively low, probably due to volatilization. Toluene and xylenes, discussed below, had similar distributions to that of benzene at PSC.

Bioaccumulation of benzene in the fat of organisms is slow because of low diffusion, but the total potential uptake is high in these tissues, because of the high lipid solubility of benzene. After absorption, benzene must undergo metabolic transformation (primarily in the liver) to exert its toxic effects. Bioaccumulation and rates of elimination of benzene differ from its metabolites (U.S. PHS 1987a). The bioaccumulation potential of benzene by aquatic organisms at pollutant concentrations anticipated in environmental waters would probably be low (U.S. EPA 1979; Clement Associates 1985).

Toluene

Toluene is a clear, colorless liquid. There is apparently little tendency for toluene to persist in the environment because it readily decomposes in soil and evaporates rapidly (U.S. PHS 1989). As for benzene, the primary means by which toluene is removed from the environment is volatilization, and (once volatilization occurs) atmospheric photooxidation of toluene generally subordinates all other fates (U.S. EPA 1979). Although it is a liquid at room temperature, toluene is sufficiently volatile that the majority of toluene in the environment exists in the vapor phase (U.S. PHS 1989).

It is likely that atmospheric toluene is precipitated in rain, since it is soluble in water (U.S. PHS 1989). Once in the water or in surface soils, toluene tends to evaporate quickly (U.S. PHS 1989). The rate of evaporation depends on whether the water is turbulent or static. In turbulent water, the half-life for toluene evaporation is five to six hours, and in static water the half-life for evaporation is up to 16 days (U.S. PHS 1989). Under average conditions, over 90 percent of the toluene in the upper soil layer volatilizes in the first 24 hours; soil half-lives of one to seven days are typical (U.S. PHS 1989). Toluene has a relatively high affinity for organic soils; it is moderately adsorbed onto soils and sediments rich in organic matter, but readily leached from soils low in organic matter.

Toluene that is not volatilized from shallow groundwater, surface water, or surface soils is typically degraded by microbial activity (U.S. PHS 1989). The rate of biodegradation depends on several factors (e.g., temperature), with biodegradation half-lives of less than one day under favorable conditions. Regardless of the conditions, volatilization remains the dominant fate process in these media (Clement Associates 1985; U.S. PHS 1989).

Toluene is moderately lipophilic, and therefore has a moderate tendency to bioaccumulate in the fatty tissues of aquatic species. The bioconcentration factor (BCF) is a measure of the degree to which an organism concentrates a compound above the levels found in the surrounding media (usually water). BCFs for toluene have been estimated to be 10.7 in fish (U.S. PHS 1989). The levels that accumulate in the flesh of aquatic species also depend on the degree to which the species metabolize toluene. Toluene can apparently be detoxified and excreted by mammals (Clement Associates 1985). Thus, bioaccumulation is probably not an important environmental fate process for toluene (Clement Associates 1985).

Xylene

Xylene is a monocyclic hydrocarbon that exists in the form of a colorless, aromatic liquid. Commercial xylene is a mixture of three isomers: o-xylene, m-xylene, and p-xylene. Xylene binds to sediments in water and to organic matter in soils, and undergoes microbial degradation in both media. Biodegradation is probably the most important fate process in sediments and in subsurface soils (Clement Associates 1985). However, biodegradation can be a slow process and xylenes have been shown to persist for up to six months in some soil types (Clement Associates 1985). Volatilization is expected to be the predominant loss mechanism from surface soils (U.S. EPA 1984a). Some xylene in subsurface soils may eventually infiltrate to groundwater if soils are low in organic carbon. Because of their low solubility in water and rapid biodegradation, however, xylenes are unlikely to leach into groundwater in high concentrations (Clement Associates 1985).

In surface water and shallow groundwater, volatilization and subsequent photooxidation by reaction with hydroxyl radicals in the atmosphere is probably the most important fate process (U.S. EPA 1979). The estimated half-life of the three xylene isomers in water ranges from 2.6 to 11 days (U.S. EPA 1984a). The half-life of xylene in air ranges from 8 hours for m-xylene to 15 hours for p-xylene (U.S. EPA 1984a).

BCFs for o-xylene, m-xylene, and p-xylene in aquatic organisms have been estimated to be 45, 105, and 95, respectively (U.S. EPA 1984a). Little research has been conducted on the effects of xylene bioaccumulation and its trophic transfer in food chains.

ECOLOGICAL CONDITIONS

Introduction

The nature and extent of site contamination, and the fate and transport mechanisms of the contaminants of concern, indicated that the primary ecological resource at risk was the aquatic system. Thus, quantitative studies of the on-site benthic and pelagic segments of the aquatic community were initiated in the spring of 1990. Since the adjacent wetlands and uplands appeared to be less contaminated, less intensive, more qualitative studies of vegetation and wildlife were considered sufficient.

A key factor to understanding the ecological effects of site contamination, the relative importance of the natural resources present at the site, and the ecological potential of the site is the evaluation of nearby areas where the ecology is relatively unaffected by contamination with hazardous substances. This was done by conducting a literature review of similar (but uncontaminated) Lake Champlain ecosystems, to document parameters that are relatively stable, temporally, such as wildlife habitats. More ephemeral or highly seasonal characteristics, such as benthic invertebrate density, were evaluated by comparing existing site conditions to those at a reference site.

A suitable reference site was defined as an uncontaminated area having similar attributes to the PSC ecosystem for certain critical factors. The direct connection to Lake Champlain, the tributary flow (water regime), and the extensive bordering wetland vegetation present at the PSC site were considered the most important factors when selecting the reference site. Based upon these factors and investigations of several potential locations, a reference site approximately 4.5 km north of PSC was selected. This site is located on Malletts Creek, which flows into Lake Champlain via Malletts Bay.

Although it is critical to compare a hazardous waste site to a similar (although uncontaminated) area, there is really no true reference site, because the physical and biological conditions present on any site are, to some degree, unique (U.S. EPA 1989c). However, such a comparison was considered an essential part of evaluating the contaminant effects and ecological potential of the PSC site for the aquatic community.

The following subsections discuss the assessment endpoints for the ecological studies, the regional ecological characteristics of the Lake Champlain ecosystem, provide a general description of the Malletts Creek reference site, outline the results of the site-specific biological field studies conducted at the PSC site, and compare the on-site aquatic community to that found at the reference site.

Objectives

The overall objective of the biological field studies was to define the ecological condition of the PSC site. There are three aspects of the ecological condition that must be addressed to meet this objective. First, significant environmental resources on-site must be identified so that they can be targeted for protection or enhancement (if impacted) as part of remediation. This is the primary reason for conducting the biological field sampling program. Secondly, ecological effects of site contamination on various segments of the ecosystem must be defined to provide input into the risk assessment. This was done by formulating assessment endpoints and evaluating the results of the biological field studies in relation to these endpoints. Finally, the ecological assessment must address the future ecological potential of the site following remediation; this is addressed at the end of this chapter.

Assessment Endpoints

Assessment endpoints concern habitat quality, which relates to "normal" or "expected" ecological functions, such as use of the site by fish and wildlife in terms of presence, diversity, and/or abundance. These expected values were derived from documented observations obtained from the reference site and other studies in the Lake Champlain region. The selection of these endpoints was based upon the premise that the on-site contamination will have a demonstrable effect on a selected species either through its avoidance of the habitat (resulting in absence or relative scarcity) or increased on-site mortality (resulting in reduced population numbers) (U.S. EPA 1989c). The assessment endpoint of "presence" is generally applied to those species whose mobility allows them to easily travel between the site and off-site areas, or for which the site represents a relatively small portion of their normal home range. It is important to note that the presence of a species on-site is not evidence of no effect, but is likely to indicate the absence of acute effects. The assessment endpoint of "diversity/abundance" is generally applied to those species (or assemblages) who are mostly sessile, and would therefore

TABLE 9-1.   Assessment and Measurement Endpoints for Selected Species

Species Group	Assessment Endpoint	Measurement Endpoint
Benthic organisms	Abundance/Diversity	Reference site/Regional abundance
Fish	Abundance/Diversity	Reference site/Regional abundance
Amphibians	Presence	Observed use of the site
Reptiles	Presence	Observed use of the site
Mink	Presence	Observed use of the site
Muskrat	Abundance	Reference site/Regional abundance
Beaver	Abundance	Reproducing population
Peromyscus mice	Presence	Observed use of the site
Ducks	Presence	Observed use of the site
Piscivorous birds	Presence	Observed use of the site

be expected to be present on-site through various life stages. It also applies to those species with home ranges less than or equal to the area of suitable habitat present on the site. These assessment endpoints meet the criteria set forth in U.S. EPA (1989c). Measurement endpoints are techniques by which assessment endpoints are evaluated. Table 9-1 lists the assessment and measurement endpoints for selected species (or species groups) for which a potential exposure risk exists.

The chosen measurement endpoints are specific to particular assessment endpoints. The assessment endpoint of "presence" is evaluated based upon observed use of the site (the measurement endpoint). The assessment endpoint of "diversity/abundance" is evaluated, except for beaver, by comparison of the abundance or diversity of the species on-site with an uncontaminated reference site, and/or the abundance or diversity of the species within the Lake Champlain region. Since the site is probably of sufficient size for only one beaver colony, the presence of a successfully reproducing population was deemed the appropriate measurement endpoint to measure abundance for this species. Field observations and existing data on the region provided the measurements by which the specific assessment endpoints were evaluated. These data are presented in the following subsections, and a comparison of site conditions with assessment endpoints is conducted as part of the ecological risk assessment.

Regional Characteristics

Certain attributes of regional habitat types can provide valuable information on the ecological conditions occurring in relatively undisturbed ecosystems. This subsection describes some of the ecological characteristics typical of Lake Champlain wetland and aquatic habitats, with emphasis on the occurrence of various floral and faunal groups in these ecosystems.

Gruendling and Bogucki (1978) provide a general description of the flora typical of Lake Champlain wetlands. Forested wetlands are usually dominated by silver maple, green ash, and occasionally by swamp white oak and red maple, with sensitive fern, wood nettle, and various shrubs typical of the understory. Emergent vegetation typically consists of bulrushes, burreed, cattail, duck potato, and/or wildrice.

Lake Champlain, in general, and Burlington Bay (adjacent to PSC), in particular, support several species of fish that depend on wetlands for some aspect of their life history. Yellow perch is the dominant fish in shallow lake waters. Other species common in Burlington Bay include walleye, brown bullhead, pumpkinseed, bowfin, northern pike, chain pickerel, largemouth bass, smallmouth bass, landlocked (Atlantic) salmon, black crappie, common carp, longnose gar, and eastern mudminnow (Gruendling and Bogucki 1978; Vermont Department of Fish and Wildlife 1978). The lake sturgeon, listed as endangered by the State of Vermont, also inhabits the lake.

Wetlands bordering Lake Champlain support a variety of semiaquatic and terrestrial wildlife, particularly avian species, which use these areas during migratory periods. Many species of birds and mammals also use these wetlands for breeding activities. Major migratory species of waterfowl that utilize Lake Champlain wetlands include American black duck, mallard, green-winged teal, blue-winged teal, ring-necked duck, common goldeneye, scaup, canvasback, merganser, wood duck, and Canada goose. The four major breeding waterfowl species are American black duck, mallard, wood duck, and blue-winged teal. Mammalian species (furbearers) include wetland-dependent species such as muskrat, beaver, mink, raccoon, and river otter, in addition to upland species such as red fox, gray fox, coyote, and bobcat (Gruendling and Bogucki 1978).

Wetland and wildlife communities at PSC are generally similar to those of other regional wetland areas bordering Lake Champlain, although the urban nature of the land uses surrounding PSC, the relatively small size of the site, the history of on-site disturbance, and on-site contamination have resulted in some differences. These will be discussed at the conclusion of this section.

Reference Site Description

A number of studies have been conducted on the waters and wetlands of Malletts Creek (e.g., Myer and Gruendling 1979). This area was selected as a reference site for comparison with PSC because of similarities in various important physiographic and ecological parameters. PSC and Malletts Creek are of similar depth and general sediment type. However, the drainage area of Malletts Creek is approximately five times greater than that of the canal, and the emergent wetlands and forested areas are more extensive. In addition, surrounding land uses at the reference site are much less urbanized than at PSC. Even with these differences, Malletts Creek was the best available site, and did provide meaningful reference for many of the aquatic characteristics potentially affected by contamination at PSC. The following is a brief summary of some of the available ecological information on Malletts Creek and adjacent Malletts Bay.

Invertebrate populations in Malletts Bay are typical of deep water lakes of the formerly glaciated regions of North America (Myer and Gruendling 1979). These populations provide an example of a typical invertebrate assemblage (in terms of species composition and diversity) in an area of Lake Champlain unaffected by contamination from hazardous substances. The inner bay, adjacent to Malletts Creek, is dominated by nonbiting midges (Chironomidae), with aquatic worms (Oligochaeta) and bivalves (Sphaeriidae) also important taxa. The oligochaete fauna of the inner bay is dominated almost exclusively by immature capilliform species (Myer and Gruendling 1979). The density of invertebrate organisms is significantly higher in the shallow areas of Malletts Bay than in deeper areas, except where the substrate is bedrock.

Wetlands bordering Malletts Creek are characterized by high spatial interspersion of emergent vegetation, forested areas, and open water. Four streams, approximately 3.5 km in total length, traverse the wetland. Vegetation bordering the eastern portion of the creek is generally dominated by deciduous trees, while the western portion is dominated by emergent vegetation. Forested areas comprise about 30 percent of the drainage basin, and are located on the levees of the tributaries and adjacent to the upland habitats (Gruendling and Bogucki 1978). Approximately 25 percent of the drainage basin consists of dispersed fields and pastures located along the edge of the wetland, creating excellent interspersion among habitat types. Emergent vegetation comprises 35 percent of the total wetland acreage (Gruendling and Bogucki 1978). Typical emergent plants include reed canary grass, burreed, cattail, and horsetail. During growing seasons with high water levels, wildrice also occurs in the emergent zone. This species significantly enhances the value of the area to wildlife, as it is an important food source for many species, particularly dabbling ducks (Gruendling and Bogucki 1978).

Wildlife use of Malletts Creek is generally similar to wildlife use described for the Lake Champlain region. Waterfowl use of the wetland is quite high during all seasons, and evidence of muskrat and beaver use is abundant. Chain pickerel is the dominant fish in the wetland (Gruendling and Bogucki 1978). According to creel surveys conducted by the Vermont Department of Fish and Wildlife, the most abundant fish species caught by anglers in Malletts Bay are, in order of decreasing abundance, yellow perch, smallmouth bass, northern pike, largemouth bass, chain pickerel, brown bullhead, black crappie, rock bass, pumpkinseed, and landlocked (Atlantic) salmon (Vermont Department of Fish and Wildlife 1984).

Wetland, aquatic, and wildlife communities at PSC are generally similar to those of the reference site, although the urban nature of the land uses surrounding PSC, the relatively small size of the site, the history of on-site disturbance, and on-site contamination have resulted in some important differences, especially among the benthos. These will be discussed at the conclusion of this section.

PSC Wildlife Habitat

A qualitative description of the wetland and upland habitats present on-site was considered a sufficient level of detail, since no overt effects of contamination on terrestrial or wetland plants were observed or expected. This description consisted of delineating the wetland-upland boundary, conducting a wetland functional assessment, compiling species lists of the dominant vegetative species present, and conducting a search for plant species listed as threatened, endangered, or of special concern.

The PSC site contains a variety of wetland and upland cover types, which support a diversity of wildlife species (Figure 9-1; Table 9-2). Due to its small size, surrounding land uses (mostly urban) influence the site's use by wildlife, as does the proximity of Lake Champlain, especially for species with large home ranges. Additionally, a railroad running along the site's west side may serve as an important travel corridor to less-developed, off-site areas for some species of mammals. Parsons et al. (1988) rated the general wildlife habitat quality of the PSC site as high (rating of 87 out of 105).

Wetlands

The hydrology of the canal and adjacent wetlands is heavily influenced by the canal's connection to Lake Champlain, and, to a lesser extent, by the inflow from a small stream to the south and several culverts connected to the Burlington sewer/storm water system. During the spring thaw, rising lake levels and snowmelt runoff combine to submerge much of the wetland area adjacent to the canal; during this period, flow is

TABLE 9-2.   Pine Street Canal habitat type summary.

Cover Type[1]	Area (hectares)
Palustrine Open Water	2.5
Palustrine Emergent Wetland	3.0
Palustrine Scrub–shrub Wetland	1.5
Palustrine Forested Wetland	_1.5_
Total Wetland	8.5
Open Grasses/Forbs	3.9
Shrub–Sapling Upland	1.3
Forested Upland	_1.6_
Total Upland	_6.8_
Total	15.3

[1]Wetland cover types follow Cowardin et al. (1979). Upland cover types were created for this evaluation, since no standardized methodology was available. Areas apply only to the region delineated in Figure 9-1.

generally from the lake into the canal. Water levels are reduced in the summer and winter periods as lake levels recede and inflows from surface runoff become more important; during this period, flow is generally from the canal into the lake. A beaver dam on the southern portion of the site influences the site's hydrology only during low-flow periods, since flood waters overtop the dam. In most years, surface ice usually covers 100 percent of the on-site water and wetland area for several months each winter. However, during brief winter warm periods, some areas of open water appear in the shallower marsh areas.

The water in the canal is generally shallow, with depths of less than 0.3 m at the southern end, to as much as 3 m in the turning basin. Most of the canal is approximately 25 to 30 m wide. Drainage on the site is poor, due to the relatively flat topography. At least three drainage swales traverse the site and receive runoff from adjacent parking lots and roads. The stream that feeds the southern portion of the site also receives runoff from adjacent parking lots and roads. Groundwater on the site both discharges to and recharges from Lake Champlain on a seasonal basis; groundwater may recharge the lake during dry periods and be recharged by the lake during wet periods.

On-site wetlands were delineated during a late May 1990 site visit, in accordance with the three-parameter approach (hydrophytes, hydric soils, and hydrology) of the Federal Manual for Identifying and Delineating Jurisdictional Wetlands (FICWD 1989). General cover types were field-mapped in February 1990, and this map was field-checked during subsequent surveys. Wetland cover types followed the classification outlined in Cowardin et al. (1979).

Palustrine emergent wetland areas are dominated by cattail, and in drier areas, by common reed. Palustrine scrub-shrub wetland areas are dominated by silver maple, red maple, speckled alder, red-osier dogwood, and willows. Dominant species in the palustrine forested wetland areas are silver maple, green ash, and cottonwood, and standing dead trees are common, especially along the canal. Open water areas contain duckweed; other submergent and floating aquatic plants appear to be uncommon, as a 1988 site visit by U.S. EPA revealed that the open portion of the canal had minimal aquatic vegetation. Wetland areas occurring near the canal and containing woody vegetation are heavily influenced by beaver foraging activity.

The wetlands on the PSC site were evaluated using the U.S. Army Corps of Engineers Wetland Evaluation Technique, Volume II (WET II) model (Adamus et al. 1987). This technique evaluates various wetland functions and values in terms of social significance, effectiveness, and opportunity. The results of this analysis indicate that on-site wetlands are of high social significance and/or effectiveness for sediment/toxicant retention, nutrient removal/transformation, production export, and wildlife and aquatic diversity/abundance. This assessment defines current conditions, which can be used as a baseline to help determine the effects of remediation activities.

Uplands

Upland forested or shrubby areas contain cottonwood, gray birch, quaking aspen, black locust, green ash, box elder, red maple, American elm, and staghorn sumac. Honeysuckle is abundant in some locations. The most mature forested areas (trees 25 to 30 m tall) occur near Maltex Pond. Trees along Pine Street and in other on-site areas are less mature, generally being less than 18 m tall. Open herbaceous areas contain a variety of grasses and annuals, including chicory, Queen Anne's Lace, black-eyed susan, clover, goldenrod, and common mullein.

In 1989, a search was conducted for Canada buffalo-berry and border meadow rue, species classified as uncommon (watch list) by the Vermont Natural Heritage Program, whose known range encompasses the PSC site. Neither species was identified on the site. Previous surveys conducted in conjunction with the I-89 Environmental Impact Statement also found no indications of unique or rare flora on the site (U.S. FHA 1977). The Fragile Area Map of Chittenden County identified no areas of special or unusual ecological interest in the site vicinity (U.S. FHA 1977). No rare, threatened, or endangered species were identified on the site by Parsons et al. (1988), and the site was not considered to contain any unique ecological resources.

PSC Wildlife Occurrence

Wildlife surveys were conducted at the PSC site during 1989-1991, and consisted of either systematic surveys, in which the entire site was surveyed for all wildlife species present (three site visits: May 1989, February 1990, and June 1990), or incidental observations where wildlife were noted concurrent with other site activities (four site visits: May 1990 (2), August 1990, and April 1991). During systematic surveys, the site was systematically walked and all observations of birds, mammals, reptiles, and amphibians, or their sign, was noted. Two avian censuses were also conducted on successive mornings during a June 1990 site visit. Small mammal trapping grids (snap traps) were also run in February and June, 1990.

As the study progressed, it became apparent that a wetland-associated mammalian species (muskrat) was potentially incurring adverse effects from site contamination or other unidentified factors, and that the information was insufficient to evaluate potential impacts to another species (beaver). Thus, several special surveys were conducted to determine the breeding status of the on-site beaver colony (June 1991) and to estimate the relative abundance of muskrat on the site (January 1991, March 1991, and June 1991).

The following subsubsections describe the results of the field surveys conducted to document wildlife occurrence at the PSC site. Information is subdivided by major taxonomic group (birds, mammals, and reptiles and amphibians) and season.

Birds

Field observations of avian species are summarized in Table 9-3. This table also identifies those species for which there is evidence of on-site breeding.

TABLE 9-3.   Wildlife species observed—Pine Street Canal.

Common Name[1]	Season Observed[2]		
	W	Sp	Su
Birds			
Double-crested cormorant		X	X
American black duck	X		X
Mallard*	X	X	X
Wood duck		X	
Herring gull	X	X	X
Ring-billed gull	X	X	X
Great black-backed gull	X		
Great blue heron		X	X
Black-crowned night heron		X	X
Green-backed heron*		X	X
Killdeer		X	X
Spotted sandpiper			X
Red-tailed hawk			X
American kestrel*		X	
Rock dove (Domestic pigeon)	X	X	X
Mourning dove		X	X
Belted kingfisher	X	X	X
Pileated woodpecker	s		
Northern flicker*		X	X
Downy woodpecker*	X	X	X
Eastern kingbird		X	X
Eastern phoebe		X	X
Oliver-sided flycatcher			X
Empidonax spp.			X
Purple martin			X
Barn swallow		X	X
Tree swallow*		X	X
Northern rough-winged swallow			X
Chimney swift		X	X
American crow	X	X	X
Blue jay	X	X	X
Black-capped chickadee*	X		X
White-breasted nuthatch	X		X
Gray catbird*	s	X	X
Northern mockingbird	X	X	X
American robin*		X	X
Cedar waxwing			X
Yellow warbler*	s	X	X
Chestnut-sided warbler			X
Common yellowthroat*		X	X
Red-winged blackbird*		X	X
Brown-headed cowbird		X	
Common grackle*		X	X
European starling	X	X	X
House sparrow*		X	X
Northern cardinal	X	X	X
House finch	X	X	X
Purple finch		X	X
American goldfinch	X	X	X
American tree sparrow	X		
Song sparrow		X	X

TABLE 9-3. (*continued*)

Common Name[1]	Season Observed[2]		
	W	Sp	Su
Mammals			
Shrew spp.	s		
Bat spp.[3]		X	
Raccoon	X		s
Weasel spp.	s		
Mink	s	s?	s
River otter		s?	
Striped skunk			X
Red fox	X	X	s
Domestic dog	s	X	s
Domestic cat	s		X
Woodchuck*[3]		X	X
Porcupine	s		
Beaver*	X	X	X
Peromyscus mice*	X		X
Vole spp.	s		
Muskrat	s	X	X
Eastern cottontail	s		
Reptiles and Amphibians			
Salamander spp.[3]		X	
Painted turtle[3]		X	X
Common snapping turtle[3]		X	
Turtle spp.[3]		X	X
Bullfrog*[3]			X
Northern leopard frog[3]			X
Frog spp.*[3]		X	X
Eastern garter snake[3]		X	X
Northern ribbon snake[3]		X	X

*Evidence of on-site breeding.
[1]See Appendix 1 for scientific names.
[2]W = Winter (21–22 February 1990; 15 January 1991; 20 March 1991); Sp = Spring (23 May 1989, 3–4 and 21 23 May 1990; 18–19 April 1991); Su = Summer (13–15 June, 6–9 August 1990; 12–13 June 1991). X = observed visually/aurally; s = observed by sign only; ? = observation uncertain.
[3]Hibernates in winter.

Winter. The PSC site provides good quality habitat for several avian groups that remain in northern Vermont during the winter months. When some open water is available, the canal and emergent wetland areas provide resting areas that are sheltered from the strong northwesterly winds off of Lake Champlain for wintering ducks, primarily mallards and American black ducks. Although several species of gulls winter in the area, their use of the site is probably minimal.

Dead trees on the site provide foraging and roosting areas for wintering species such as woodpeckers and chickadees. Although shrub species that bear fruits into the winter period are not abundant on the site, some fruits and seeds are available for wintering finches, mockingbirds, and other species. The dense shrubby areas may also serve as winter cover for some passerines.

Spring/Summer.   The PSC site provides good quality breeding and foraging habitat for a variety of bird species. While the small size of the site precludes it from supporting large numbers of breeding birds of any particular species, the site does provide breeding habitat which is locally rare, such as emergent wetlands, due to the urban nature of the surrounding area. Upland, ground-nesting birds are unlikely to breed on the site, however, due to high predation rates from domestic animals and other predators.

The canal and emergent wetlands provide nesting and brood-rearing habitat for American black ducks and mallards; a brood of six mallards was observed during wildlife surveys. This area, and the forested wetlands on-site, may also provide breeding and brood-rearing habitat for wood ducks; dead trees with suitably large cavities for nest sites were observed along the western edge of the canal. The large tree cavities in this area may also be used for nesting by American kestrels; although raptors were rarely observed on-site, a possible on-site nest of this species was reported. Parsons et al. (1988) rated the waterfowl habitat quality as moderate on the PSC site (rating of 45 for ground-nesters and 52 for cavity-nesters out of 105).

Wetland and open water areas also provide sheltered areas for aquatic foragers such as gulls, cormorants, kingfishers, and herons. The canal shoreline, in particular, is littered with downed trees and rocks, providing resting and foraging locations for these aquatic birds, as well as for aquatic and semiaquatic mammals, fish, reptiles, and amphibians. The surrounding wooded areas on-site may also provide potential nesting habitat for solitary nesting herons, such as the green-backed heron or the black-crowned night heron. Open water areas also provide foraging habitat for aerial insectivores, such as swallows. Shorebird use of the area is probably minimal and limited to upland foraging species such as killdeer.

The palustrine emergent wetlands provide nesting areas for passerines such as red-winged blackbirds, which may be the most abundant birds present on-site during the breeding season. The dense shrubby areas near these emergent wetlands provide good quality nesting habitat for such species as yellow warblers, common yellowthroats, and gray catbirds.

Upland wooded areas provide breeding habitat for common species such as common grackles, northern cardinals, American robins, blue jays, American goldfinches, and American crows. Upland herbaceous areas are important foraging areas for granivores such as finches, as well as insectivores such as swallows, flycatchers, and some warblers.

Specialized habitat features existing on-site include many dead trees along the canal, which provide suitable nesting and feeding substrates for primary excavating cavity-nesting species, such as downy woodpeckers and northern flickers. Cavities created and abandoned by these species may be used by other species (secondary cavity-nesters) for nesting and shelter. Examples include tree swallows and European starlings. These dead trees, and power lines running adjacent to the site by the railroad tracks, serve as roosting and perching locations for many bird species, particularly swallows and blackbirds. Spoil piles near Pine Street contain steep dirt banks, which may provide suitable locations for bank nesters such as northern rough-winged swallows and belted kingfishers. Buildings surrounding the site provide nesting locations for bird species associated with structures, such as barn swallows and house sparrows. These species, while nesting off-site, likely use the site to forage.

During spring and fall migrations, bird species that are not permanent residents of the region may utilize the site temporarily to forage or rest before continuing their migration. Although observations were not conducted in the fall, no species that were obviously migrants (i.e., northern Vermont was outside of their breeding range) were observed during spring surveys. This may have been due to the limited sampling effort, however.

In summary, the avifauna observed at the PSC site was generally typical of that expected based upon geographical location, habitat present, and surrounding land uses. No overt effects of site contamination were observed.

Mammals
Field observations of mammals, or their sign, are summarized in Table 9-3.

Winter. The winter mammalian fauna inhabiting the PSC site is generally typical of a small wild area in an urban landscape. Species present are, at a minimum, tolerant of some urban activities and may actively use the surrounding areas to meet certain of their life requisite needs, such as foraging or breeding sites.

As of June 1991, the PSC site supported only a single colony of beaver, whose dam, numerous lodges, and foraging activities make them the most visible mammalian residents of the site. Shoreline woody vegetation is heavily impacted by beaver foraging activity in the southern portion of the site, and less so in the central and northern portions of the site near the canal. As the study progressed, observed foraging activity was reduced in the southern portion of the site and increased in the central and northern portions. During the last site visit (June 1991), fresh foraging sign was solely confined to the northern portion of the site. This included the wooded area near the turning basin, which contains a high density of preferred shrub species, the area of forested wetland on the western edge of the canal, and within the fenced area north of the turning basin. Since beaver interact dynamically with their environment, these existing habitat use patterns are likely to change as the animals breed, feed, and continue to use the area.

The beaver dam appears to have a minimal influence on the hydrology of the site, especially during high flow conditions. Highly variable water levels in the canal, due to rising and falling Lake Champlain water levels and storm water inflow, could explain the large number of lodges observed on the site. As these lodges either flood (as water levels rise) or become high and dry (as water levels recede), the beaver may shift their activities to another lodge. Winter activity was observed during 1990 at the lodge on the southern input stream (see Figure 9-1); this was also the only area with open water during the February 1990 site visit. During the winter of 1991, another lodge about midway up the canal and on the eastern shore was active. This northward movement up the canal in lodge utilization is consistent with the pattern observed for foraging areas.

Only minimal muskrat sign was observed during the winter period, although the habitat appeared to be suitable. Evidence of muskrat use was confined to several scat piles observed during Winter 1990; the remains of what appeared to be winter (1991) feeding lodges were found in two locations during the June 1991 survey. These lodges were not observed during the brief winter survey conducted in January 1991, although the ice conditions near these locations were unsafe during that winter survey and the area could not be thoroughly searched. Active muskrat lodges were observed on the Malletts Creek reference site, however, during the January 1991 survey.

A variety of small mammals that are active in winter also use the site. These include mice, voles, shrews and, although not observed, probably moles as well. Winter small mammal trapping only yielded one *Peromyscus* mouse (either a deer or white-footed mouse), although conditions were poor for trapping. Numerous small runs in the snow, created by voles, shrews, and mice, were observed, however. Small mammal species, in addition to *Peromyscus* mice, which may possibly use the site, include the masked shrew, the northern short-tailed shrew, the hairy-tailed and star-nosed moles, the meadow vole, and the southern red-backed vole (DeGraaf and Rudis 1983).

Although not observed, eastern gray squirrels probably use the site to some degree. This species is adaptable to urban landscapes (Williamson 1983), but the scarcity of mast- (e.g., acorns) producing trees reduces the likelihood of winter use. The eastern cottontail rabbit is another terrestrial herbivore that can adapt, to some degree, to man-dominated landscapes. Rabbit tracks and browse sign were observed in the dense shrubby area near the turning basin. This area provides excellent cover and forage for this species. Porcupine feeding sign and tracks were also observed in this area during the winter period.

Two terrestrial omnivores adaptable to urban areas, the raccoon and the striped skunk, use the site during the winter period. Raccoon tracks were observed during the February 1990 survey and two partially decomposed raccoon carcasses were found (together) in the canal during the March 1991 site visit. Signs of skunk were not observed during the winter period although a dead skunk was found on-site in the summer. A third terrestrial omnivore, the red fox, was observed on-site. This species is less likely to utilize habitats near urban areas, but signs (scats and tracks) were commonly observed. These fox probably use the railroad tracks as a travel corridor to off-site areas, and use the PSC site to forage. Red fox are subordinate to coyotes in territorial encounters (Major and Sherburne 1987), and high coyote densities south of Burlington (Person and Hirth 1991) may push these fox into suboptimal areas such as the PSC site. The two dead fox discovered on-site may indicate that these animals are being forced into less favorable habitat, where mortality rates would likely be higher relative to more optimal habitat.

Tracks of a number of semiaquatic carnivorous mammals were observed on-site. Although snow conditions were generally poor, several clearly-defined tracks of mink and weasel (probably ermine or long-tailed weasel) were found on a thin layer of snow atop the ice on the canal. River otter is another species that may utilize the site during the winter period. Based upon home range size and/or the relative scarcity of observations, these species are likely to spend significant time off-site, travelling to and from the site via the canal's connection with Burlington Bay.

Being surrounded by an urban landscape, the site is also utilized by domestic dogs and cats. During the February 1990 survey, tracks of both of these species were observed, with dog tracks the more common. Domestic animals can affect wildlife populations in these situations, as these species are predators and can kill, injure, or stress susceptible wildlife.

Spring/Summer. The mammalian fauna utilizing the site during the spring and summer seasons is generally similar to that described for the winter period, with a few exceptions. Woodchucks, which hibernate on-site during the winter period, were commonly observed on the PSC site in spring and summer, and likely breed there in fair numbers. Freshly dug burrows were observed in all upland areas in the spring and summer, and young were observed on one occasion. Two carcasses (the cause of death was unknown) were found during the April 1991 site visit in separate locations within the tank farm area north of the turning basin. Woodchuck burrows, when abandoned, may be used by other species, such as red fox and several species of snakes, for reproduction or shelter.

Another species that hibernates during the winter, the eastern chipmunk, is also likely to use the site in the spring and summer, although this species was not observed. Bats, which hibernate or migrate during the winter, also are likely to use the site. A dead bat (species unknown due to decomposition) was found on-site. The two species most likely to use this area are the little brown myotis and the big brown bat; these species would likely use the open water, emergent wetland, and field habitats to forage for insects.

A total of seven *Peromyscus* mice were caught during June 1990 trapping. Most were caught in shrubby areas or woody/herbaceous ecotones, particularly in the area around Maltex Pond. The presence of several immature mice, as well as a lactating female, in the sample indicate successful breeding activity. All mice appeared healthy, with no evidence of any external morphological abnormalities (e.g., skin lesions).

Although beavers were observed during every site visit from May 1989 through April 1991, no more than two beavers were ever observed concurrently during these surveys. Thus, these data were insufficient to determine whether or not the on-site beaver colony was successfully reproducing; lack of successful reproduction, if confirmed, could possibly be related to site contamination. To correct this deficiency, a special survey was conducted on PSC in June 1991. Based on concurrent observations from two teams of observers during the June 1991 survey, the minimum 1991 beaver colony size on the PSC site was five individuals, which likely consisted of the resident pair and three subadults. Young-of-the-year, which are easily distinguished (by size) early in the breeding season from adult and subadult beaver, were not observed outside of the lodge nor were young heard within the active lodge (probably because the lodge walls were thick and it was windy). Beaver in northern areas generally breed in January–March, and the gestation period is approximately 105 to 107 days (Hill 1982). Assuming the midpoint of the breeding period (15 February), young would be born in the beginning of June and would have been less than two weeks old at the time of the survey. Although born fully furred and with their eyes open, young beaver would be expected to spend very little time outside of the lodge at this young age. Thus, it is not surprising that young-of-the-year were not observed at this time. The presence of the subadults, however, confirms that successful reproduction occurred during 1990. It should be noted that the active lodge emanated a petroleum-like odor, probably from PAHs and BTX volatilizing from the mud used to construct the lodge.

As indicated earlier, the beaver colony has consistently moved northward in the canal system over the course of the study. This movement is likely due to the effects of changing water levels and the availability of forage. The Winter 1991 active lodge, for example, was partly out of the water in June, due to reduced water levels in the canal, and one of the entrances into the lodge was above the existing water level. This is the likely reason the lodge was abandoned. The currently-active lodge (as of June 1991) is at the edge of the canal and turning basin, which is about as far north as suitable habitat exists for lodge building on the site.

Since only minimal muskrat sign was observed on PSC during the May 1989, February 1990, and June 1990 wildlife surveys, and no live muskrats were observed during these site visits or during a January 1991 special survey, additional site visits (in March and June 1991, plus observations during other site work in April 1991) were scheduled in an attempt to determine the status of this species on the PSC site. This was deemed important, since the most suitable on-site habitat for muskrat, the cattail marsh on the southern end of the site, was among the most contaminated areas, and because muskrats will often construct bank burrows, exposing them to contaminated soils. The lack of muskrat sightings in what appeared to be structurally suitable habitat suggested a possible effect of site contamination that needed further investigation.

Despite this additional effort, only small numbers of muskrats were observed on-site, including single sightings during the March and April 1991 site visits, and visual observations of two, possibly three, individuals during the June 1991 survey. All of these sightings occurred within or near the cattail marsh on the southern portion of the site, which, in addition to being the most suitable habitat present on-site, also has some of the highest observed soil contamination levels, as mentioned earlier. Tracks were also observed in the mud within the nearly dry areas of the marsh on the eastern side of the canal during June 1991. Several bank burrows, none apparently active, were also found along the canal edge. The only active bank burrow observed on-site was found in March 1991 on the bank of the small stream on the extreme southern end of the site. No breeding lodges were observed on-site. While these additional surveys did document muskrat use of the PSC site, observed numbers were much lower than expected based upon the quality and extent of the habitat present.

Several possible hypotheses can be advanced to explain the lower-than-expected use of the site by muskrat: (1) muskrat populations may be reduced in the Champlain Valley due to a cyclic decline, (2) water levels on-site may be too variable, as this species prefers stable water levels, or (3) contaminants in sediments and wetland soils are having deleterious effects on this species. Muskrat populations do not appear to be at a cyclic low (Myers 1991) and muskrats were commonly observed at the Malletts Creek reference site. Parsons et al. (1988) reported an unspecified amount of muskrat use of the PSC site, and rated the suitability of the existing habitat as fair to good (score of 30 on a scale of 50). The possible effects of site contaminants on this species are addressed in the ecological risk assessment section.

Tracks and/or scats of raccoon, striped skunk, red fox, and domestic dogs and cats were commonly observed on-site during the spring and summer periods. Mink sign was consistently observed as well, indicating that this species is often present on-site. Possible feeding sign of river otter was found near Maltex Pond. As discussed earlier, carcasses of striped skunk, raccoon, and red fox were found on the PSC site. The skunk may have been hit by a car on Pine Street and crawled onto the site, based upon the type and extent of its injuries; cause of death of the raccoons and foxes was not apparent due to decomposition.

Reptiles and Amphibians

Field observations of reptiles and amphibians are summarized in Table 9-3.

Winter.   Reptiles and amphibians, being cold-blooded, hibernate during the winter months, and thus were not observed during winter surveys. Salamanders, newts, and frogs may hibernate in aquatic (bottom mud) or terrestrial (typically under litter) habitats, depending upon species. Turtles typically winter under mud in ponds or other water bodies. Snakes generally hibernate on land, usually in the burrows of other species or in rock or slash piles (DeGraaf and Rudis 1983).

Spring/Summer.   Relative to birds and mammals, most reptiles and amphibians are much more difficult to observe. The most visible and numerous group observed on-site was frogs. Although only two species were identified, bullfrog and northern leopard frog, other species that are likely to be found on-site include eastern American toad, northern spring peeper, green frog, wood frog, and pickerel frog (DeGraaf and

Rudis 1983). Some tadpoles were caught during fish electroshocking and dip-net sampling, indicating that some species are actively (and successfully) reproducing on-site. The tadpoles caught with dip-nets were probably bullfrogs, based upon size.

Most frogs require permanent, relatively shallow and stagnant water to breed; Maltex Pond appears to provided excellent breeding habitat for this group. A large number of the observations of frogs were at Maltex Pond. The canal and associated wetlands are also excellent habitat for immature forms, but predation rates on tadpoles by fish and birds are probably quite significant in these areas. Bullfrogs were also heard calling from the deeper water in the marsh by the railroad tracks.

Painted turtles were often observed basking on logs in Maltex Pond or on rocks along the edge of the turning basin. A single common snapping turtle was also observed on-site in the spring. Another species common to the Lake Champlain area, but not observed, is the map turtle (DeGraaf and Rudis 1983). A dead turtle (species and cause of death unknown) was found on the eastern bank of the canal during aquatic surveys.

Newts and salamanders are common to aquatic habitats such as those that exist on the PSC site. Salamanders (species unknown) were abundant in Maltex Pond during late spring/early summer. Species likely to be present on-site include northern dusky salamander, spotted salamander, northern two-lined salamander, and red-spotted newt (DeGraaf and Rudis 1983).

Two species of snakes were observed on-site. Eastern garter snakes were common, and a single northern ribbon snake was also observed. Other species common to the area, which may be present at the site, include northern water snake, northern brown snake, eastern smooth green snake, northern ringneck snake, and eastern milk snake (DeGraaf and Rudis 1983).

In summary, while data on the amphibians and reptiles present at the PSC site are limited, at least one group (frogs) is actively and successfully reproducing on-site. This is important, as immature forms (tadpoles) will overwinter for one or more seasons in sediments prior to becoming adults (DeGraaf and Rudis 1983), and the canal sediments and adjacent wetland soils are among the most contaminated media present on-site. No overt effects of site contamination on frogs, or on other reptiles and amphibians, were observed however.

PSC Aquatic Community

The aquatic environment was surveyed at the PSC site and the Malletts Creek reference site in order to document the existing ecological conditions at PSC, and to determine if there were any apparent impacts from the on-site contamination, relative to the reference site. Segments of the aquatic community examined included benthos, fish, and zooplankton. Water and sediment quality measurements were also taken during the aquatic surveys.

Benthic

The benthic segment of the aquatic community (living in or on the sediments or other submerged substrate) of any water body is generally indicative of the long-term environmental health of these ecosystems. The benthos is primarily made up of sessile organisms, which must generally be able to tolerate the range of sediment and water quality conditions on-site, as they cannot easily escape stressful conditions. Thus, they are generally good indicators of the typical conditions at a particular sampling point.

The benthos of the PSC and reference sites were surveyed by taking quantitative grab and dip-net samples; qualitative dip-net sampling was also conducted. In May 1990, a reconnaissance survey was conducted at each site in order to determine sampling locations for full-scale sampling in June and August 1990. Based on the results of the reconnaissance survey, two benthic grab stations were established at the reference site (RF-1 and RF-2), and three stations were established at the PSC site (PS-1, PS-2, and PS-3; Figure 9-1) for June and August benthic grab sampling. During each survey, three replicate grabs were taken from a boat at each station using a 930 square cm (144 square inch) Ekman grab sampler. All samples were sieved through a No. 30 (500 μm mesh) sieve bucket, transferred in to labelled polypropylene bottles, and preserved in an ethyl alcohol solution for later identification. Sediment samples were also collected in June 1990 at benthic grab stations and analyzed for volatile organics, semivolatile organics, total organic carbon, and grain size.

Quantitative dip-net sampling was also conducted, in June and August 1990, at locations that could not be readily sampled by the Ekman dredge (i.e., they were inaccessible by boat and/or the substrate was not suitable). This sampling occurred at two locations in the canal (PS-4 and PS-5; Figure 9-1) and at one location at the reference site (RF-3). Quantitative dip-net sampling consisted of netting for 10 minutes over 9 m of stream. This effort was repeated three times at each station. Samples were processed as described for grab samples.

Qualitative dip-net sampling was also conducted along the eastern shore of the canal and at selected portions of the reference site during May, June and August 1990. During these efforts, the dip net was swept along the sediment and vegetative surfaces and all sampled organisms were field-identified to the lowest possible taxonomic level.

Physical/Chemical Environment.   The physical and chemical conditions of the sediments vary within the Pine Street Canal. Sediment types range from a thick tar substance to silty sand. Based on the results of the May reconnaissance survey, where individual grab samples were taken at the center of transects spaced throughout the canal, the distribution of these sediment types is generally patchy throughout the canal with no consistent trends, although sediments in the turning basin appear to be composed mainly of silt, while more sand was observed further to the south. Total organic carbon content, however, was higher in the southern portion of the canal (28 percent at PS-3) than near the turning basin (about 5 percent at PS-1).

Species Composition.   The benthos is generally composed of two distinct groups (Smith 1980). One group (the infauna) inhabits the sediments, and is thus directly influenced by the physical and chemical properties of the sediments. The second group (the epifauna) consists of those species more closely associated with the sediment/water interface and the surfaces of submergent or emergent vegetation. These species generally have more direct contact with the water than with the sediments.

Although the density of organisms varies throughout the canal, the benthic infauna inhabiting the sediments is consistently dominated by tubicifid oligochaetes, as indicated by the results of benthic grab sampling (Tables 9-4 and 9-5). Tubicifids comprised 50 to 96 percent of the total benthic infauna in the canal and turning basin during the May reconnaissance survey. Results of the June and August benthic grab sampling showed similar trends, although station PS-1 (near the turning basin) had relatively low (8 to 17) percentages of these organisms. Tubicifids are considered to be opportunistic species, highly tolerant of stressed or organically enriched conditions (Aston

TABLE 9-4.   Summary of mean density of benthic organisms from benthic grab sampling, June and August 1990.

| | Mean Density (Per Square Meter) | | | | | | | | | |
| | June | | | | | August | | | | |
Taxon	RF-1	RF-2	PS-1	PS-2	PS-3	RF-1	RF-2	PS-1	PS-2	PS-3
Insects	1.0	11.3	1.6	4.3	3.7	38.3	10.8	1.2	1.1	0.2
Crustaceans	0.0	0.2	0.2	0.1	13.0	2.0	1.6	5.5	5.6	0.2
Oligochaetes	3.3	0.6	0.5	71.6	27.8	1.1	0.7	0.6	6.1	3.8
Gastropods	0.1	0.2	0.0	0.0	0.3	0.4	0.1	0.0	0.0	0.0
Bivalves	3.3	1.9	0.0	0.0	2.2	4.3	1.7	0.0	0.1	0.0
Other	0.5	1.8	0.0	0.5	0.3	0.7	1.0	0.1	0.0	0.3
All Organisms	8.2	16.0	2.3	76.5	47.3	46.8	15.9	7.4	12.9	4.5

1973). When found in relatively high numbers, they are considered to be indicative of stressed conditions. They are adapted to burrowing in soft sediments, and derive most of their nutrition from bacteria. They ingest large volumes of sediments continuously in order to extract the small fraction of nutrient material contained within. Tubicifids are very tolerant of oxygen depletion and organic enrichment (Aston 1973; Brinkhurst and Cook 1974). Tubicifidae were found at the reference site in much lower relative numbers, comprising 2 to 45 percent of the total benthic infauna. It is likely that the high numbers of tubicifids at PSC are directly related to the contaminants in the sediments, since other parameters (such as dissolved oxygen and nutrient concentrations) were similar between the two sites. Maltex Pond, which was remediated in 1985, had only 0.2 percent of its benthic infauna comprised of tubificid oligochaetes during the May reconnaissance survey.

The invertebrates more closely associated with the water column (at the sediment surface and on vegetation) were generally more diverse than those associated with the sediments. Crustaceans (Cladocera, *Simocephalus serrulatus*) and gastropods (Plan-

TABLE 9-5.   Summary of density of benthic organisms from benthic grab sampling, May 1990.

	Density (Per Square Meter)								
	Reference Site			Pine Street Canal					
					Canal				
Taxon	RF-1	RF-2*	RF-3	Turning Basin	N	C	S	Emergent Wetland	Maltex Pond
Insects	25.5	14.9	4.6	0.0	3.5	2.7	7.2	16.4	2.8
Crustaceans	6.0	34.2	0.3	0.0	0.0	0.0	1.2	0.7	37.3
Oligochaetes	15.7	19.5	18.2	1.6	4.0	77.8	169.2	227.7	0.1
Gastropods	2.0	11.2	7.9	0.1	0.0	0.3	8.6	21.4	12.7
Bivalves	4.6	10.8	0.4	0.0	0.0	0.0	5.1	6.5	2.0
Other	0.4	1.2	0.2	0.0	0.0	0.2	8.2	0.7	0.0
All Organisms	54.2	91.8	31.6	1.7	7.5	81.0	199.5	273.4	54.9

*Average of two samples.

orbidae and Physidae) were the most abundant epifaunal taxa at PSC in June. In August, gastropods (Planorbidae and Viviparidae) were abundant in the canal. At the reference site, crustaceans (Isopoda, *Caecidotea* spp.) and bivalves (Sphaeriidae) were the most abundant taxa in June, while in August, crustaceans (Amphipoda, *Hyalella azteca*) and insects (Hemiptera, *Neoplea* spp.) were abundant. Odonates (insects) were also common during all sampling periods. Table 9-6 summarizes the species composition, by major taxonomic group, observed during quantitative dip-net surveys.

In general, PSC had a higher percentage of gastropods and a lower percentage of crustaceans than the reference site. The lower relative abundance of crustaceans at PSC is likely due to the fact that crustaceans are considered to be pollution sensitive organisms (Hart and Fuller 1974). It is not clear if the higher relative abundance of gastropods at PSC is related to site contamination. There is some limited evidence that gastropods are pollution tolerant. However, it is also known that the physical substrate type influences gastropod distribution (Hart and Fuller 1974). It is likely that differences in the physical substrate type between the PSC and reference sites affected the relative abundance of gastropods.

Greater percentages of sensitive species, such as amphipods and odonates (Hart and Fuller 1974), occurred at the reference site than at PSC, while pollution-tolerant species, such as oligochaetes, were more abundant at PSC. Invertebrate assemblages associated with the water column in the canal do, however, appear to be healthier than those living in the sediments.

Some spatial differences were revealed in the canal as a result of qualitative dip-net sampling. From the southern end of the canal to the filled barge slip, isopods dominate, with amphipods, damsel fly larvae, gastropods, molluscs, and chironomids also occurring. Adjacent to the filled barge slip, there is a scarcity of any benthos. From the barge slip to the turning basin, the benthic segment of the aquatic community is similar to that occurring in the southern portion of the canal. The scarcity of benthos near the barge slip, as indicated from qualitative dip-net sampling, suggests that the pure coal tar product at the sediment surface in this location is adversely affecting the epifaunal benthos (present at the vegetation-water or sediment-water interfaces), as well as the infaunal benthos (see following). North and south of the barge slip, qualitative dip-net sampling indicates a relatively healthy benthic epifauna along the canal shoreline. Dip-net sampling on the turning basin shoreline was not conducted.

The benthic segment of the community at Maltex Pond, which is isolated from the canal itself, is very healthy and diverse. Invertebrate densities were high, and the most

TABLE 9-6.   **Summary of percent occurrence of major taxonomic groups from quantitative dip-net sampling.**

| | Mean Percent Occurrence in Samples | | | | | |
| | June | | | August | | |
Taxon	PS-4	PS-5	RF-3	PS-4	PS-5	RF-3
Insects	24	23	18	40	24	50
Crustaceans	36	25	50	15	6	48
Oligochaetes	17	6	3	5	4	1
Gastropods	16	38	12	27	41	1
Bivalves	4	1	14	5	0.5	0.3
Other	2	6	3	7	24	0

TABLE 9-7.   Summary of benthic community indices from benthic grab sampling.

	June			August		
Station	Number of Taxa	Shannon-Wiener Diversity	Simpson's Dominance	Number of Taxa	Shannon-Wiener Diversity	Simpson's Dominance
RF-1	12	1.60	0.28	32	1.40	0.46
RF-2	17	1.61	0.29	23	1.66	0.33
PS-1	13	2.08	0.20	7	1.49	0.31
PS-2	9	0.10	0.97	10	0.38	0.87
PS-3	25	1.27	0.43	8	0.63	0.75

commonly occurring groups were amphipods, chironomids, culex (mosquito) larvae, and gastropods.

Community Indices.   A summary of several ecological community indices of the benthic infauna is presented in Table 9-7. Diversity and dominance varied considerably at PSC for the benthic segment of the aquatic community. Diversity was highest near the turning basin (PS-1), however, the density of organisms was relatively low. Diversity was lowest near the filled barge slip (towards the center of the canal; PS-2) and densities there were highest. At the reference site, diversities were in the same range or higher than at PSC. The variability in diversity, as well as densities, at PSC is likely due to the interspersion of substrate types, which ranged from silty mud to thick coal tar product. It should be noted that, in areas of the canal where pure coal tar occurred, conditions were abiotic (no living benthic infauna). Samples from these areas could not, however, be processed, since the material could not be sieved and these data are not reported in the tables of this chapter. Areas of pure coal tar were common in the vicinity of the filled barge slip.

Pelagic

The pelagic segment of the aquatic community refers to those species inhabiting the water column, and includes zooplankton and fish. This community segment is generally more influenced by the conditions in the water column, as opposed to those in the sediments. In addition, fish move in and out of the canal seasonally, and are also capable of leaving if conditions become unsuitable (e.g., low dissolved oxygen concentrations). The water quality conditions at PSC are outlined in this subsubsection, followed by a characterization of the zooplankton and fish assemblages that inhabit the water column.

The fish assemblage at the PSC and reference sites was sampled in June and August 1990. In order to obtain representative samples, electroshocking and gill netting were conducted. During the August survey, 22 fish from the reference site and 37 fish and one tadpole from the PSC site were preserved with dry ice and shipped to the laboratory for tissue analysis. Whole body or muscle tissue (fillet) samples were analyzed for individual PAH compounds, but not for PAH metabolites or BTX.

Zooplankton samples were taken at two stations at the reference site (RF-1 and RF-2) and at three stations at the PSC site (PS-1, PS-2, and PS-3) during June and August 1990 surveys. Samples were collected by using a submersible pump to filter 40 liters of water through a No. 20 (76 μm mesh) plankton net. Each concentrated sample was preserved with formalin and sent to the laboratory for enumeration and identification to the lowest possible taxon.

A Hydrolab was used to measure standard water quality parameters (temperature, pH, and dissolved oxygen). During June and August surveys, water samples were taken and analyzed for volatile organics, semivolatile organics, total phosphorus, orthophosphate, nitrate-nitrite, and ammonia.

Water Quality.   Water quality at PSC was relatively good, with levels of volatile and semivolatile contaminants rarely exceeding detection limits. In terms of standard water quality parameters and nutrients, the water quality in the canal is similar to that of the reference site for those parameters measured. The waters of the reference site are considered to be clean and generally nonstressed, as there are no known sources of contaminants entering Malletts Creek and the surrounding land use is mostly agricultural or undeveloped.

Concentrations of nitrate/nitrite, orthophosphate, and total phosphorous at the PSC and reference sites were generally higher in August than in June. Ammonia levels were highest in the southern portion of the canal, relative to the northern portion of the canal and the reference site, in both June and August. Dissolved oxygen concentrations (at a depth of 1 m) were consistently lower at the reference site (range 0.6 to 2.3 mg/1) than at PSC (range 3.5 to 11.5 mg/1). This is likely due to the high productivity of the dense submergent vegetation at the reference site; submergent aquatic vegetation was uncommon at PSC. The dissolved oxygen concentration at the reference site was especially low in August (0.6 to 0.9 mg/1), probably due to a lack of flushing.

Zooplankton.   A total of 58 zooplankton taxa were identified at PSC during the two aquatic surveys. Fifty-four taxa were identified at the reference site. Twenty-six taxa were common to both sites. The most abundant taxa at both sites were rotifers and copepods.

Densities ranged from 130 to 222 organisms per liter in June, and from 8 to 34 organisms per liter in August at PSC. At the reference site, densities ranged from 6 to 15 organisms per liter in June and 27 to 62 organisms per liter in August. The difference in densities of zooplankton between the two sites is likely due to the larger drainage area at the reference site, which results in greater flushing and a more riverine type of environment, conditions that are less suitable for supporting a self-sustaining zooplankton assemblage. In addition, the canal is more influenced by Lake Champlain, due to its more proximate connection. In general, the zooplankton assemblage at PSC appears fairly healthy and diverse. This is consistent with the generally good water quality conditions observed on-site.

In studies summarized by Myer and Gruendling (1979), over 50 taxa of zooplankton (not including protozoa) have been identified in the main body of Lake Champlain, with Copepods and Cladocera dominating. Seventeen taxa identified in the lake also occurred at PSC. Mean monthly densities in the lake range from 10 to 600 organisms per liter. The differences in species composition and densities between the lake and the canal are likely due to differences in depth and other physical factors.

Fish.   The most abundant species of fish collected at both the PSC and reference sites were golden shiners and yellow perch (Table 9-8). High densities of small fish (less than 3 cm in length) were present in the canal in May, but were not present

TABLE 9-8. Summary of the number of fish caught during electroshocking and gill-netting at the Pine Street Canal and reference sites, June and August 1990.

| | Reference Site | | | | Pine Street Canal | | | |
| | Electroshocking | | Gill-Netting | | Electroshocking | | Gill-Netting | |
Species	June	August	June	August	June	August	June	August
Golden shiner	2	2	1	0	1	20	87	0
Yellow perch	26	0	2	13	2	27	26	0
Pumpkinseed	0	1	0	1	2	23	0	0
Bowfin	0	2	0	3	0	0	0	0
Minnow spp.	0	0	0	0	0	4	0	0
Smallmouth bass	3	0	1	0	0	0	0	0
Sunfish spp.	3	0	0	0	0	0	1	0
Rock bass	0	0	0	0	1	2	0	0
Banded killifish	0	0	0	0	2	0	0	0
Northern pike	0	0	0	0	0	0	2	0
Chain pickerel	1	0	2	0	0	0	0	0
Largemouth bass	0	0	2	0	0	1	0	0
White perch	0	0	0	0	0	0	1	0
White sucker	0	0	0	0	1	0	0	0
Emerald shiner	0	0	0	0	1	0	0	0
Blacknose shiner	0	0	0	0	1	0	0	0
Brown bullhead	0	0	0	0	0	1	0	0
Channel catfish	0	0	0	0	0	1	0	0
Unidentified	0	0	0	0	0	2	0	0
Total Fish	35	5	8	17	11	81	117	0
Tadpole	0	0	0	0	0	1	0	0

during the June and August sampling. These small fish were likely larval fish using the canal as a nursery prior to dispersal to deeper waters in other areas. Alternatively, consumption by bass and other predatory fish, which were abundant in the canal during spring, may have decimated these larval fish. A total of 13 fish species were collected at PSC, while 7 species were collected at the reference site. Far fewer fish were collected by gill-netting at the entrance to the canal in August than in June, although significantly more fish were collected during August electroshocking, relative to June. Only two bottom-dwelling fish, a brown bullhead and a channel catfish, were collected in the canal; no bottom-dwelling fish were collected at the reference site. The sampling methodology employed, however, did not effectively sample bottom-dwelling fish.

Yellow perch are abundant throughout Lake Champlain. They generally spawn in weedy or brushy areas such as those present at PSC. Being slow swimmers and travelling in large schools, they are prime food for predaceous fish. They generally seek protection in weeds and brush, as occurs at PSC, and feed on small zooplankton and larval insects (McClane 1974) common in the canal. Golden shiners are also a very common Lake Champlain species commonly found in or near weed beds. They spawn over submergent vegetation and feed on planktonic crustaceans, aquatic insects, molluscs, and algae (McClane 1974). Thus, PSC provides both spawning and foraging habitat for this species. Game fish such as smallmouth bass and northern pike utilize the canal for foraging. The bass feeds on crustaceans, insect larvae, crayfish, and fish, while the northern pike feeds largely on other fish, principally perch (McClane 1974).

Fish that feed in the water column are relatively abundant in the canal, due to the generally good water quality and resulting forage. It can be speculated that species requiring sediment habitat for spawning will not be able to successfully reproduce in the canal, due to the toxicity of the sediments, which could influence egg survival.

Concentrations of 16 individual PAH compounds were measured in muscle tissue (fillets) and whole body samples of fish and a single tadpole collected in August at the PSC and reference sites. Since most of the fish sampled were bait fish, which are likely to be consumed whole by predators, only data from whole body samples were used in ecological analyses. The results show that, of the 29 samples analyzed from PSC, concentrations were below detection limits for all individuals with the exception of a tadpole, two golden shiners, and two pumpkinseeds. PAHs detected in fish samples were fluorene, pyrene, and benzo(a)anthracene. In addition to these three PAHs, phenanthrene, fluoranthene, chrysene, benzo(b)fluoranthene, benzo(k)fluoranthene, and benzo(a)pyrene were detected in the tadpole sample. However, PAH metabolites were not tested for. None of the 13 fish from the reference site had detectable levels of PAHs in their tissues. It must be pointed out, however, that problems were discovered when the fish tissue data were validated. Due to laboratory analysis problems, including exceeding method-specified holding times, the likelihood of false negative results (i.e., the compound is reported as below detection limits when it was actually present at levels above detection limits) is increased. Thus, these data must be interpreted with caution, as the results are likely to underestimate the actual tissue concentrations of PAHs present in the organisms.

Summary of Ecological Conditions

Wetland, wildlife, and aquatic communities at PSC are generally similar to those described for other regional wetlands, although the urban nature of the land uses surrounding PSC and the history of on-site disturbance has resulted in some differences. Forested wetland species composition is similar to regional communities, although emergent areas tend to be more monotypic at PSC, being dominated by cattail and common reed, which is likely due to past disturbance. Fish use of PSC appears similar to that at other shallow regional wetlands, while differences between zooplankton assemblages can generally be attributed to physical factors. PSC avian species composition is generally typical of regional conditions, although the site is generally too isolated and shallow to be used by diving ducks and contains few, if any, exposed areas of sediment attractive to foraging shorebirds. Wetland-associated mammalian species composition appears consistent with regional conditions, although muskrats are present at lower-than-expected numbers at PSC. Some upland wildlife species, such as coyote and bobcat, are absent from the PSC site, due to the small size of the site (relative to home range requirements) and the urban nature of surrounding land uses.

Wetland, aquatic, and wildlife communities at PSC are generally similar to those at the reference site, although effects attributable to on-site contamination have resulted in some important differences, especially among the benthic infauna. The benthic infauna at PSC (except Maltex Pond) is dominated by tubicifid oligochaetes, which are pollution-tolerant organisms, while the reference site is dominated by pollution-sensitive species such as amphipods. In highly contaminated areas of the canal, where coal tar product is present at the sediment surface, no living benthic organisms are present. Zooplankton and fish assemblages are similar between sites, or differences are attributable to physical factors (as opposed to contaminant effects), although some PSC fish had detectable levels of PAHs in their tissues, while reference site fish were clean. Differences between the PSC and reference sites for wetland and wildlife communities are generally as described above for regional conditions.

No overt effects on upland and wetland vegetation from site contaminants were apparent, although some wooded areas along the canal are heavily impacted by beaver foraging activity and the effects of periodically high water levels. The avifauna observed at the PSC site was generally typical of that expected based upon geographical location, habitat present, and surrounding land uses. No overt effects of site contamination were observed.

The mammalian fauna inhabiting the PSC site is generally typical of a small wild area in an urban landscape. Species present are, at a minimum, tolerant of some urban activities, and may actively use the surrounding areas to meet certain of their life requisite needs, such as foraging or breeding sites. Most species potentially at risk of exposure to site contaminants, such as mice, woodchucks, and beavers, showed no overt effects and were determined to be successfully reproducing on-site. However, muskrat populations were lower than expected, which may be due to the effects of site contamination, although this could not be conclusively demonstrated.

While data on the amphibians and reptiles present at the PSC site are limited, at least one group, frogs, is successfully reproducing on-site. This is important, as immature forms (tadpoles) will overwinter for one or more seasons in sediments prior to becoming adults, and the canal sediments and adjacent wetland soils are the most contaminated media present on-site. No overt effects of site contamination on frogs, or other reptiles and amphibians, were observed.

In general, the portion of the aquatic community closely associated with the canal sediments (the benthic infauna) appears to be significantly impacted by sediment contamination, resulting (in some areas) in abiotic conditions. In contrast, the benthic invertebrate assemblage in Maltex Pond is healthy and diverse; sediments in Maltex Pond were remediated in 1985. The portion of the community more closely associated with the water column, including zooplankton, fish, and invertebrates, appears to be less affected by contamination, since the water column is relatively clean. However, some fish sampled from the canal did have detectable levels of PAHs in their tissues.

ECOLOGICAL RISK ASSESSMENT

Introduction

An ecological risk assessment has been defined as a qualitative and/or quantitative appraisal of the actual or potential effects of a hazardous waste site on plants and animals other than humans or domesticated animals (U.S. EPA 1989d). Ecological risk has also been defined as a function of hazard and exposure. Hazard relates to the intrinsic ability of a contaminant to cause an adverse effect under a particular set of circumstances. Exposure is a function of two components: (1) the estimated amount of the contaminant present in an environmental medium that is available to a receptor at any given time, and (2) the number, dynamics, and characteristics of receptors that come into contact with the contaminant (U.S. EPA 1989a).

The baseline ecological risk assessment for the PSC site consists of three distinct parts. The first is an exposure assessment in which exposure pathways, species or species groups potentially at risk of exposure, and assessment and measurement endpoints are identified. Second, a toxicity assessment was performed, in which the effects of the contaminants of concern on aquatic and terrestrial flora and fauna were quantified for the various exposure pathways. The third task involved conducting a risk characterization, in which baseline risks were quantified and media-specific ecological effect levels were developed. Ecological effect levels were defined as the concentration of a particular contaminant in a particular medium below which no demonstrable reproductive effects to ecological receptors are likely to occur. Ecological effect levels were developed based on established numerical criteria (e.g., interim sediment quality criteria), or on exposure pathway modeling using site-specific data, as well as information obtained from the literature. These effect levels can be used as a guide to determine the necessity for remediation, and to determine the areal extent of remediation activities.

Following the baseline risk assessment, potential ecological endpoints for site remediation are outlined. The decision to remediate, and the development of potential ecological endpoints for the PSC site following remediation, represents the classical ecological conflict at hazardous waste sites (U.S. EPA 1989c). The site currently has significant ecological value, and only a few readily demonstrable effects of site contamination were apparent. There is the real possibility that the removal of contaminated media could destroy some of the site's existing ecological value. Some of the features that contribute to the value, such as mature forested wetlands, would be difficult to recreate within a reasonable time-frame. However, there is also the possibility that remediation will ultimately enhance the site's overall ecological value. Preserving the site's unique ecological attributes, while at the same time enhancing the site's overall ecological value through remediation is a difficult task, and will usually involve various trade-offs to achieve the proper balance.

Information from the first two parts of this chapter (nature and extent and ecological conditions) were essential for conducting the baseline ecological risk assessment. Nature and extent identified the contaminants and media of concern. Ecological conditions provided data on the types of organisms and habitats present on-site, which allowed potential sensitive receptors at risk of exposure to be identified.

Objectives

The objectives of this ecological risk assessment are to quantify existing (baseline) risks to ecological resources from exposure to site contaminants, and to develop media-specific ecological effect levels for aquatic and terrestrial organisms consistent with assessment endpoints.

Exposure Assessment

This subsection analyzes the potential exposure pathways and effects on terrestrial and aquatic fauna at the PSC site, using data on the existing site conditions, coupled with additional information on the life histories of biota utilizing the site. For those species, or species groups, with a perceived potential risk, ecological assessment endpoints are identified and evaluated against measurement endpoints using empirical data obtained from biological field investigations.

Terrestrial and aquatic plants are not considered at risk, since toxic effects in plants have rarely been demonstrated from exposure to benzene, toluene, xylene, or PAHs. In addition, the on-site flora did not exhibit any signs of stress that could be attributed to the contamination of the site by these compounds. Thus, plants are addressed in this assessment only as a potential medium, taking up contaminants from soil or sediments and transferring them to herbivorous animals, who consume their tissues.

Exposure Pathways

There are four major ways in which fauna might be exposed to contaminants: (1) direct ingestion of contaminated abiotic media, (2) the consumption of contaminated animal or plant tissues, (3) direct inhalation, and (4) absorption through skin or gill surfaces. These pathways are discussed, in general terms, below.

Only species or assemblages whose activities frequently bring them into direct contact with the canal sediments, wetland or upland soils, or canal surface water; directly consume plants growing on or in these media; or feed upon species possessing one or both of these characteristics, are evaluated for contaminant effects. These species or assemblages were selected based on their potential for exposure via one or more of the pathways listed above. For terrestrial and vegetated wetland areas, the selection process also incorporated the concept of representative species, where only one of a group of species exhibiting similar habitat requirements and life history traits was selected for evaluation. The other species within the group were considered to have a similar risk of exposure and would be expected to exhibit similar effects (U.S. EPA 1989b). For aquatic areas, the selection of a single species was more problematic, so entire assemblages (e.g., benthic invertebrates) were evaluated. Other species or assemblages that are present on the site, but do not meet the exposure criteria, were screened out during the exposure assessment, because they were not considered to be at risk of significant exposure.

Direct Ingestion of Contaminated Abiotic Media.   Most of the bioavailable on-site contamination (of PAHs and BTX) lies in the canal surface sediments and wetland surface soils, with lower levels found in upland surface soils; surface waters were generally uncontaminated. Direct ingestion of contaminated soil or sediment could occur while animals grub for food, feed on plant matter covered with contaminated soil, filter feed in areas where sediments have been resuspended in the water column, or while preening or grooming themselves. In addition, aquatic deposit feeders directly ingest large quantities of bulk sediment in order to obtain the energy-rich fraction; these organisms would likely have a significant exposure from this pathway.

Ingestion of Contaminated Tissues.   It is likely that some terrestrial, wetland, and aquatic plants rooted in contaminated soils or sediments would uptake some of the contaminants and incorporate these compounds into their tissues, thereby presenting a possible risk to animals feeding upon those plants. Although data are extremely limited, there is little evidence of biomagnification within the food chain from consumption of plant tissues contaminated with PAHs, suggesting that this pathway is of minimal importance. No data on plant uptake and bioaccumulation were found for BTX.

Predatory organisms (secondary consumers) may be at risk when feeding upon prey containing elevated levels of contaminants in their ingesta (undigested or partially digested material in the organism's gastrointestinal tract) or tissues. The ability of contaminants to be passed from lower to higher trophic levels is dependent upon their fate inside the prey animals. If the contaminant is solely ingested and excreted by the prey, without absorption from the gastrointestinal (GI) tract and storage in its tissues, then the risk to higher trophic levels is minimized; the dose received by the predator would be dependent upon the contaminant concentration in the prey's ingesta, its ingestion rate, the passage of time since consumption, and the amount of the contaminated ingesta consumed. In this scenario, the exposure to the predator would be from consuming the undigested material in the prey's stomach, not the prey's tissues.

The risk of exposure to predatory organisms is greater if the prey assimilates the ingested contaminants, or directly bioaccumulates contaminants present in the media, into its tissues. The risk of exposure to predators in this case would then be dependent upon the concentration of contaminants in the particular tissues consumed, and the rate of food consumption. The dose received would also depend upon the rate of assimilation from the GI tract of the contaminants (or toxic metabolites resulting from chemical changes in the compounds) during digestion, and the rate of metabolism.

Various common food species (e.g., earthworms, amphibians, and small fish) for upper-level consumers are known to bioaccumulate PAHs and BTX to levels above those found in abiotic media. Although no site-specific data on tissue concentrations of contaminants exist for earthworms, these organisms have been shown to bioaccumulate certain PAHs up to (but generally not above) levels found in soils (Marquenie and Simmers 1984; Marquenie et al. 1987). A whole-body sample of a single tadpole collected from the canal contained elevated levels of PAHs (but was not analyzed for BTX), and a small percentage of sampled fish also had detectable levels of some PAHs in their (whole body) tissues. It should be noted that the tissue analysis did not test for the presence of PAH metabolites. Thus, prey items on-site may pose a potential exposure risk to predators that consume them.

Inhalation Exposure.   Volatile organic compounds and, to a lesser extent, lower molecular-weight PAHs, tend to volatilize from surface soils or surface water. In vapor form, these compounds may become bioavailable to organisms during respiration, and pose an important exposure route. The lungs, with their large surface area for gas exchange, readily absorb many chemicals and pass them directly into the bloodstream. Most hazards from volatile organic compounds are associated with inhalation exposure (Clement Associates 1985); PAHs, especially compounds with higher molecular weights, have a reduced exposure risk from inhalation relative to BTX, due to their tendency to remain adhered to surface particles. Absorption of BTX from air, through the lung surface, into the bloodstream depends to some degree on the concentration of the compound in the air, and ranges from 10 to 50 percent in rodents for benzene (U.S. PHS 1987a). For toluene and xylene, about 60 percent of the compound present in the air inhaled by rodents was absorbed and retained (U.S. EPA 1984a; 1984b).

Exposure via inhalation would be most important to organisms that burrow in contaminated soils, especially those that bear their young in below-ground burrows (e.g., woodchucks). Contaminants present in the soil would volatilize into the confined airspace of the burrow, where they could reach relatively high concentrations. This exposure would be most severe for prenatal and post-natal animals, as they would have continuous exposure during the early period of their life, prior to their exit from the burrows, when they may be most susceptible to the effects of the contaminants, due to the high rate of growth experienced at this life stage (U.S. EPA 1989d). Exposure of the pregnant mother to contaminants in the burrow air may also have effects on the adult animal and the developing fetus.

Dermal Exposure.   Direct exposure to the soils or sediments is another pathway important to some species. This could result from direct dermal contact with the soils or sediments on unprotected surfaces (e.g., gill membranes [from suspended sediments], exposed skin, and exposed mucosal membranes). Although less important at the PSC site, due to the low observed concentrations, contact with surface waters is another potential dermal exposure for aquatic and semiaquatic organisms.

Dermal exposure is likely to be most important to burrowing mammals (such as woodchucks and muskrats), amphibians and reptiles that hibernate in the canal sediments (such as frogs and turtles), benthic invertebrates, and bottom-dwelling fish. These taxa all have extensive contact with sediments and soils (the most contaminated media on-site) during all or most life stages. For burrowing mammals, this exposure is heightened in young animals, which are usually born hairless, due to direct contact of unprotected skin with the soil surface.

Exposure to wetland soils and canal sediments is likely to pose a greater dermal exposure risk than exposure to upland soils, due to the higher moisture content of the former media. This would allow higher levels of contaminants to desorb from the soil particles into the pore water, increasing the bioavailability to the exposed organisms. Exposure via pore water would more closely parallel laboratory experiments where the compound is applied in an aqueous solution, thus allowing more confident extrapolation of laboratory results to organisms exposed in the field.

Potential Ecological Receptors

This subsection discusses the various species groups considered to be potentially at risk of exposure to site contaminants, based on their life history attributes. The potential for off-site transfer of contaminants is also assessed.

Benthic Organisms.   Benthic organisms living in the canal sediments would be at risk due to their direct contact with this contaminated medium. Exposure would result from direct contact with exposed outer membranes and respiratory surfaces, the direct ingestion of sediments during feeding activities, and the consumption of contaminated prey or organic matter (depending upon the species' feeding habits).

Since most benthic organisms are relatively sedentary, off-site transfer of contaminants by those species closely associated with the sediments is considered minimal. Benthic organisms more closely associated with the water column may be carried off-site by currents, but this transfer mechanism is not likely to be significant.

Fish.   Certain species of fish (e.g., brown bullhead and channel catfish) are at risk, as they have considerable dermal exposure to contaminated sediments. These species are associated with the bottom sediments for feeding, spawning, or nursery areas, and may burrow into sediments to rest and hide from predators (McClane 1974). In addition to dermal exposure, those species that bottom-feed may also directly ingest sediments as well as contaminated prey items living in or on the sediments. Sediments placed in solution by feeding activities may also come in contact with gill membranes, providing additional exposure. Bottom-dwelling fish were rarely caught (only two individuals) in the canal during aquatic surveys, although the sampling methodology used did not efficiently sample bottom-dwelling fish; these species were equally rare in reference site samples.

Some sampled fish did exhibit detectable levels of several PAHs in their tissues during whole-body chemical analysis. Since these organisms are mobile, and are likely to utilize Lake Champlain during some portions of the year, some minimal off-site transfer of contaminants is possible. Since most PAHs and BTX are rapidly metabolized by many organisms, this transfer is considered insignificant.

Amphibians.   Salamanders, newts, and frogs are at risk of exposure because of their close association with the canal sediments, wetland soils, or upland soils, either as immature forms, adults, or both. Most newts and salamanders are terrestrial hibernators, whereas most species of frogs hibernate under water in mud (DeGraaf and Rudis 1983). Thus, exposure to contaminants in sediments or soils continues even during hibernation (although metabolism is greatly slowed), because of direct absorption through their relatively unprotected membranous skin. This exposure is expected to be greatest for some species of frogs, since they would be in contact with the most contaminated media (the canal sediments). It is significant that these organisms conduct considerable metabolite exchange directly through their skin (Schmidt-Nielsen 1983).

Salamanders and newts consume earthworms, aquatic insects, and small fish or tadpoles (DeGraaf and Rudis 1983). These prey items are among the most likely to contain elevated levels of contaminants in their tissues. Due to their relatively small home ranges and short dispersal distances (DeGraaf and Rudis 1983), there is little potential for these organisms to transfer contaminants off-site.

Reptiles.   Turtles and, to a lesser degree, snakes are at risk of exposure, based upon their life history characteristics. Turtles are mostly aquatic and spend considerable time on the bottom sediments of waterbodies, the medium with the highest levels of contaminants. The species of turtles observed on-site also hibernate in the mud on pond or lake bottoms (DeGraaf and Rudis 1983), and would be exposed to contaminants in the same manner as frogs during this period. Many snakes are sensitive to pollutants and have frequent contact with soil or sediment (Hall 1980; DeGraaf and Rudis 1983).

Turtles consume tadpoles, small fish, crustaceans, and some carrion (DeGraaf and Rudis 1983). Semiaquatic snakes also consume fish, frogs, aquatic insects, and salamanders, while more terrestrially oriented species may consume large numbers of soil invertebrates, especially earthworms (DeGraaf and Rudis 1983). These food items are likely to contain the highest levels of contaminants of the available food items present at the site. Turtles and snakes are generally sedentary, minimizing the possibility of them transferring contaminants to off-site areas.

Mammals.   Several mammalian primary consumers are potentially at risk of exposure to site contaminants. These include, for wetland and aquatic areas, muskrats and beavers, and for upland areas, woodchucks and other small rodents (e.g., mice, moles, and voles). Because of the lower levels of contamination in the upland soils, relative to the wetland soils and canal sediments, these upland species probably have a smaller risk of significant exposure.

Both muskrats and beavers spend considerable time in potentially contaminated water, and in contact with the contaminated sediments and wetland soils. Beavers carry sediments (mud) in their forefeet for the purpose of lodge and dam building (Novak 1987). Muskrats may also build lodges utilizing emergent vegetation (Allen and Hoffman 1984), but (based upon on-site observations) are more likely to utilize bank burrows dug directly into the contaminated soils. For both species, the floors of the lodge or burrow are at least partly composed of bare soils and/or sediments. Young animals (especially muskrats, which are born hairless) would have considerable dermal exposure to these sediments before emerging from the lodge or burrow. Both adults and young would be exposed to contaminants volatilizing from the soils or sediments into the confined airspace of the burrow or lodge.

These two species may directly ingest contaminated soils or sediments in the course of dam, lodge, or burrow construction, during grooming activities, and as they forage. Contaminants could be absorbed directly through the skin and mouth membranes (dermal exposure), and from the digestive tract after ingestion. Exposure to contaminated water could also occur from these routes; the canal sediments and wetland soils often discharge contaminants (evidenced by an oily sheen) into the water when disturbed.

Muskrats may be exposed to contaminants from the plants they eat. Muskrats prefer emergent vegetation (especially cattails), which root directly on some of the most contaminated soils on-site (Allen and Hoffman 1984). While some species of plants are known to bioaccumulate PAHs from the growth medium, there is little evidence of trophic transfer from plants to higher-level consumers, so this exposure is likely to be relatively small. The ingestion of soils or sediments while digging and consuming plant tubers is likely to be a more important exposure pathway. Similarly, beaver exposure via contaminated plant tissues is also likely to be minimal. This species forages mainly on woody plants growing on upland areas (Allen 1983), where soil contamination levels are lower. Sediments deposited on the bark surface during transport and storage of the cut stems near the lodge, however, may be ingested during foraging activities. While beavers do consume soft plant tissues (primarily leaves) in the spring and summer (Allen 1983), this exposure is not likely to be considerable.

Beavers and muskrats may transfer contaminants to off-site locations when they disperse. Muskrats disperse annually, whereas beavers usually disperse in their second year (Perry 1982; Hill 1982). Although PAHs and BTX may bioaccumulate in fatty tissues, they tend not to persist for long periods of time at significant concentrations, due to metabolic breakdown. Thus, when removed from exposure, contaminant concentrations in the tissues of these organisms are expected to decrease relatively quickly.

Upland, burrowing mammals, such as the woodchuck and *Peromyscus* mice, could be at some risk of exposure due to dermal contact with the soil. These animals, particularly young animals born in burrows, could also be at risk due to inhalation exposure to contaminants volatilizing from the soil surface into the confined airspace of the burrow. Soil could also be directly ingested by these species during feeding and grooming activities.

Mammal secondary consumers with a possible risk of exposure include mink and raccoon. Mink, and to a lesser degree raccoon, preferentially feed upon aquatic animals (e.g., fish, frogs, and tadpoles) and small mammals, depending upon relative prey abundances (Linscombe et al. 1982; Kaufmann 1982). These prey species are among the organisms most likely to contain significant levels of contaminants in their tissues. Mink and raccoon will also consume carrion, particularly if it is not extensively decomposed (Linscombe et al. 1982; Kaufmann 1982). If such animals died due to acute effects of contaminants, consumption of the carcasses could result in a significant exposure.

Since mink and raccoon have large home ranges, relative to the size of the site, the percentage of time spent on-site and the percentage of food obtained on-site would influence the potential exposure. Raccoon home ranges vary between one and three km in diameter (160 to 480 ha) while mink home ranges vary between one and five km of river, or up to 800 ha (Linscombe et al. 1982; Kaufmann 1982; DeGraaf and Rudis 1983; Eagle and Whitman 1987). Raccoons are particularly tolerant of human activities (DeGraaf and Rudis 1983), and likely utilize the surrounding habitats. Mink are somewhat tolerant of human activities (Allen 1986) and would likely utilize the shoreline of Lake Champlain. Utilization of these off-site areas would limit the potential exposure of these species to site contamination, but would also allow for off-site transport of contaminants.

Birds.   Among avian primary consumers, certain species of dabbling ducks may be at risk of exposure to site contaminants. The species observed on-site, the mallard, American black duck, and wood duck, all (to varying degrees) directly ingest wetland soils or canal sediments as they forage. The consumption of bottom sediments during foraging is well documented for dabbling ducks (e.g., Eisler 1988). These species primarily consume submergent vegetation (relatively rare on-site), floating vegetation (e.g., duckweed), and aquatic invertebrates (Bellrose 1980). Additional exposure may occur if these food items contain elevated levels of contamination in their tissues, and from consumption of contaminated water (due to release of contaminants from the disturbed sediments). Contaminated media adhering to the feathers of the animal may also be ingested during preening activities.

There is some evidence that mallards nest on-site (an observed brood), and suitable nesting cavities exist along the canal for wood duck nesting as well. Breeding activity would increase the time spend on-site by these species, and therefore increase their exposure, as well as exposing young to contaminated media.

Ducks may be dermally exposed to contamination through direct contact with water and sediments; ducks will often rest on upland areas with direct belly contact with the ground. While the waterproof coating of the feathers would likely prevent direct exposure of the skin surface under most circumstances, skin contact is likely for female ducks during the breeding season, when they have no protective feathering over their brood patches (located on the abdomen) (Berger 1961). In addition, this brood patch is highly vascularized, to allow more efficient heat transfer (Berger 1961; Pettingill 1970), increasing the likelihood of absorption of contaminants into the blood stream. Contamination of the brood patch area with water or sediments may be transferred to eggs or young during incubation or brooding activities (Albers 1978). Bird eggs are porous and allow many toxic substances, such as oil, to diffuse in (Ehrlich et al. 1988); small amounts of some PAHs (e.g., chrysene and benzo(a)pyrene) and BTX (e.g., xylene) are known to be deleterious when applied to eggs, causing egg failure (David 1982; Hoffman and Albers 1984; Eisler 1987).

Avian secondary consumers at potential risk of exposure include belted kingfishers, double-crested cormorants, several species of herons, and American robins. Kingfishers and cormorants consume small fish, while herons eat small fish, as well as amphibians and small mammals (DeGraaf and Rudis 1983; Ehrlich et al. 1988). American robins consume earthworms (DeGraaf and Rudis 1983; Ehrlich et al. 1988). These prey species are among the most likely to contain contaminants in their tissues. These secondary consumers are unlikely to be exposed along any other exposure route. Limited data (from mallards) suggest that birds are not very sensitive to PAHs in their diet (Eisler 1987); their tolerance to BTX is largely unknown.

Comparison to Assessment Endpoints

Assessment endpoints, and the measurement endpoints by which they are evaluated, were previous described at the beginning of the Ecological Conditions section, and were summarized in Table 9-1. The species (or species groups) to be evaluated in this subsubsection are those for which a potential for exposure was identified in the previous subsubsection. Field observations at PSC and the reference site, as well as existing data on the ecological resources in the Lake Champlain region, provide the measurements by which the specific assessment endpoints are evaluated.

Benthic Organisms.   The benthic segment of the aquatic community inhabiting a water body is generally a good indicator of the long-term environmental condition and health of the aquatic system. The benthos is primarily composed of slow-moving, relatively sessile organisms, which cannot easily escape stressful conditions, such as changes in the quality of the sediments and/or water. As some species are more sensitive to these types of stresses than others, the species composition and relative abundance of various organisms can serve as a good indicator of the typical conditions at a particular sampling point.

The benthic infauna inhabiting the sediments at the PSC site appears to be greatly affected by the sediment conditions. The sediments in the canal are spatially heterogeneous and the diversity and density of organisms vary correspondingly. In areas of heavy contamination, indicated by the presence of free coal tar product on the sediment surface, abiotic conditions (no living benthos) prevailed. In other canal locations, pollution-tolerant species such as tubicifed oligochaetes (Brinkhurst and Cook 1974) were much more common (representing 50 to 96 percent of the benthic infauna) than at the reference site (2 to 45 percent) or Maltex Pond (0.2 percent). The benthic segment of the aquatic community at the reference site, correspondingly, was composed of a greater percentage of pollution-sensitive species, such as amphipods and odonates (Hart and Fuller 1974).

These observed patterns appear to be directly related to contamination of the sediments. The U.S. EPA (1988) has drafted interim sediment quality criteria for several PAHs (but not for BTX). These criteria were exceeded in sampled canal sediments for three (acenaphthene, phenanthrene, and pyrene) of the six PAHs for which criteria exist. These results suggest that exposure to contaminated canal sediments has had detrimental effects on the benthic infauna.

Benthic invertebrates more closely associated with the water column do not appear to be as stressed as those associated with the sediments. This is evidenced by the greater relative abundance (compared with the on-site infauna) of pollution-sensitive species such as amphipods and copepods observed at the PSC site. It is also important to note that the benthic fauna in Maltex Pond is relatively healthy and diverse. This pond was dredged of contaminated sediments in 1985, and has naturally recolonized to a healthy and functioning aquatic community, which is isolated from the canal.

In summary, the benthic infauna, but not the epifauna, appears to be affected by contaminants in the canal sediments. The benthic infauna inhabiting the canal sediments failed to meet the measurement endpoint, showing reduced diversity and/or abundance (the assessment endpoints) relative to the reference site. The health of the epifaunal assemblage at PSC indicates that the assessment endpoint may be achievable for the infauna if there is uncontaminated substrate available.

Fish.   Lake Champlain, in general, and Burlington Bay (adjacent to the PSC site), in particular, support several species of fish that depend upon wetlands for some aspect of their life history. Yellow perch is the dominant fish in shallow lake waters (Gruendling and Bogucki 1978). Many other fish species are common in this region, including walleye, brown bullhead, pumpkinseed, bowfin, northern pike, chain pickerel, largemouth and smallmouth bass, carp, and eastern mudminnow (Gruendling and Bogucki 1978; Vermont Department of Fish and Wildlife 1978). Observations and sampling confirm that these regional patterns are generally true of the PSC site as well.

The U.S. EPA (1986) has set Ambient Water Quality Criteria for some compounds, which are designed to protect most aquatic organisms from the toxic effects of contaminants dissolved in surface waters. Although data are currently considered insufficient by the U.S. EPA to develop criteria for PAHs and most volatile organics, acute and chronic Lowest Observed Effect Level (LOEL) values for acenaphthene, naphthalene, fluoranthene, benzene, and toluene are available. The surface waters of the canal did not exceed acute or chronic LOEL values for those volatile organics and PAHs with existing values. PAHs, in particular, very rarely exceeded detection limits in surface water samples. Thus, it is not surprising that fish and other aquatic organisms associated with the water column appear to be relatively unaffected by site contaminants. This is evidenced by the healthy and diverse zooplankton and fish populations (including larval fish) observed during aquatic studies at the PSC site. Compared with the reference site, the fish assemblage at PSC had as many, or more, species present, and fish were consistently more abundant at the PSC site as well. The abundance of larval fish (of species which spawn over submergent vegetation) at PSC in the spring indicates that the site is also being used as a nursery area. There were few bottom-dwelling fish (which would be associated with the contaminated sediments) caught at PSC, although these species were equally rare in samples from the reference site. This is likely due to the sampling methodology used, which did not efficiently sample bottom-dwelling fish.

Some fish sampled at PSC did exhibit detectable levels of some PAHs in their (whole body) tissues, whereas no analyzed fish obtained from the reference site contained detectable levels of PAHs. The one bottom-dwelling fish (a channel catfish) caught and analyzed for contaminants at the PSC site did not, however, show detectable levels of PAHs in its tissues. Interim sediment quality criteria for several PAHs (primarily phenanthrene) were exceeded at the PSC site, however, indicating that aquatic species coming into contact with these sediments have the potential to be adversely affected. However, phenanthrene is metabolized by many species of aquatic organisms, including fish (Eisler 1987), so the general lack of significant bioaccumulation shown in the analysis of fish tissue was not unexpected. Unfortunately, PAH metabolites were not tested for during tissue analyses.

In summary, the fish and zooplankton assemblages associated with the water column show no evidence of acute adverse effects from site contaminants. Low levels of PAHs in the tissues of some of these organisms, however, suggests that more subtle, chronic effects may be possible. The relative lack of bottom-dwelling fish at the PSC site cannot be definitively related to contamination of the sediments, due to the equal scarcity of these species in reference site samples, although disruption of the benthic infauna by site contaminants would limit the foraging opportunities for these species in the canal. For fish, the measurement endpoint (comparison of the assessment endpoints of diversity and abundance with conditions on the reference site) was generally satisfied.

Amphibians.   A number of species of frogs are common in the Champlain Valley, and are usually associated with areas containing semipermanent or permanent water (DeGraaf and Rudis 1983). Other important habitat characteristics include the presence of aquatic vegetation, overhead cover, and densely vegetated banks; food items include insects, crustaceans, and fish. These conditions are present at the PSC site.

Several species of frogs were observed on-site during the field surveys. Elevated levels of PAHs detected in a single tadpole sample (which was not analyzed for BTX or PAH metabolites) indicates that frogs, at least in the larval stage, are being exposed to, and bioaccumulating, PAHs. Although many species of aquatic amphibians are especially sensitive to pollution (Hall 1980), both adult and juvenile (tadpole) frog life stages were fairly abundant on-site, suggesting that the contamination was not grossly affecting reproductive success. Based upon qualitative observations, however, salamanders, and to a lesser extent frogs, were generally more abundant in Maltex Pond (where sediments and soils were relatively uncontaminated) than in the wetlands directly connected to the canal, suggesting that these species may be avoiding the most grossly contaminated soils and sediments to some degree. Newts were not observed during biological field surveys. Thus, while acute effects of on-site contamination were not evident, some more subtle and chronic effects are suggested. The measurement endpoint (observed use of the site) was, however, generally met.

Reptiles.   Several species of turtles are common in the Lake Champlain region, and the PSC site provides suitable habitat for several, including snapping and painted turtles (DeGraaf and Rudis 1983). Snapping turtles were observed on-site only once, and painted turtles were observed relatively infrequently. However, the quiet and unobtrusive behavior of these species may account for the scarcity of observations; almost all observations were of basking individuals, when they are most visible and thus amenable to observation.

As with frogs, painted turtles were more frequently observed in Maltex Pond (where sediments and soils were relatively uncontaminated) than in the wetlands directly associated with the canal. It can be speculated that these organisms may be avoiding the most grossly contaminated soils and sediments present on-site to some degree. A dead turtle, with no external signs of trauma, was found in one of the most contaminated areas of the canal during aquatic surveys, although the cause of death could not be determined.

The distribution of turtle observations may also be influenced by habitat use patterns, as Maltex Pond is shallow, isolated, and provides suitable sites for basking. Turtles were also observed along the shores of the turning basin, basking on the rocks and logs along the shore. Basking sites were less common in the southern portions of the site (where contaminant levels were also higher) and fewer turtles were observed in this area. Thus, while acute effects were not demonstrable, more subtle, chronic effects are possible, but their detection is confounded by habitat use patterns. The measurement endpoint (observed use of the site) was, however, generally met. Data on snakes were too limited to allow for an evaluation.

Mink.   The PSC site provides suitable habitat for mink, which is characterized by permanent or semipermanent water bodies bordered by woody and emergent wetland vegetation and other shoreline cover (such as downed logs and rocks) (Allen 1986). Favored prey items, including fish, amphibians, muskrats, and small mammals are present. Mink sign (including tracks and feeding sign) confirm that this species does use the site, at least sporadically. Since mink movements are normally restricted to waterways, most reported home range sizes are in linear rather than areal measurements

(Eagle and Whitman 1987). The large home range of this species [1 to 5 km] (Linscombe et al. 1982; DeGraaf and Rudis 1983; Eagle and Whitman 1987), the relatively small size of the site (canal length less than 1 km), and the presence of suitable habitat along the shores of nearby Lake Champlain, suggest that this species may be a transitory visitor to the site.

The primary exposure risk to mink at the PSC site is the consumption of contaminated prey. Due to the likelihood that this species is not permanently confined to the site (and thus not receiving continual exposure), and the levels of PAH contamination found in the tissues of the aquatic species sampled, acute effects are not considered likely. Data are insufficient to evaluate the possibility of more subtle, chronic effects to this species. However, the assessment endpoint (presence) is satisfied as the measurement endpoint (observed use of the site) was met.

Muskrat.   Muskrats are common inhabitants along the shores and backwaters of Lake Champlain (Gruendling and Bogucki 1978). The PSC and reference sites both provide the emergent wetland habitat, characterized by vegetation such as cattail, preferred by this species. Parsons et al. (1988) rated the PSC site as moderately good muskrat habitat (score of 30 out of 50). Muskrat populations often undergo cyclic fluctuations in numbers; currently muskrat populations levels are "about average" in the Lake Champlain region (Myers 1991).

Muskrats were observed on both the PSC and reference sites, although data are insufficient to compare the relative abundances at the two sites, due to limited observations at the reference site. Although muskrats, or their sign, were observed less frequently than expected, based upon the habitat present on-site, this cannot be definitively attributed to site contamination. An alternate explanation, equally plausible, is that fluctuating water levels in the emergent wetlands on the southern portion of the site (the highest quality habitat present) discourage use by muskrats. Stable water levels are one of the most important habitat features for this species (Allen and Hoffman 1984) and large portions of the emergent wetland area were devoid of standing water on at least one site visit. Also, high water levels (also observed frequently on-site) may flood bank burrows, causing their abandonment.

Of the wildlife species present at the PSC site, muskrats are among the species with the highest potential exposure to site contaminants. Their possibly reduced abundance at the PSC site may be the result of habitat factors, site contamination, or a combination of both. There are no quantitative data to establish a causal relationship between the contamination of the site and the reduced abundance of this species, but this possibility cannot be discounted. Thus, insufficient data exist for a complete evaluation of the measurement endpoint for this species.

Beaver.   Beavers are also common mammals in the aquatic environments of the Lake Champlain region (Gruendling and Bogucki 1978). The PSC site provides good quality habitat for this species in the form of adequate water depth and sufficient woody forage on adjacent upland areas.

Evidence of beaver use is prevalent throughout the site. Numerous upland areas all along the canal exhibit signs of beaver feeding activity. A minimum of seven lodges (plus a dam) have also been found at various locations along the shores of the canal and beavers were frequently sighted during aquatic and terrestrial field studies. Based upon field observations conducted in June 1991, the on-site beaver colony consists of a minimum of five individuals, likely composed of the resident pair and three subadults.

The presence of the subadults indicates that the beavers successfully reproduced in 1990. The status of the 1991 litter could not be determined since young-of-the-year were probably too young to be outside of the lodge for extended periods at the time of the survey (mid-June), and were thus not observed.

Despite the potential exposure to site contaminants incurred by beavers on this site, there is no evidence of any overt or acute effects on the beaver colony. The colony is successfully reproducing, thus meeting the criteria set forth in the measurement endpoint for this species.

Peromyscus Mice.   *Peromyscus* mice are common rodents in New England, utilizing a variety of wooded or shrubby habitat types (DeGraaf and Rudis 1983). Suitable habitat for these species occurs in the noninundated woody areas along both sides of the canal and near the turning basin.

A total of eight mice were caught during small mammal trapping efforts at the PSC site. None of the mice exhibited any externally evident abnormalities, although tissue samples were not taken. The presence of a lactating female and several immature mice in the sample shows that these organisms are also successfully reproducing on the site. While no obvious effects from site contaminants were evident, trapping data and information from surface soil sampling were difficult to correlate, since trapping and soil samples were collected at different times, and sampling points were not co-located, making it difficult to determine if this species was avoiding the more contaminated on-site areas. A single mouse was caught near the location of the former coal gasification plant, which is one of the most contaminated upland areas on-site, suggesting that observed levels in surface soils were not acutely impacting these organisms. The measurement endpoint (observed use of the site) was exceeded, based upon the evidence of successful reproduction.

Ducks.   The PSC site provides suitable nesting, foraging, and resting habitat for several species of dabbling ducks. The site is sheltered from winds blowing off of Lake Champlain and the canal and adjacent wetlands contain a variety of floating and emergent plants (although relatively few submergent aquatic plants), as well as aquatic invertebrates, for foraging ducks. Suitable nesting substrates exist within the emergent wetlands for mallards and American black ducks, and suitable cavities exist in dead trees along the canal for nesting wood ducks.

These three species were each observed at the site during biological field sampling. In addition, a brood of mallards was observed in the emergent vegetation on the southern portion of the site, suggesting that successful reproduction took place on-site. Thus, there is no evidence of any acute or chronic effect from site contaminants to these species and the measurement endpoint of observed use of the site is satisfied. Although an observation of an unusually lethargic mallard did occur on-site, a number of possible explanations for this behavior are possible including disease (e.g., botulism), off-site lead poisoning (from ingestion of spent lead shot), and habituation to humans (likely through being fed at parks and other areas). The limited amount of time most ducks likely spend on the site, in addition to the relatively high tolerance of ducks to ingestion of PAHs (Eisler 1987), also suggest that effects are unlikely to occur at the site.

Piscivorous Birds.   The PSC site provides excellent foraging habitat for a variety of piscivorous birds, including kingfishers, cormorants, and herons, due to the abundant supply of small fish and other aquatic organisms (e.g., tadpoles) present in the canal and turning basin. The primary risk to these species would be ingesting contaminated prey items, as tissue samples from fish and a single tadpole did contain detectable levels of PAHs in whole body samples.

Belted kingfishers, double-crested cormorants, and three species of herons (great blue, green-backed, and black-crowned night herons) were observed on-site during biological field surveys. Since these species likely spend a considerable amount of time off-site and limited data (from mallards) suggest that birds are not very sensitive to PAHs in their diet (Eisler 1987), their exposure to site contaminants is not likely to result in observable effects. The frequent observation of these species on the site (meeting the measurement endpoint), and no observations of any abnormal behaviors or carcasses, supports a conclusion that these birds are not incurring observable effects from site contaminants.

Toxicity Assessment

This subsection summarizes the available information from the literature on acute and chronic effects of PAHs, benzene, toluene, and xylene (the contaminants of concern) on aquatic and terrestrial flora and fauna. Little toxicity data relating the concentrations of these contaminants to effects on terrestrial wildlife exist, since the direct effects of contaminants on many wild mammal and other terrestrial or semiaquatic wildlife species are difficult to observe. These organisms frequently occur at relatively low densities, are secretive, often range widely, and dead and sick individuals are seldom found in the wild (Wren 1987). An extensive bibliography on the effects of environmental contaminants on *Peromyscus* mice (Buttler 1988) contained no data or references relating to PAH or BTX effects. Similarly, Schafer and Bowles (1985) tested the acute toxicity of 933 chemicals to deer and mouse mice; PAHs and BTX were not among the compounds tested.

Because of their widespread use for human health assessments, there exists a large database of toxicity studies on laboratory rodents, primarily rats and mice. Therefore, these species are often used as surrogates for the estimation of the toxicity of contaminants to other (wild) mammals (U.S. EPA 1989d). This may, however, yield conservative values, since limited comparisons have shown that laboratory rodents are generally more sensitive to dietary exposures than wild rodents (Hoffman et al. 1990).

The literature on the effects of PAHs and BTX on aquatic biota is also somewhat limited, and largely focuses on the effects of contaminants dissolved in water on these species. Almost all studies relating PAH concentrations in sediments to effects on biota have been done for saltwater systems (NOAA 1990).

Polynuclear Aromatic Hydrocarbons
Because the toxic effects of the various PAHs differ among compounds (generally as a function of molecular weight), a total of 11 selected PAHs are addressed here individually. These compounds are naphthalene, phenanthrene, anthracene, fluoranthene, benzo(a)anthracene, chrysene, benzo(b)fluoranthene, benzo(k)fluoranthene, benzo(a)pyrene, dibenz(a,h)anthracene, and indeno(1,2,3-cd)pyrene. These particular compounds were selected since they represent a range of molecular sizes (from two-ring compounds, such as naphthalene, to six-ring compounds, such as indeno(1,2,3-cd)pyrene), they were analyzed for and detected in samples of on-site media, and at least a limited amount of information on their toxicity was available. Two additional

compounds, acenaphthene and pyrene, are also used, to a limited degree, since interim sediment quality criteria exist for these compounds.

PAHs are moderately persistent in the environment, and may potentially cause adverse effects to vegetation, fish, and wildlife. A variety of adverse biological effects have been reported in numerous species of organisms under laboratory conditions, including carcinogenic effects, as well as effects on survival, growth, and metabolism (Eisler 1987).

The carcinogenicity of individual PAHs differs. Unsubstituted lower molecular weight compounds containing two–three rings (e.g., naphthalene, fluorenes, phenanthrenes, and anthracenes) exhibit acute toxicity and other adverse effects to some organisms, but are noncarcinogenic. In contrast, the higher molecular weight compounds (four–seven rings) are significantly less acutely toxic, but many are demonstrably carcinogenic, mutagenic, or teratogenic (causing fetal malformations or disturbance to fetal growth) to a wide variety of organisms, including fish and other aquatic life, amphibians, birds, and mammals (Eisler 1987). Past studies indicate that interspecific and intraspecific responses to carcinogenic PAHs are quite variable, and are significantly modified by many chemicals, including other PAHs that are weakly carcinogenic or noncarcinogenic. Differences in responses may be attributable to differences in the ability to absorb and assimilate PAHs from food. For example, crustaceans and fish readily assimilate PAHs from contaminated food, whereas assimilation by molluscs and polycheate annelids is limited. In all cases, however, metabolism and excretion is quite rapid following ingestion (Eisler 1987).

Some PAHs are transformed to intermediates that are highly toxic, mutagenic, or carcinogenic. Most authorities agree that metabolic activation by the liver mixed-function oxidase system is a necessary prerequisite for PAH-induced carcinogenesis and mutagenesis (Eisler 1987). Rodents, for example, can convert PAHs to various derivatives (phenols, quinones, and epoxides), and can also activate PAHs to produce carcinogenic metabolites. These intermediate metabolites, and not the PAHs themselves, have been identified as the mutagenic, carcinogenic, and teratogenic agents (Eisler 1987). PAHs are poorly absorbed from the gastrointestinal (GI) tract; the main routes of elimination of PAHs and their metabolites include the heptobilary system and the GI tract (Eisler 1987).

The potential effects of PAHs on aquatic biota include reduced survival, decreased foot uptake, carcinogenesis, inhibited reproduction, decreased heart rate and respiration, increased weight of body organs in fish, and photosynthetic inhibition in algae and macrophytes (Eisler 1987). PAHs vary substantially in their toxicity to aquatic organisms. In general, toxicity increases as molecular weight increases, although high molecular weight PAHs have low acute toxicity, perhaps due to their low solubility in water (Eisler 1987). Many PAHs, especially lower molecular weight compounds, are acutely toxic at concentrations between 50 and 1,000 $\mu g/l$ and sublethal effects have been documented at concentrations as low as 0.1 to 5.0 $\mu g/l$ (Eisler 1987). Lowest Observed Effect Level (LOEL) values for acute toxicity are 1,700 $\mu g/l$ for acenaphthene (520 $\mu g/l$ for chronic toxicity), 3,980 $\mu g/l$ for fluoranthene, and 2,300 for naphthalene (620 $\mu g/l$ for chronic toxicity) (U.S. EPA 1986).

Most species of aquatic organisms studied to date accumulate PAHs from low concentrations in the ambient medium (water and sediment). Uptake is highly species specific, being highest in species incapable of metabolizing PAHs. Bioaccumulation factors tend to increase as the molecular weight of the PAH compound increases, with increases in the amount of organic matter in the medium, and with increases in the lipid content of the organism (Eisler 1987). When placed in clean water, organisms depurated higher molecular weight PAHs more slowly than lower molecular weight compounds, probably due to the lower solubility of the higher weight compounds in water. Depuration rates vary by species, but are usually rapid, except in some species of invertebrates (Eisler 1987). The role of sediment in PAH uptake can be important. When sediment PAH levels are elevated, benthos obtain a majority of their PAHs from sediments through their ability to mobilize PAHs from the sediment/pore water matrix (Eisler 1987). The elevated levels in the tissues of these organisms could provide a significant source of PAHs to predatory fish. However, fish have the ability to efficiently metabolize and degrade PAHs (Eisler 1987).

Some evidence links PAHs to cancer in wild fish populations, however, especially to bottom-dwelling fish from areas heavily contaminated with PAHs. For example, sediments heavily contaminated with industrial PAH wastes have directly caused elevated PAH body burdens and increased frequency of liver tumors in fish (Eisler 1987). In one study, sediments and sediment extracts from the Buffalo River, New York, contained elevated levels of carcinogenic PAHs (1,000 to 16,000 μg/kg). Brown bullheads, in response to repeated applications of Buffalo River sediment extracts, showed higher frequencies of epidermal tumors when compared to controls (Eisler 1987). In a separate study, a positive relationship was established between sediment PAH levels and liver tumors in fish from the Black River, Ohio. Sediment PAH concentrations ranged from 50 to 100 mg/kg for some individual compounds. Brown bullheads exposed to the sediment contained from 1.1 to 5.7 mg/kg of several PAH compounds in their tissues and exhibited a 33 percent higher frequency of liver tumors than controls (Eisler 1987). In a third study, from the Niagara River in New York, brown bullheads had significantly higher total lesion incidences at a site heavily contaminated with PAHs, when compared with a reference site (Hickey et al. 1990).

Noncarcinogenic PAHs vary substantially in their toxicity to aquatic organisms. Toxicity is most pronounced in crustaceans, and least among teleosts (bony fishes). In all but a few cases, PAH concentrations that are acutely toxic to aquatic organisms are several orders of magnitude higher than concentrations in the most heavily polluted waters (Eisler 1987). The EPA does not consider available data adequate to establish freshwater acute or chronic Ambient Water Quality Criteria for PAHs, although LOEL values are available for acute and chronic exposure to acenaphthene and naphthalene, and for acute exposure to fluoranthene (U.S. EPA 1986). The U.S. EPA (1988) has developed interim sediment quality criteria for six PAH compounds: acenaphthene (732 mg/kg carbon), phenanthrene (139 mg/kg carbon), fluoranthene (1,883 mg/kg carbon), pyrene (1,311 mg/kg carbon), benzo(a)pyrene (1,063 mg/kg carbon), and benzo(a)anthracene (1,317 mg/kg carbon).

Only limited data are available on the potential effects of PAHs on amphibians and reptiles. In amphibians and reptiles, as in mammals, the mixed-function oxidase system acts to detoxify PAHs, although the rate of metabolism tends to be slower than in mammals. However, amphibians are quite resistant to PAH carcinogenesis, when compared to mammals (Eisler 1987).

PAHs can be taken into the mammalian body by inhalation, ingestion, or skin contact. PAHs have been shown to have carcinogenic, toxic, and sublethal effects on laboratory mammals. The effects on mammals can be divided between two groups of PAHs: carcinogenic PAHs and noncarcinogenic PAHs.

Several PAHs are among the most potent carcinogens known to exist, producing tumors in some laboratory animals through single exposures to microgram quantities

(Eisler 1987). PAHs that are known carcinogens (mostly four–six ring structures) have been shown to cause mammalian tumors both at the point of application and systemically; their effects have been demonstrated in nearly every tissue tested, regardless of the route of administration (Clement Associates 1985; Eisler 1987).

In mammals, numerous carcinogenic PAH compounds are distinct in their ability to produce tumors in skin and most epithelial tissues. Topically applied PAHs can also pass through mammalian skin and cause carcinogenesis in many internal organs. Exposure to carcinogenic PAHs has been shown to cause tumors in the stomach, lung, and skin of laboratory animals (mice, rats, and hamsters) (Clement Associates 1985). In most cases, there are often no overt effects until the dose is high enough to produce a high tumor incidence (Eisler 1987). PAHs have also been shown to cause destruction of blood and lymph tissues, ovatoxicity, antispermatogenesis, changes in the intestinal and respiratory epithelia, mutagenesis, and immunosuppression (Eisler 1987).

PAH carcinogens transform cells through genetic injury involving metabolism of the parent compound into a reactive diol epoxide, which alters nucleic DNA and RNA (Eisler 1987). Many chemicals are known to modify the action of carcinogenic PAHs by altering the metabolism of the carcinogen, preventing carcinogens from reaching their critical target sites, and causing competitive antagonism with carcinogens (Eisler 1987).

The environmental effects of most noncarcinogenic PAHs are poorly understood. Available information suggests that these PAHs are not very potent teratogens or reproductive toxins. Documented noncarcinogenic internal effects include damage to the liver and kidney, and noncarcinogenic external effects include destruction of sebaceous glands, hyperkeratosis (hardening of the skin), and ulceration (Clement Associates 1985; Eisler 1987).

Little research has been conducted on the effects of PAHs on species other than laboratory mammals. The database on the acute and chronic effects of PAHs on rodents (mice and rats) under laboratory conditions is extensive for some compounds. Tables 9-9 and 9-10 summarize the acute and chronic toxicity values, respectively, for the 11 PAH compounds under consideration (as well as for BTX) to laboratory mammals. In Table 9-9, acute effects have been limited to LD_{50} (the dose lethal to 50 percent of a population) or LC_{50} (the concentration lethal to 50 percent of a population) values reported in the literature. Acute effect data are relatively unambiguous. LD_{50} values are generally based upon single oral or dermal doses. LC_{50} values are for single, short-term (generally four-hour) inhalation exposures.

On the other hand, chronic effect data (Table 9-10) were much more difficult to synthesize. Exposure durations were quite variable among studies, different exposure methods and/or vehicles were used, as were different strains of mice or rats, and results were reported in different and sometimes unstandardized units (e.g., not standardized to body weight). In general, oral exposure studies were most similar to one another and were usually based upon daily doses or daily administration of compounds in food or water. Exposure periods were often limited, rarely exceeding one year; studies of maternal/fetal effects limited exposures to the period of pregnancy. Dermal application rates, exposure durations, and vehicles used to apply the compound (in aqueous solution) were highly variable among studies. Applications generally ranged from three to seven times per week (most frequently five), but overall study durations were usually long-term (six months to lifetime exposures). Inhalation exposure studies also had widely varying exposure periods, generally 6–18 hours per day and five days per week; continuous exposures were never reported in the chronic studies evaluated. Study durations were also variable (1–24 months); studies of maternal/fetal effects were limited to the period of pregnancy.

Chronic effect levels in Table 9-10 stress reproductive impairment (e.g., reduced fertility and fetal or embryonic effects) or effects that significantly reduce the ability of young animals to survive to reproductive age. These other effects include carcinogenesis (e.g., skin and lung cancers) since tumors may directly effect reproductive systems (Eisler 1987). It should be noted that not all of the values in Table 9-10 are LOEL.

Numerous studies have shown that unsubstituted PAHs do not accumulate in mammalian adipose tissue, despite their high lipid solubility, probably because they tend to be rapidly and extensively metabolized (Eisler 1987). Thus, long-term storage and biomagnification through food chains is not likely to occur for PAHs.

Little information on the toxicity of PAHs to birds has been collected. Two studies have been conducted on the toxicity of PAHs to mallards. When fed 4,000 mg total PAH (mostly as lower molecular weight compounds) per kg body weight for seven months, no mortality or visible signs of toxicity resulted. Other effects were noted, however, including an average increase in liver size of 25 percent, and increased blood flow to the liver of 30 percent.

In the second study, adverse sublethal effects were noted at concentrations of between 0.036 and 0.18 μg PAH per egg following application of various PAHs (e.g., chrysene and benzo(a)pyrene) to the surface of mallard eggs (Eisler 1987). It has been suggested that the presence of PAHs in petroleum may confer many of the well-documented adverse biological effects reported after eggs have been exposed to polluting oils (Albers and Gay 1982; Hoffman and Albers 1984).

Schafer et al. (1983) reported acute oral effect levels for two passerine bird species (red-winged blackbird and house sparrow) and three PAH compounds. LD_{50} values

TABLE 9-9.   Acute toxicity values for laboratory mammals by exposure pathway.

Compound	Ingestion		Dermal		Inhalation	
	Value mg/kg	Effect	Value mg/kg	Effect	Value mg/m³	Effect
Benzene	930	LD_{50}	ND[1]	—	32000	LC_{50}
Toluene	2000	LD_{50}	12124	LD_{50}	15000	LC_{50}
Xylene	2000	LD_{50}	ND	—	13000	LC_{50}
Naphthalene	1780	LD_{50}	ND	—	ND	—
Phenanthrene	700	LD_{50}	ND	—	ND	—
Anthracene	ND	—	ND	—	ND	—
Fluoranthene	2000	LD_{50}	3180	LD_{50}	ND	—
Benzo(a)anthracene	ND	—	ND	—	ND	—
Chrysene	ND	—	ND	—	ND	—
Benzo(b)fluoranthene	ND	—	ND	—	ND	—
Benzo(k)fluoranthene	ND	—	ND	—	ND	—
Benzo(a)pyrene	50	LD_{50}	ND	—	ND	—
Dibenz(a,h)anthracene	ND	—	ND	—	ND	—
Indeno(1,2,3-cd)pyrene	ND	—	ND	—	ND	—

[1]No data.

Note: LD_{50} values are for single doses. LC_{50} values are for single, short-term (generally 4-hour) exposures.

References: U.S. EPA (1984a; 1984b; 1984c; 1984d); U.S. PHS (1987a; 1987b; 1987c; 1987d; 1987e; 1987f; 1989); Clement Associates (1985); Eisler (1987); IRIS (1991); RTECS (1991); HSDB (1991).

TABLE 9-10.   Chronic toxicity values for laboratory mammals by exposure pathway.

Compound	Ingestion Value mg/kg/day	Effect	Dermal Value mg/kg	Effect	Inhalation Value mg/m³	Effect
Benzene	270	Fetal	1.20	Cancer	160	Fetal
Toluene	260	Fetal	ND[1]	—	320	Fertility
Xylene	21	Fetal	ND	—	50	Fetal
Naphthalene	533	Behavioral	ND	—	32	Cancer
Phenanthrene	ND	—	71	Cancer	ND	—
Anthracene	3300	Cancer	2500	Cancer	11	Fetal
Fluoranthene	250	Maternal	280	Cancer	ND	—
Benzo(a)anthracene	500	Cancer	0.34	Cancer	ND	—
Chrysene	99	Cancer	3.60	Cancer	ND	—
Benzo(b)fluoranthene	40	Cancer	2.90	Cancer	ND	—
Benzo(k)fluoranthene	72	Cancer	2820	Cancer	ND	—
Benzo(a)pyrene	10	Fetal	0.05	Cancer	0.05	Cancer
Dibenz(a,h)anthracene	38	Cancer	0.03	Cancer	ND	—
Indeno(1,2,3-cd)pyrene	72	Cancer	40	Cancer	ND	—

[1]No data.

References: U.S. EPA (1984a; 1984b; 1984c; 1984d); U.S. PHS (1987a; 1987b; 1987c; 1987d; 1987e; 1987f, 1989); Clement Associates (1985); Eisler (1987); IRIS (1991); RTECS (1991); HSDB (1991).

for acenaphthene, phenanthrene, and anthracene exceeded 100 mg/kg body weight (the highest dose tested) for these species.

Lower molecular weight PAHs are absorbed by plants more readily than higher molecular weights PAHs, probably due to their higher solubility in water. Documented phytotoxic effects of PAH exposure are rare. Most plants can catabolize benzo(a)pyrene, and possibly other PAHs, but pathways are not clearly defined (Eisler 1987). Some plants contain chemicals known to protect against PAH effects. In other plants, PAHs may act as growth hormones (Eisler 1987).

Benzene

Benzene appears to be of low acute toxicity when administered by various routes to laboratory animals (Clement Associates 1985). Most benzene hazards are associated with inhalation exposure; dermal exposure to liquid benzene may also produce various effects (Clement Associates 1985). Several adverse health effects from benzene exposure have been demonstrated for small mammals (rats and mice), however. Adverse effects include chromosomal damage (although benzene has not been found to be mutagenic in microorganisms), fetotoxicity, embryotoxicity and carcinogenicity. The most significant noncarcinogenic health effects of benzene are hematotoxicity (damage to the blood-forming mechanisms in bone marrow), immunotoxicity, and neurotoxicity (U.S. PHS 1987a). Hematotoxic effects have been correlated more strongly with chronic, long-term exposure than with acute, short-term exposure (U.S. PHS 1987a). Immunotoxic effects include damage to the lymphatic cells responsible for antibody production (B-cells) and self-mediated immunity (T-cells). Neurotoxic effects of inhalation exposure include disturbed neuronal transport characteristics, narcosis (relaxation), loss of reflex actions, and general behavioral alterations (U.S. PHS 1987a).

After absorption, benzene must undergo metabolic transformation to exert its toxic effects. Metabolism occurs primarily in the liver, where benzene is converted to benzene oxide and phenol. The bioaccumulation and rates of elimination of benzene differ from its metabolites. Excretion occurs primarily in the urine, in the form of phenols, or directly from the lungs, as benzene (U.S. PHS 1987a).

A considerable amount of data are available on specific dose limits and effects of benzene on laboratory animals (U.S. PHS 1987a). The LC_{50} value for a four-hour inhalation exposure to benzene was estimated to be 32,000 mg/m^3 (Table 9-9). For oral exposure, the LD_{50} was 930 mg/kg. Estimates of critical doses for chronic exposures are given in Table 9-10.

Some specific information is also available for aquatic organisms. The EC_{50} (concentration at which half the experimental animals show an effect) values for benzene in a variety of invertebrate and vertebrate freshwater aquatic species range from 5,300 μg/l to 386,000 μg/l (Clement Associates 1985). The concentration of 5,300 μg/l is listed in the U.S. EPA Quality Criteria for Water (U.S. EPA 1986) as the LOEL for acute toxicity of benzene. Because of the lack of sufficient data, no formal criteria have been established for acute or chronic benzene toxicity to aquatic life in freshwater (U.S. EPA 1986).

Toluene

Toluene is readily absorbed from the respiratory and GI tracts and, to a lesser degree, through the skin. Animals given toluene orally or by inhalation had high concentrations of toluene in their adipose tissue and bone marrow (U.S. PHS 1989). Because of its lipophilic properties, toluene has a moderate tendency to bioaccumulate, with the level of accumulation dependent on the degree to which the organism metabolizes toluene. Excretion occurs in some organisms within 12 hours of exposure (U.S. PHS 1989).

Although toluene has not been shown to be carcinogenic or mutagenic in animals or humans (U.S. EPA 1984b), several toxic and sublethal effects have been demonstrated at relatively high doses in laboratory animals. The most significant are embryotoxic and fetotoxic effects (Clement Associates 1985). Other effects include hearing impairment, decreased body and brain weight, lung and kidney damage, and growth inhibition (U.S. PHS 1989).

Synergistic effects have been documented for toluene and other toxic compounds. For example, co-administration of toluene along with benzene or styrene has been shown to suppress the metabolism of benzene or styrene in rats (Clement Associates 1985).

The embryotoxic and fetotoxic effects of toluene are perhaps the most significant. Oral administration of toluene at doses as low as 260 mg/kg produced a significant increase in embryonic death in mice (Table 9-10). Other, sublethal fetotoxic effects on mice include decreased fetal weight at doses of 434 mg/kg and increased incidence of cleft palate at 867 mg/kg (Clement Associates 1985). The oral and dermal LD_{50} values and inhalation LC_{50} values are estimated to be 2,000 mg/kg, 12,124 mg/kg, and 15,000 mg/m^3, respectively (Table 9-9).

Aquatic organisms are relatively insensitive to toluene. The EC_{50} and LC_{50} values for five freshwater species tested with toluene ranged from 12,700 to 313,000 μg/l (Clement Associates 1985). Two freshwater algal species tested with toluene were relatively insensitive, with EC_{50} values of at least 245,000 μg/l reported (Clement Associates 1985). The U.S. EPA (1986) lists 17,500 μg/l as the freshwater acute LOEL for toluene. No formal freshwater acute or chronic criteria have been established, however, because of the lack of sufficient data (U.S. EPA 1986).

Xylene

Although no carcinogenic, mutagenic, or teratogenic effects of xylene have been identified in rats and mice, xylene has been shown to be fetotoxic in both species. Acute exposure to high levels of xylene can also cause sublethal effects, including central nervous system damage and irritation of mucous membranes in adult rats and mice (Clement Associates 1985). The oral LD_{50} value for xylene in rodents is 2,000 mg/kg, and the LC_{50} for inhalation exposure is 13,000 mg/m^3 (Table 9-9). Critical doses for chronic exposure are given in Table 9-10.

Information on the toxicity of xylene to terrestrial wildlife or domestic animals is extremely limited. Because of the generally low acute toxicity of xylene observed in laboratory animals, it is unlikely that xylene would be highly toxic to wild or domestic birds and mammals (Clement Associates 1985). However, quail eggs exposed to an aqueous solution of xylene applied to the egg surface showed decreased hatch rates and embryo viability at concentrations greater than 0.05 percent (David 1982).

Some studies suggest that xylene adversely affects growth and survival of aquatic species. Xylene adversely affected adult trout at concentrations as low as 3.6 mg/l in a continuous flow system, and trout fry avoided xylene at concentrations greater than 0.1 mg/l (Clement Associates 1985). The LD_{50} value for adult trout was determined to be 13.5 mg/l (Clement Associates 1985). No Ambient Water Quality Criteria have been established for acute or chronic freshwater exposure to xylene.

Risk Characterization

The subsubsection Comparison to Assessment Endpoints qualitatively characterized the baseline risk to on-site biota from site contamination based upon the results of field studies. The results of that assessment reveal that certain on-site media pose a substantial risk and/or are having observable effects on specific biota. In this subsection, the baseline risk is quantified by comparing calculated ecological effect levels with observed contaminant concentrations in the media of concern. As defined earlier, ecological effect levels are the concentrations of particular contaminants (the contaminants of concern) in particular media (the media of concern) below which no demonstrable reproductive effects to ecological receptors are likely to occur. Ecological effect levels were developed based on established numerical criteria (e.g., interim sediment quality criteria) or on exposure pathway modeling using site-specific data, as well as information obtained from the literature. These effect levels can be used as input into the decisionmaking process as remediation goals and criteria are established. Thus, this analysis can be used by decisionmakers to evaluate the possible ecological implications of various remediation strategies. To further aid this process, possible ecological endpoints for restoring important fish and wildlife habitat are also outlined.

Modeling Methodology

Models relating contaminant exposure to effects on ecological receptors are generally unavailable (U.S. EPA 1989e). The estimation of exposure to aquatic receptors is more advanced than that of terrestrial systems and components (U.S. EPA 1989a). However, only a few aquatic exposure assessment models have been constructed to date. There are virtually no validated models for predicting multimedia exposure to terrestrial receptors. The U.S. EPA (1989a) estimates the state of the art in terrestrial modeling as poor for inhalation exposures and poor to fair for ingestion, dermal contact, and plant uptake.

Historically, exposure modeling in support of ecological risk assessments for organic toxicants has concentrated on aquatic exposures. Only limited work has been done to address the problem of estimating chemical exposures in terrestrial ecosystems. One such model (U.S. EPA 1989f) has been developed to predict pesticide exposure to small- and medium-sized birds in agricultural settings. Unfortunately, models such as this have not been expanded to deal with other organic compounds (such as PAHs and BTX) or other taxa. Other models (e.g., Newell et al. 1987) have been developed to predict trophic transfer within food chains but, unfortunately, do not relate the initial concentrations to levels of contamination in physical media. Because of these factors, it was necessary to construct simple exposure models for use in this study.

Wetland and Upland Soils.   The calculation of ecological effect levels for wetland and upland areas used selected mammals (mostly rodents) as representative organisms, since most of the toxicity data for PAHs and BTX (from laboratory studies utilizing mice and rats) are limited to this faunal group. Also, these species have direct contact with the soils, and are thus likely to represent organisms with direct exposure pathways to the most contaminated media on-site. However, it should be recognized that these organisms may not be the most sensitive species present at PSC to the particular site contaminants. For example, certain reptiles and amphibians are very sensitive to some toxic compounds (Hall 1980), but sufficient toxicity information was not available for PAHs and BTX to quantitatively characterize the risk to these organisms.

Ecological effect levels were intended to protect the reproductive viability of wildlife populations. Reproductive effects include maternal or paternal effects (e.g., reduced fertility), fetal effects (e.g., birth defects or low birth weight), and effects on young animals that would significantly decrease their chances of surviving to reproductive age (e.g., skin and lung cancers). The lowest doses (termed critical doses) known to produce adverse reproductive effects for dermal exposure (of young in burrows), inhalation exposure (in confined burrows), and ingestion exposure (from incidental ingestion of contaminated soils and consumption of contaminated food sources) were obtained from the literature, and are summarized in Table 9-10. These critical doses were then converted, for each organism and exposure pathway, to the soil concentration necessary to cause toxic effects using simple models. Insufficient data on some model inputs necessitated making assumptions on the values of some parameters. These assumptions were based upon the best available information and, when possible, the most conservative values were generally used. Ecological effect levels for a particular media were then based, on a compound-by-compound basis, on the pathway and animal whose model output resulted in the lowest soil concentration. This approach was taken because available information was insufficient to model total risk from multiple exposure pathways and simultaneous exposures to multiple contaminants.

Ecological effect levels (mg/kg soil) for ingestion exposure were calculated using the following formula:

$$\text{Effect Level} = \text{CD} \times \text{BW} \ / \ \text{F}$$

where:

CD = Critical dose of compound (mg/kg body weight/day)

BW = Body weight (kg)

F   = Food/Soil consumption per day (kg)

Critical doses were obtained from the literature, and are summarized in Table 9-10 for chronic effects; acute doses are also available (Table 9-9). Body weights represent average weights (of both sexes combined) for adult animals, and were obtained from Burt and Grossenheider (1976) and Chapman and Feldhamer (1982). Food consumption was obtained from Chapman and Feldhamer (1982), or if unavailable, calculated from an animal with a known rate of food consumption, using formulas in Robbins (1983) for basal metabolic rate. Since contaminant levels in food items were assumed to be in equilibrium with the soil (see following), the proportion of food and soil consumed is irrelevant.

The use of this ingestion formula involves several important assumptions:

1. Plant or animal tissues consumed have contaminant levels equal to the concentration in soil. This assumes that vegetation or prey species bioaccumulate contaminants to (but not above) soil levels, and that there is no biomagnification. This is likely to be a reasonable or even conservative assumption, especially for herbivorous rodents (of which four of the five representative species chosen are), based on the known fate and transport mechanisms for PAHs and BTX.
2. Absorption in the GI tract of contaminants from soil ingested incidental to feeding activities is the same as that for food items.
3. All food or media ingested is uniformly contaminated, that is, exposure is constant and at a consistent level.

Ecological effect levels (mg/kg soil) for dermal exposure were calculated using the following formula:

$$\text{Effect Level} = CD \times BW \times K \; / \; (CV \times SD \times RAF)$$

where:

$$CD \; = \; \text{Critical dose of compound (mg/kg body weight)}$$

$$BW \; = \; \text{Body weight (kg) of young}$$

$$K \quad = \; \text{Unit conversion constant } (1 \times 10^6)$$

$$CV \; = \; \text{Contact volume (cm}^3)$$

$$SD \; = \; \text{Soil density (kg/m}^3)$$

$$RAF = \; \text{Relative adsorption factor (\%)}$$

Critical doses were obtained from the literature and are summarized in Table 9-10 for chronic effects. Body weights represent average weights for young animals, and were obtained from Burt and Grossenheider (1976) and Chapman and Feldhamer (1982). Soil densities were 2047.5 kg/m^3 for wetland soils, and 1811.3 kg/m^3 for upland soils, based upon average moisture and organic matter levels. Relative Absorption Factors (RAFs), developed for human health risk assessments (U.S. EPA 1989g), describe the percentage of the compound that will be absorbed from soil in direct contact with skin. RAFs are five percent for PAHs and 50 percent for BTXs. While these values are probably not applicable to most adult mammals, due to the presence of a dense coat of fur, human-derived RAFs can probably be safely extrapolated to young mammals, as was done here, since the young are normally born hairless.

Contact volume was calculated using the following formula:

$$CV = 10 \times BW^{0.67} \times PC$$

where:

$$BW = \text{Body weight (g)}$$

$$PC = \text{Proportion in contact with soil (\%)}$$

Total surface areas were calculated based upon the relationship of total surface area (in cm^2) to the 0.67 power of body weight multiplied by 10 (Robbins 1983). The proportion of the total surface area located on the underside of an animal (PC) was calculated to be 0.222 (22.2%), based upon a *Peromyscus* mouse; this proportion was assumed to be similar for the other mammals. Underside surface areas were converted into contact volumes by assuming the animal was in contact with the top 1 cm of soil (the effective interface). It should be noted that, while alternative methods exist for calculating dermal exposure (U.S. EPA 1989h), this approach was considered the most suitable for dermal exposure of burrowing mammals.

Ecological effect levels (mg/kg soil) for inhalation exposure were calculated using the following formula:

$$\text{Effect Level} = CD \times (C1 \ / \ MW) \times R \times (K \ / \ VP) \times C2 \times CT$$

where:

$$CD = \text{Critical concentration of compound (mg/m}^3\text{)}$$

$$C1 = \text{Unit conversion constant (g/1000 mg)}$$

$$MW = \text{Molecular weight of coal tar (g/mole)}$$

$$R = \text{Gas constant } (8.2 \times 10^{-5} \text{ atm m}^3\text{/mole K)}$$

$$K = \text{Temperature (}^\circ\text{K)}$$

$$VP = \text{Vapor pressure (mm Hg)}$$

$$C2 = \text{Unit conversion constant (760 mm Hg/atm)}$$

$$CT = \text{Coal tar concentration (3500 mg/kg)}$$

This equation is based upon Raoult's Law and the Ideal Gas Law, and assumes that the concentrations of contaminants are small relative to the concentration of coal tar. Critical doses were obtained from the literature, and are summarized in Table 9-10 for chronic effects; acute doses are also available for BTX (Table 9-9). Doses were assumed to be independent of body weight, since all laboratory studies reported only the air concentration, without adjustment for the body weight of the test animals. The temperature was assumed to be 280°K (7°C). This equation assumes that the contaminants are volatilizing off of coal tar, as opposed to pore water or directly from soil particles. Coal tar concentrations of 2,000 to 5,000 mg/kg are indicative of heavily contaminated soils (Fitchko 1989); the midpoint of this range (3,500 mg/kg) was used for this calculation. The model also assumes that the calculated value is the equilibrium air concentration in a confined space (e.g., a burrow), which remains constant (i.e., any fresh air entering from outside of the burrow quickly reaches the equilibrium concentration for the contaminant). This model is somewhat simplified, in that it does not take into account such factors as soil moisture levels and relative humidity (for which representative values were difficult to obtain and factor into the equation), and thus gives equivalent values for both upland and wetland soils. It is also based on conservative assumptions (as are the ingestion and dermal models), which are likely to yield conservative effect levels.

Canal Sediments.   Based upon an analysis of the benthic data from ecological field studies, the sediments at PSC are having demonstrable, adverse effects on the benthic infauna. Sediment types are highly interspersed within the canal, and range from pure coal tar product, containing no benthic organisms, to silty clay, dominated by opportunistic benthic species. Because of the observed adverse effects to the benthic infauna, toxicity testing with on-site sediments was not necessary to determine baseline risk. Rather, interim sediment quality criteria (U.S. EPA 1988) were used to develop ecological effect levels for aquatic areas with sediments and permanent overlying water. These criteria were developed by the EPA utilizing the equilibrium partitioning approach, which uses chronic water quality criteria and back-calculates the sediment concentration needed to obtain this concentration in water. Since chronic water quality criteria are not currently available for PAH compounds, predicted fathead minnow chronic toxicity values are used (U.S. EPA 1988). This calculation of the partitioning between water and sediment is based upon the quantity of the sorbents in the sediment and the appropriate sorbent coefficient. Since the primary sorbent is assumed to be the organic carbon in the sediments, interim sediment quality criteria values must be normalized to the total organic carbon (TOC) content of the sediments. Thus, interim sediment quality criteria, which are available for six PAHs but not for benzene, toluene, or xylene, are designed to protect benthic organisms from exposure pathways involving contact with the interstitial (pore) water, and not the sediments directly.

Since PAHs are readily metabolized by many organisms and tend not to bioaccumulate in the long term, food-chain effects are not of major concern. Thus, it is assumed that ecological effect levels based upon the protection of primary consumers inhabiting the benthos will also protect higher-level organisms that may feed on them.

The final ecological effect level was based upon the PAH with the lowest interim sediment quality criteria (of the six compounds for which criteria have been developed), provided that the existing levels observed in the canal violated the criteria for that compound. A TOC level of five percent was assumed when calculating the effect level, since this was approximately the lowest value observed during sediment sampling in the canal. The lowest observed TOC value was used since the bioavailability of the contaminants decreases as TOC increases, since more organic carbon is available for adsorbing the contaminants. This five percent level is also consistent with levels observed at the reference site. The ecological effect level would simply be the criteria value multiplied by 0.05. PAH effect levels for the individual compounds were converted to equivalent total PAH (tPAH) values based upon empirical ratios obtained from site-specific data (M&E 1992).

Ecological Effect Levels For Wetland Soils
Of the three wetland species considered (muskrats, beavers, and mink), the muskrat yielded the lowest ecological effect level. This is not surprising, since dermal and ingestion exposure effects are highly dependent upon body weight, and the muskrat weighs the least of these three species. Inhalation exposure, assumed to be independent of body weight, yielded equivalent results for each of the three species considered. Thus, the muskrat was used to derive the ecological effect level for wetland soils in emergent wetland areas. Since muskrats are not likely to extensively use forested or

scrub–shrub wetlands, as these areas lack suitable food supplies, effect levels based upon muskrat exposure were deemed unsuitable for these habitat types. Since beavers will preferentially utilize these habitats, ecological effect levels for wooded wetland types were based upon beaver exposure to contaminated soils.

Emergent Wetlands.   Table 9-11 shows the emergent wetland ecological effect levels for the critical exposure pathway for each compound, based upon chronic effects. Of the volatile organics, benzene had the lowest ecological effect level (0.286 mg/kg soil), based upon an inhalation exposure. This ecological effect level was less than both the mean and maximum soil concentration observed in the PSC emergent wetland soils (Table 9-11). Of the PAHs, dibenz(a,h)anthracene had the lowest effect level (0.568 mg/kg soil), based upon a dermal exposure. Benzo(a)pyrene had a similar level (0.947 mg/kg soil), also based upon a dermal exposure. The ecological effect levels for both compounds were less than the mean and maximum concentrations observed in PSC emergent wetland soils (Table 9-11). PAH effect levels for the individual compounds were converted to equivalent total PAH (tPAH) values based upon empirical ratios obtained from site-specific surface soil data (M&E 1992). When converted, benzo(a)pyrene yielded the lowest ecological effect level (13.7 mg tPAH/kg soil).

Although not considered when setting ecological effect levels, soil concentrations necessary to cause acute effects (death) were also calculated for compounds and pathways with sufficient data. Xylene was the only compound whose mean or maximum observed soil concentration exceeded the acute effect level.

TABLE 9-11.   Pine Street Canal emergent wetland areas—Ecological effect levels.

Compound	Animal	Critical Pathway	Effect	Effect Level (mg/ kg soil)	Existing Site Conc.[1]
Benzene	Muskrat	Inhalation	Reproductive	0.286	**
Toluene	Muskrat	Inhalation	Reproductive	1.900	*
Xylene	Muskrat	Inhalation	Reproductive	1.110	**
Naphthalene	Muskrat	Inhalation	Lung Cancer[3]	30.9	**
Phenanthrene	Muskrat	Dermal[2]	Skin Cancer[3]	1344.1	
Anthracene	Muskrat	Inhalation	Reproductive	12835.4	
Fluoranthene	Muskrat	Ingestion	Reproductive	750.6	
Benzo(a)anthracene	Muskrat	Dermal	Skin Cancer	6.4	**
Chrysene	Muskrat	Dermal	Skin Cancer	68.2	.
Benzo(b)fluoranthene	Muskrat	Dermal	Skin Cancer	54.9	
Benzo(k)fluoranthene	Muskrat	Ingestion	Cancer	216.2	
Benzo(a)pyrene	Muskrat	Dermal	Skin Cancer	0.947	**
Dibenz(a,h)anthracene	Muskrat	Dermal	Skin Cancer	0.568	**
Indeno(1,2,3-cd)pyrene	Muskrat	Ingestion	Cancer	216.2	

[1]*—Effect level less than maximum observed soil concentration.
**—Effect level less than mean and maximum observed soil concentration.
—Effect level greater than maximum observed soil concentration.
[2]Only pathway with available toxicity information.
[3]Skin and lung cancers are considered reproductive effects as they influence the survival to reproductive age of young in burrows.

TABLE 9-12.   Pine Street Canal wooded wetland areas—Ecological effect levels.

Compound	Animal	Critical Pathway	Effect	Effect Level (mg/ kg soil)	Existing Site Conc.[1]
Benzene	Beaver	Inhalation	Reproductive	0.286	**
Toluene	Beaver	Inhalation	Reproductive	1.900	
Xylene	Beaver	Inhalation	Reproductive	1.110	
Naphthalene	Beaver	Inhalation	Lung Cancer[3]	30.9	
Phenanthrene	Beaver	Dermal[2]	Skin Cancer[3]	2428.7	
Anthracene	Beaver	Inhalation	Reproductive	12835.4	
Fluoranthene	Beaver	Ingestion	Reproductive	1515.2	
Benzo(a)anthracene	Beaver	Dermal	Skin Cancer	11.6	*
Chrysene	Beaver	Dermal	Skin Cancer	123.1	
Benzo(b)fluoranthcne	Beaver	Dermal	Skin Cancer	99.2	
Benzo(k)fluoranthene	Beaver	Ingestion	Cancer	436.4	
Benzo(a)pyrene	Beaver	Dermal	Skin Cancer	1.71	**
	Beaver	Ingestion	Reproductive	60.6	
Dibenz(a,h)anthracene	Beaver	Dermal	Skin Cancer	1.03	**
Indeno(1,2,3-cd)pyrene	Beaver	Ingestion	Cancer	436.4	

[1]*—Effect level less than maximum observed soil concentration.
**   Effect level less than mean and maximum observed soil concentration.
—Effect level greater than maximum observed soil concentration.
[2]Only pathway with available toxicity information.
[3]Skin and lung cancers are considered reproductive effects as they influence the survival to reproductive age of young in lodges.

Wooded Wetlands.   Table 9-12 shows the wooded wetland ecological effect levels for the critical exposure pathway for each compound, based upon chronic effects. Of the volatile organics, benzene had the lowest effect level (0.286 mg/kg soil), based upon an inhalation exposure. This ecological effect level was less than both the mean and maximum soil concentration observed in the PSC wooded wetland soils (Table 9-12).

Since beavers are likely to build their lodges on the canal edge adjacent to wooded wetland habitat types, a more restrictive ecological effect level for PAHs was calculated for wooded wetland areas within 3 m of the canal's edge. This level was based upon individual effects (cancer) from a dermal exposure of beavers to benzo(a)pyrene (BaP), similar to the approach taken with muskrats. Effect levels were 1.71 mg BaP/kg soil or 24.8 mg tPAH/kg soil. The ecological effect level for benzo(a)pyrene was less than the mean and maximum concentrations observed in PSC wooded wetland soils (Table 9-12).

Since beavers are likely to utilize wooded wetlands more than 3 m from the canal bank solely for foraging (as opposed to lodge construction), PAH effect levels for these habitats were based upon an ingestion exposure. Benzo(a)pyrene had the lowest ecological effect level. Effect levels were 60.6 mg BaP/kg soil or 878.4 mg tPAH/kg soil. This ecological effect level for benzo(a)pyrene exceeded both the mean and maximum concentration observed in PSC wooded wetland soils (Table 9-12).

Although not considered when setting ecological effect levels, soil concentrations necessary to cause acute effects (death) were also calculated for compounds and pathways with sufficient data. No compound had a mean or maximum observed soil concentration that exceeded the acute effect level.

Ecological Effect Levels for Upland Soils

Of the two upland species considered (*Peromyscus* mice and woodchuck), the mouse yielded the lowest ecological effect levels. Table 9-13 shows the ecological effect levels for the critical exposure pathway for each compound, based upon chronic effects.

Of the volatile organics, benzene had the lowest ecological effect level (0.286 mg/kg soil), based upon an inhalation exposure. This level was less than the maximum soil concentration observed in PSC upland soils but exceeded the mean soil concentration (Table 9-13). Of the PAHs, dibenz(a,h)anthracene had the lowest ecological effect level (0.20 mg/kg soil), based upon a dermal exposure. Benzo(a)pyrene had a similar level (0.33 mg/kg soil), also based upon a dermal exposure. The ecological effect levels for both compounds were less than the mean and maximum concentrations observed in PSC upland soils (Table 9-13). When converted to equivalent total PAHs, benzo(a)pyrene yielded the lowest ecological effect level, 4.8 mg tPAH/kg soil.

Although not considered when setting ecological effect levels, soil concentrations necessary to cause acute effects (death) were also calculated for compounds and pathways with sufficient data. No compound had a mean or maximum observed soil concentration that exceeded the acute effect level.

The upland ecological effect level for PAHs appears to be conservative. More mice were caught in the shrubby area near Maltex Pond, including a lactating female and an immature, than in any other location on-site during small mammal trapping efforts. This suggests that the mouse population was successfully reproducing in this area. Surface soil data from this area (M&E 1992) show benzo(a)pyrene levels at about twice the calculated ecological effect level. The uncertainty in the analysis, which is very difficult to quantify (U.S. EPA 1989c), probably accounts for this discrepancy.

TABLE 9-13.   Pine Street Canal upland areas—Ecological effect levels.

Compound	Animal	Critical Pathway	Effect	Effect Level (mg/ kg soil)	Existing Site Conc.[1]
Benzene	*Peromyscus*	Inhalation	Reproductive	0.286	*
Toluene	*Peromyscus*	Inhalation	Reproductive	1.900	*
Xylene	*Peromyscus*	Inhalation	Reproductive	1.110	*
Naphthalene	*Peromyscus*	Inhalation	Lung Cancer[3]	30.9	
Phenanthrene	*Peromyscus*	Dermal[2]	Skin Cancer[3]	466.0	
Anthracene	*Peromyscus*	Ingestion	Cancer	3657.6	
Fluoranthene	*Peromyscus*	Ingestion	Reproductive	277.1	
Benzo(a)anthracene	*Peromyscus*	Dermal	Skin Cancer	2.2	*
Chrysene	*Peromyscus*	Dermal	Skin Cancer	23.6	
Benzo(b)fluoranthene	*Peromyscus*	Dermal	Skin Cancer	19.0	
Benzo(k)fluoranthene	*Peromyscus*	Ingestion	Cancer	79.8	
Benzo(a)pyrene	*Peromyscus*	Dermal	Skin Cancer	0.3	**
	Peromyscus	Ingestion	Reproductive	11.1	*
Dibenz(a,h)anthracene	*Peromyscus*	Dermal	Skin Cancer	0.2	**
Indeno(1,2,3-cd)pyrene	*Peromyscus*	Ingestion	Cancer	79.8	

[1]*—Effect level less than the maximum observed soil concentration.

**—Effect level less than the mean and maximum observed soil concentration.

—Effect level greater than the maximum observed soil concentration.

[2]Only pathway with available toxicity information.

[3]Skin and lung cancers are considered reproductive effects as they influence the survival to reproductive age of young in burrows.

While it is difficult to correlate trap data with soil sampling data, these results suggest that the calculated ecological effect level is conservative by at least a factor of two. This may be due to the assumption that dermal effect levels (cancer) would impact mouse reproduction due to decreased survival of young to reproductive age. Because mice tend to have naturally short life-spans, and early litters may reproduce the same year they are born (age to sexual maturity is about two months; DeGraaf and Rudis 1983), this assumption may not be totally valid for these organisms at the observed contaminant levels. Also, upland soils tend to have low moisture levels, which may influence the amount of contaminant that is transferred from the soil to the skin of the organism, decreasing the exposure to below the calculated level (which is based upon laboratory studies that use an aqueous solution during application), even after adjustment using relative absorption factors. Another factor that may influence the exposure is that mice may make their nests in litter, logs, rocks, and other debris (DeGraaf and Rudis 1983), not necessarily always in soil burrows, which would decrease their direct exposure to bare soil. Using the next highest effect level from an oral exposure to benzo(a)pyrene (10 mg/kg body weight) with demonstrated direct effects (fetal) on reproducing mice, the calculated ecological effect level would be 11.1 mg/kg soil (or 160.6 mg tPAH/kg soil). The maximum observed soil concentration for benzo(a)pyrene (13 mg/kg) in upland soils still exceeded this revised effect level.

Ecological Effect Levels for Canal Sediments

Since interim sediment quality criteria have not yet been developed for benzene, toluene, or xylene, ecological effect levels were calculated only for PAHs, based on existing criteria for six PAH compounds. Phenanthrene has the lowest criteria value of these six compounds (139 mg/kg carbon), and the observed levels in the canal exceeded the criteria value by four (at 15 percent TOC) to 33 (at five percent TOC) times (Table 9-14). Thus, effect levels were based upon phenanthrene and assumed a minimum TOC content of five percent, making the ecological effect level 6.95 mg/kg sediments ($139 \times 0.05 = 6.95$) (Table 9-15). When converted to total equivalent PAH, the effect level is calculated to be 42.4 mg tPAH/kg sediment.

Summary of Ecological Effect Levels

The ecological effect levels (Table 9-16) are as follows:

1. Effect levels for total PAHs in emergent wetland soils were 13.7 mg/kg soil (based upon a dermal exposure of muskrats to benzo(a)pyrene); for BTX, the level was 0.286 mg/kg (based upon an inhalation exposure of muskrats to benzene).
2. Effect levels for total PAHs in wooded wetland soils within three m of the canal bank were 24.8 mg/kg soil (based upon a dermal exposure of beavers to benzo(a)pyrene). For wooded wetland soils more than three m from the canal bank, total PAH effect levels were 878.4 mg/kg soil (based upon an ingestion exposure of beavers to benzo(a)pyrene). For BTX (both areas), the level was 0.286 mg/kg (based upon an inhalation exposure of beavers to benzene).
3. Effect levels (effects on individual animals) for total PAHs in upland soils were 4.8 mg/kg soil (based upon a dermal exposure of *Peromyscus* mice to benzo(a)pyrene). Total PAH effect levels based upon reproductive population effects resulted in a soil level of 160.6 mg/kg soil (based upon an ingestion exposure of mice to benzo(a)pyrene). For BTX, the level was 0.286 mg/kg (based upon an inhalation exposure of mice to benzene).

TABLE 9-14.   Pine Street Canal surface sediment and wetland surface soil quality.

Station	Total Organic Carbon (mg/kg)	Acenaphthene Measured (mg/kg)	Acenaphthene Normalized (mg/kg C)	Phenanthrene Measured (mg/kg)	Phenanthrene Normalized (mg/kg C)	Fluoranthene Measured (mg/kg)	Fluoranthene Normalized (mg/kg C)
PS-1	43000[a]	14[b]	326	5.5	128	9.2	214
PS-2	100000[a]	43[b]	430	9.4	94	15	150
PS-3	280000[a]	9	32	25	89	19	68
1989	50000[d]	130[e]	2600[c]	230[e]	4600[c]	86[e]	1720
Canal	150000	130[e]	867[c]	230[e]	1533[c]	86[e]	573
	350000	130[e]	371	230[e]	657[c]	86[e]	246
Wetland	50000[d]	330[e]	6600[c]	480[e]	9600[c]	120[e]	2400[c]
	150000	330[e]	2200[c]	480[e]	3200[c]	120[e]	800
	350000	330[e]	943[c]	480[e]	1371[c]	120[e]	343
Criteria			732		139		1883

Station	Total Organic Carbon (mg/kg)	Pyrene Measured (mg/kg)	Pyrene Normalized (mg/kg C)	Benzo(a)pyrene Measured (mg/kg)	Benzo(a)pyrene Normalized (mg/kg C)	Benzo(a)anthracene Measured (mg/kg)	Benzo(a)anthracene Normalized (mg/kg C)
PS-1	43000[a]	13	302	4.1	95	3.4	79
PS-2	100000[a]	19	190	8.2	82	8.6	86
PS-3	280000[a]	33	118	12	43	14	50
1989	50000[d]	140[e]	2800[c]	32[e]	640	43[e]	860
Canal	150000	140[e]	933	32[e]	213	43[e]	287
	350000	140[e]	400	32[e]	91	43[e]	123
Wetland	50000[d]	210[e]	4200[c]	17.5[e]	350	57[e]	1140
	150000	210[e]	1400[c]	17.5[e]	117	57[e]	380
	350000	210[e]	600	17.5[e]	50	57[e]	163
Criteria			1311		1063		1317

[a]Measured value—August 1990.
[b]Detection limit.
[c]Exceeds U.S. EPA interim sediment quality criteria (mean value).
[d]TOC data unavailable; a representative range of values is used.
[e]Maximum observed concentration.

TABLE 9-15.   Pine Street Canal aquatic areas—Ecological effect levels.

Compound	Interim Sediment Criteria (mg/kg C)[a]	Normalized Effect Level (mg/kg)[b]
Acenaphthene	732	36.60
Phenanthrene	139	6.95
Fluoranthene	1883	94.15[c]
Pyrene	1311	65.55[c]
Benzo(a)anthracene	1317	65.85[c]
Benzo(a)pyrene (Phenanthrene)	1063	53.15[c]
Effect Level		6.95

[a]U.S. EPA (1988).
[b]Effect levels are based on a minimum total organic carbon content of 5 percent.
[c]Maximum observed value less than effect level.

TABLE 9-16.   Summary of ecological effect levels for PAHs—Pine Street Canal.

Area	Species	Effect Levels (mg/kg) Actual	Total PAH Factor	Total PAH	Criteria
Emergent wetland	Muskrat	0.95	0.069	13.7	Individual effect (cancer). Dermal exposure: benzo(a)pyrene.
Wooded wetland < 3 m from bank	Beaver	1.71	0.069	24.8	Individual effect (cancer). Dermal exposure: benzo(a)pyrene.
Wooded wetland > 3 m from bank	Beaver	60.61	0.069	878.4	Population effect (reproduction). Ingestion exposure: benzo(a)pyrene.
Upland	Mouse	11.08	0.069	160.6	Population effect (reproduction). Ingestion exposure: benzo(a)pyrene.
Canal	—	6.95	0.164	42.4	Based upon interim sediment quality criterion for phenanthrene at a TOC of 5 percent.

4. Effect levels for total PAHs in canal sediments were 42.4 mg/kg sediment (based on the interim sediment quality criterion for phenanthrene). This level is based on a five percent total organic carbon content in the sediment.

In both media (soils and sediments), the minimum excavation depth for preventing ecological exposure to contaminated soils or sediments is approximately 1.5 m, provided that clean fill is used to return the areas to their former grade following excavation. This is the approximate maximum depth of muskrat and woodchuck burrows (Perry 1982; Lee and Funderburg 1982), the species most likely to dig deep burrows in wetland and upland areas, respectively, on the PSC site and should also be sufficient for the reestablishment of benthic organisms in canal sediments.

The ecological effect levels from the models agree closely with the few published soil criteria (Fitchko 1989; Beyer 1990). Benzene soil effect levels reported in the literature vary from 0.04 to 1.0 mg/kg; 0.3 mg/kg was the most common value (two studies). This latter value is quite close to the model result of 0.286 mg/kg. Benzo(a)pyrene effect levels found in the literature range widely from 0.004 to 1.0 mg/kg; a value of 0.02 was considered protective of biological resources (Beyer 1990). Model results from PSC (based upon benzo(a)pyrene levels) were consistently higher than the literature values, with effect levels of 0.947 mg/kg in emergent wetland soils, 1.71–60.61 mg/kg in wooded wetland soils, and 0.33–11.1 mg/kg for upland soils.

Characterization of Baseline Risk

Baseline risks to ecological receptors from exposure to the contaminants of concern can be characterized by comparing the calculated ecological effect levels to the observed contaminant concentrations in the media of concern. If ecological effect levels are higher than the observed maximum concentration of a contaminant, no adverse effects are likely to occur and the risks are minimal. If effect levels are less than maximum concentrations, some adverse effects are possible in isolated areas of the site containing these elevated concentrations. If effect levels are less than the mean concentration of a contaminant, this suggests that potential adverse impacts could be widespread over the site.

In emergent wetland areas, PAH (based on benzo(a)pyrene) and volatile organic (based upon benzene) effect levels were both less than their respective mean and maximum observed site concentrations in soils, suggesting potential adverse effects. In wooded wetland areas, volatile organic (based upon benzene) effect levels were also less than the observed mean and maximum observed soil concentration. For PAHs, effect levels for wooded wetland areas within three meters of the canal bank were less than the observed mean and maximum observed soil concentration, indicating the potential for adverse effects. For wooded wetland areas more than three meters from the canal bank, the effect level exceeded the maximum observed soil concentration, suggesting that risks in these areas are likely to be minimal.

Observed PAH concentrations in the canal sediments exceeded interim sediment quality criteria for three of the six compounds with interim criteria values. Phenanthrene concentrations exceeded criteria values by four to 33 times, depending upon the level of total organic carbon present. Thus, the potential for adverse effects from exposure to canal sediments is relatively high. This is supported by field observations of adverse effects to the benthic infauna inhabiting the canal.

In upland areas, PAH (based on benzo(a)pyrene) and volatile organic (based upon benzene) effect levels were both less than their maximum observed site concentrations in soils but greater than the observed mean soil concentrations. This suggests that potential adverse effects would be limited to relatively small areas containing relatively high concentrations of contaminants. More widespread effects would not be expected.

Uncertainty
The ecological risk assessment analysis has an unquantifiable uncertainty associated with it. This uncertainty is attributable to four main factors:

1. Insufficient data on some model inputs necessitated assumptions that could not be tested. Since, where possible, conservative assumptions were made, this may possibly lead to overstatement of the potential risks to ecological receptors.
2. Extrapolating toxic effect data from laboratory studies of mice and rats to mammalian wildlife species may overstate the risk to ecological receptors, as limited studies have shown that laboratory animals tend to be more sensitive than wild mammals.
3. Insufficient data on exposure frequencies and the effects of chemical interactions on toxicity did not allow total risks from all exposure pathways and chemicals to be modeled simultaneously, resulting in a possible understatement of the risk to ecological receptors.
4. Species other than mammals (e.g., reptiles and amphibians), for which effect data were unavailable and which spend much of their life in contact with the canal sediments, may be adversely affected at concentrations below the effect levels calculated for the modeled mammals. This would potentially understate the risk to ecological receptors.

Protection of Ecological Values
The following is a general discussion of the existing ecological values of the PSC site to be considered for protection during remediation or enhancing/restoring following

remediation, as well as potential impacts of remedial activities on these values. In order to accomplish the preservation or restoration of important ecological values, a plan outlining the goals (termed ecological endpoints) and methodology needs to be formulated. The guidelines in this subsubsection provide one possible scenario for the creation of such a plan. The example scenario addresses restoration on-site, but the same goals and methodology would be applicable to off-site restoration or habitat creation.

Terrestrial Habitats.   The wildlife community inhabiting the upland areas of the PSC site appears to be generally healthy and diverse. No readily demonstrable effects could be linked with site contamination of upland soils, although soil invertebrates were not studied. The baseline risk assessment results suggest that potential adverse effects would be limited to relatively small areas with high concentrations of contamination. These "hot spots" do not contain any critical wildlife habitat but rather consist of easily replaceable species (such as shrub-stage aspen and various herbaceous plants) which could be replanted following remediation.

The activity with the greatest potential to adversely impact terrestrial habitats is the potential disturbance of large areas of upland habitat associated with remediation equipment, storage areas, and other facilities, and as a result of remediation activities on adjacent wetland or aquatic areas (such as the construction of graded slopes or access corridors). The potential disturbance would vary depending upon the remediation technology chosen, how much of the processing will occur on-site, and the extent of remediation on adjacent wetland and aquatic areas. For example, composting contaminated media on-site will require large amounts of land, while incinerating contaminated media off-site will result in relatively little disturbance to the upland areas. Deep excavation of canal sediments may require graded side slopes, which would significantly encroach upon upland areas.

Most of the shrubby or herbaceous upland areas on-site could be easily, effectively, and relatively inexpensively returned to their original conditions if disturbed during remediation. It would be desirable, from an ecological perspective, to protect from disturbance upland areas containing relatively mature trees, as these resources would be difficult to replace in a reasonable time frame. Thus, one potential ecological endpoint for upland areas may be the preservation of mature forested areas, and the restoration of disturbed shrubby and herbaceous areas to existing conditions.

Vegetated Wetland Habitats.   The comparison to assessment endpoints did not clearly demonstrate adverse effects that could be definitively attributed to contamination of the wetland soils, although the apparently reduced abundance of muskrats (and possibly frogs and turtles) in the more contaminated vegetated wetland areas suggests a possible effect. The baseline risk assessment results suggest that contaminant levels in the wetland soils, especially in the southern portion of the site, are likely to cause chronic (reproductive), but not acute, adverse effects to wildlife species.

A potential ecological endpoint for vegetated wetland habitats may be the protection and perpetuation of the present wetland ecosystem, including both the wooded and emergent wetlands, since this habitat type is generally uncommon in the site vicinity, due to the urban nature of surrounding land uses. The on-site beaver colony is a key component of this wetland system. While it is probably not feasible to protect the colony during remediation activities, protection and enhancement of beaver habitat would allow the site to be recolonized (naturally or by introduction) by this species following remediation. The maintenance of this wetland system would provide a benefit to a multitude of species which live in and depend upon this habitat. These species include, in addition to the beaver, amphibians, reptiles, fish, aquatic invertebrates, wading birds, songbirds, ducks, muskrats, and mink. Of particular ecological importance is the preservation of the mature forested wetlands, since the live and dead trees in these areas are important habitat resources that are difficult to replace in a reasonable time frame. The results of the wetland functional assessment, previously described, could also be used as a guide to protect or enhance other wetland functions and values.

Protection of the critical habitat components of the beaver involves the maintenance of adequate water levels and the establishment of an adequate supply of forage and building materials in the form of deciduous trees and shrubs within 100 meters of the low water edge (Allen 1983). Deciduous forested wetlands and uplands, dominated by alder, cottonwood, aspen, maple, and willow are a preferred source of winter food, as well as building materials to construct lodges and dams.

Preserving or restoring the hydrology and water depth on the site would also be necessary to provide suitable beaver habitat. At the water's edge, it would be desirable to create a combination of steep banks interspersed with areas with more gradual banks. Steep banks, lined with downed trees and rocks (which currently exist on-site), provide perching locations for wading birds, basking sites for turtles, cover for fish, and many other ecological values. The more gradual banks would provide beavers access to land areas to forage, and provide suitable locations for muskrats to construct bank burrows.

If remediation requires the destruction of forested areas within 100 meters of the water's edge, the areas could be replanted with woody, deciduous species preferred by beavers. If snags, or live mature trees with the potential to become snags are removed during remediation, bird nest boxes, in various sizes to accommodate species such as wood ducks and tree swallows, could be erected to replace some of the ecological values lost. Emergent wetland areas that are disturbed during remediation could be replanted with suitable plant species in such a manner as to ensure good water/vegetation interspersion. This will replace critical habitat for muskrats, ducks, and various other aquatic and semiaquatic species that feed or nest in these areas.

Aquatic Habitats.   The existing condition of the aquatic sediments at the PSC site does not meet the assessment endpoints, as the sediments in the canal are having demonstrable toxic effects on benthic organisms. The baseline risk assessment results also suggest that the potential for adverse effects from exposure to canal sediments is relatively high.

One possible ecological endpoint for PSC aquatic habitats is the restoration of a healthy and functioning benthic infauna. One possible way to meet this ecological endpoint would be to remove the existing canal sediments.

If the sediments are removed, one alternative is to return them to the canal following treatment. Possible sediment treatment alternatives currently available include incineration and solvent extraction. If incineration is used, the end-products would not be suitable for replacement into the canal, at least within the top meter, since incineration will produce ash. In order for the canal sediments to be recolonized by indigenous benthic species, the substrate should be composed of fine material such as silts and clays, with a suitable component of organic matter, to allow organisms to burrow and feed.

If solvent extraction is used as the treatment process, the end-product could be suitable for retention. However, it would be prudent to conduct toxicity tests on the treated sediments before placing them back into the canal. The U.S. EPA has established protocols for aquatic toxicity testing, which are available for conducting freshwater sediment toxicity tests on chironomids, daphnids, fathead minnows, and amphipods. Treated sediments may need to be augmented with organic matter (to a minimum level of five percent by weight) to provide a suitable substrate.

ACKNOWLEDGMENTS

This case study is based upon the Supplemental Remedial Investigation and Baseline Risk Assessment reports for the Pine Street Canal Superfund Site prepared by Metcalf & Eddy for the U.S. EPA Region I. This material is used here with U.S. EPA permission. However, the opinions expressed herein in no way reflect the policies or opinions of the U.S. EPA.

S. Petron assisted in all aspects of the exposure modeling, and P. Gwinn, M. Doyle, and J. Young contributed to the development of the inhalation exposure model. M. Doyle and J. Young of Metcalf & Eddy, and R. Gilleland of the U.S. EPA, provided comments on draft versions of this chapter. K. Goodwin prepared the figure.

REFERENCES

Adamas, P. R., E. J. Chairain, Jr., R. D. Smith, and R. E. Young. 1987. *Wetland Evaluation Technique (WET), Volume II: Methodology*. Wetlands Research Program, U.S. Army Corps of Engineers, Environmental Laboratory, Operational Draft. USACOE/USDOT, Washington, D.C.

Albers, P. H. 1978. The effects of petroleum on different stages of incubation in bird eggs. *Bulletin of Environmental Contamination and Toxicology* 19:624–630.

Albers, P. H. and M. L. Gay. 1982. Unweathered and weathered aviation kerosene: chemical characterization and effects on hatching success of duck eggs. *Bulletin of Environmental Contamination and Toxicology* 28:430–434.

Allen, A. W. 1983. Habitat suitability index models: Beaver. U.S. Department of the Interior, Fish and Wildlife Service. FWS/OBS-82/10.30 Revised. 20 pp.

Allen, A. W. 1986. Habitat suitability index models: Mink. U.S. Department of the Interior, Fish and Wildlife Service. FWS/OBS-82/10.127 Revised. 23 pp.

Allen, A. W. and R. D. Hoffman. 1984. Habitat suitability index models: Muskrat. U.S. Department of the Interior, Fish and Wildlife Service. FWS/OBS-82/10.46. 27 pp.

Aston, R. J. 1973. Tubificids and water quality: A review. *Environmental Pollution* 5:1–10.

Bellrose, F. C. 1980. *Ducks, Geese, and Swans of North America*. Harrisburg, PA: Stackpole Books, 540 pp.

Berger, A. J. 1961. *Bird Study*. New York: Dover Publications, Inc., 389 pp.

Beyer, W. N. 1990. Evaluating soil contamination. U.S. Fish and Wildlife Service Biological Report 90(2). 25 pp.

Brinkhurst, R. O. and D. G. Cook. 1974. Aquatic earthworms: Annelida, Oligochaeta. In *Pollution Ecology of Freshwater Invertebrates*, C. W. Hart and S. H. Fuller (eds.), pp. 143–156. New York: Academic Press.

Burt, W. H. and R. P. Grossenheider. 1976. *A Field Guide to the Mammals, Third Edition*. Boston, MA: Houghton Mifflin Co., 289 pp.

Buttler, B. 1988. Peromyscus (Rodentia) as Environmental Monitors: A Bibliography. Presented at the 67th Annual Meeting of the American Society of Mammalogists, 21–25 June 1987.

Chapman, J. A. and G. A. Feldhamer (eds.). 1982. *Wild Mammals of North America: Biology, Management, and Economics*. Baltimore, MD: John Hopkins University Press, 1147 pp.

Clement Associates. 1985. Chemical, Physical, and Biological Properties of Compounds Present at Hazardous Waste Sites, Final Report. Prepared under subcontract to GCA Corporation for the U.S. EPA. Subcontract No. 1-625-999-222-003.

Cowardin, L. M., V. Carter, F. C. Golet, and E. T. LaRoe. 1979. Classification of Wetlands and Deepwater Habitats of the United States. U.S. Department of the Interior, Fish and Wildlife Service. FWS/OBS-79/31. 103 pp.

David, D. 1982. Cited in HSDB (1991).

DeGraaf, R. M. and D. D. Rudis. 1983. New England Wildlife: Habitat, Natural History, and Distribution. USDA Forest Service General Technical Report NE-108.

Eagle, T. C. and J. S. Whitman. 1987. Mink. In *Wild Furbearer Management and Conservation in North America*, M. Novak, J. A. Baker, M. E. Obbard, and B. Malloch (eds.), pp. 615–624. Ontario: Ministry of Natural Resources.

Edwards, N. T. 1983. Polycyclic aromatic hydrocarbons (PAHs) in the terrestrial environment—A review. *Journal of Environmental Quality* 12:427–441.

Ehrlich, P. R., D. S. Dobkin, and D. Wheye. 1988. *The Birder's Handbook: A Field Guide to the Natural History of North American Birds*. New York: Simon and Schuster, Inc., 785 pp.

Eisler, R., 1987. Polycyclic Aromatic Hydrocarbon Hazards to Fish, Wildlife, and Invertebrates: A Synoptic Review. U.S. Fish and Wildlife Service Biological Report 85(1.11).

Eisler, R. 1988. Lead Hazards to Fish, Wildlife, and Invertebrates: A Synoptic Review. U.S. Fish and Wildlife Service Biological Report 85(1.14).

FICWD (Federal Interagency Committee for Wetland Delineation). 1989. Federal Manual for Identifying and Delineating Jurisdictional Wetlands. U.S. Army Corps of Engineers, U.S. Environmental Protection Agency, U.S. Fish and Wildlife Service, and U.S.D.A. Soil Conservation Service, Washington D.C. Cooperative Technical Publication. 76 pp. plus appendices.

Fitchko, J. 1989. *Criteria for Contaminated Soil/Sediment Cleanup*. Northbrook, IL: Pudvan Publishing Co.

Gruendling, G. K. and D. J. Bogucki. 1978. Assessment of the Physical and Biological Characteristics of the Major Lake Champlain Wetlands. Prepared for the Wetlands Task Force, Lake Champlain Basin Study, New England River Basins Commission. Technical Report 5.

Hall, R. J. 1980. Effects of Environmental Contaminants on Reptiles: A Review. U.S. Fish and Wildlife Service Special Scientific Report—Wildlife No. 228. 12 pp.

Hart, C. W., Jr. and S. L. H. Fuller (eds.). 1974. *Pollution Ecology of Freshwater Invertebrates*. New York: Academic Press, Inc., 389 pp.

Hazardous Substances Data Bank (HSDB). 1991. Computerized database on the toxicology of hazardous chemicals.

Hickey, J. T., R. O. Bennett, and C. Merckel. 1990. Biological Indicators of Environmental Contaminants in the Niagara River: Histological Evaluation of Tissues From Brown Bullheads at the Love Canal-102nd Street Dump Site Compared to the Black Creek Reference Site. U.S. Fish and Wildlife Service, Cortland, N.Y. 124 pp.

Hill, E. P. 1982. Beaver. In *Wild Mammals of North America: Biology, Management, Economics*, J. A. Chapman and G. A. Feldhamer (eds.), pp. 256–281. Baltimore: Johns Hopkins Univ. Press.

Hoffman, D. J. and P. H. Albers. 1984. Evaluation of potential embryotoxicity and teratogenicity of 42 herbicides, insecticides, and petroleum contaminants to mallard eggs. *Archives of Environmental Contamination and Toxicology* 13:15–27.

Hoffman, D. J., B. A. Rattner, and R. J. Hall. 1990. Wildlife toxicology. *Environmental Science and Technology* 24:276–283.

Integrated Risk Information System (IRIS). 1991. Computerized database on health risks of chemicals, sponsored by the U.S. Environmental Protection Agency.

Kaufmann, J. H. 1982. Raccoons and Allies. In *Wild Mammals of North America: Biology, Management, Economics*, J. A. Chapman and G. A. Feldhamer (eds.), pp. 567–585. Baltimore: Johns Hopkins Univ. Press.

Lee, D. S. and J. B. Funderburg. 1982. Marmots. In *Wild Mammals of North America: Biology, Management, Economics*, J. A. Chapman and G. A. Feldhamer (eds.), pp. 176–191. Baltimore: Johns Hopkins Univ. Press.

Linscombe, G., N. Kinler, and R. J. Aulerich. 1982. Mink *(Mustela vison)*. In *Wild Mammals of North America: Biology, Management, Economics*, J. A. Chapman and G. A. Feldhamer (eds.), pp. 629–643. Baltimore: Johns Hopkins Univ. Press.

Major, J. T. and J. A. Sherburne. 1987. Interspecific relationships of coyotes, bobcats, and red foxes in western Maine. *Journal of Wildlife Management* 51:606–616.

Marquenie, J. M. and J. W. Simmers. 1984. Use of a Bioassay to Evaluate the Bioaccumulation of Contaminants by Animals Colonizing a Wetland Created With Contaminated Dredged Material. In The Third United States–Netherlands Meeting on Dredging and Related Technology, 10–14 September 1984, pp. 131–134. U.S. Army Corps of Engineers, Water Resources Support Center AD-A182 670.

Marquenie, J. M., J. W. Simmers, and S. H. Kay. 1987. Preliminary Assessment of Bioaccumulation of Metals and Organic Contaminants at the Times Beach Confined Disposal Site, Buffalo, N.Y. U.S. Army Corps of Engineers Miscellaneous Paper EL-87-6. 67 pp.

McClane, A. J. (ed.). 1974. *McClane's Field Guide to Freshwater Fishes of North America*. New York: Henry Holt and Co.

Metcalf & Eddy, Inc. (M&E). 1992. Supplemental Remedial Investigation Report. Pine Street Canal Site, Burlington, Vermont. Report prepared for U.S. EPA Region I under contract 68-W9-0036.

Myer, G. E. and G. K. Gruendling. 1979. Limnology of Lake Champlain. Prepared for the Eutrophication Task Force, Lake Champlain Basin Study, New England River Basins Commission. Technical Report 30.

Myers, T. 1991. Vermont Department of Fish and Wildlife. Personal Conversation.

National Oceanic and Atmospheric Administration (NOAA). 1990. The Potential for Biological Effects of Sediment-Sorbed Contaminants Tested in the National Status and Trends Program. NOAA Technical Memorandum NOS OMA 52.

Newell, A. J., D. W. Johnson, and L. K. Allen. 1987. Niagara River Biota Contamination Project: Fish Flesh Criteria for Piscivorous Wildlife. N.Y. State Department of Environmental Conservation Technical Report 87-3. 182 pp.

Novak, M. 1987. Beaver. In *Wild Furbearer Management and Conservation in North America*, M. Novak, J. A. Baker, M. E. Obbard, and B. Malloch (eds.), pp. 283–312. Ontario: Ministry of Natural Resources.

Parsons, J., E. Thompson, and T. Hudspeth. 1988. The Identification and Characterization of Burlington, Vermont's Wetlands and Significant Natural Areas, with Recommendations for Management. Prepared for the City of Burlington Community and Economic Development Office. 87 pp.

Perry, H. R., Jr. 1982. Muskrats. In *Wild Mammals of North America: Biology, Management, Economics*, J. A. Chapman and G. A. Feldhamer (eds.), pp. 282–325. Baltimore: Johns Hopkins Univ. Press.

Person, D. K. and D. H. Hirth. 1991. Home range and habitat use of coyotes in a farm region of Vermont. *Journal of Wildlife Management* 55:433–441.

Pettingill, O. S., Jr. 1970. *Ornithology in Laboratory and Field, Fourth Edition*. Minneapolis, MN: Burgess Publishing Company, 524 pp.

Registry of Toxic Effects of Chemical Substances (RTECS). 1991. Computerized database on the toxic effects of chemicals, maintained by the National Institute for Occupational Safety and Health (NIOSH).

Robbins, C. T. 1983. *Wildlife Feeding and Nutrition*. New York: Academic Press, 343 pp.

Schafer, E. W., Jr. and W. A. Bowles, Jr. 1985. Acute oral toxicity and repellency of 933 chemicals to house and deer mice. *Archives of Environmental Contamination and Toxicology* 14:111–129.

Schafer, E. W., Jr., W. A. Bowles, Jr., and J. Hurlbut. 1983. The acute oral toxicity, repellency, and hazard potential of 998 chemicals to one or more species of wild and domestic birds. *Archives of Environmental Contamination and Toxicology* 12:355–382.

Schmidt-Nielsen, K. 1983. *Animal Physiology: Adaptation and Environment.* London: Cambridge University Press, 619 pp.

Sims, R. C. and M. R. Overcash. 1983. Fate of polynuclear aromatic compounds (PNAs) in soil-plant systems. *Residue Reviews* 88:1–68.

Smith, R. L. 1980. *Ecology and Field Biology, Third Edition.* New York: Harper & Row Publishers, 835 pp.

U.S. EPA. 1979. Water-Related Environmental Fate of 129 Priority Pollutants, Volumes I and II. EPA-4401/479-029b.

U.S. EPA. 1984a. Health Effects Assessment for Xylene. EPA/540/1-86/006. 29 pp.

U.S. EPA. 1984b. Health Effects Assessment for Toluene. EPA/540/1-86/033. 33 pp.

U.S. EPA. 1984c. Health Effects Assessment for Benzo(a)pyrene. EPA/540/1-86/022. 31 pp.

U.S. EPA. 1984d. Health Effects Assessment for Dibenz(a,h)anthracene. EPA/540/1-86.

U.S. EPA. 1986. Quality Criteria for Water, 1986. Office of Water Regulation and Standards. EPA 440-5-86-001.

U.S. EPA. 1988. Interim Sediment Criteria Values for Nonpolar Hydrophobic Organic Compounds. Office of Water, Criteria and Standards Division. 34 pp.

U.S. EPA. 1989a. Superfund Exposure Assessment Manual Technical Appendix: Exposure Analysis of Ecological Receptors. Unpublished Draft Report.

U.S. EPA. 1989b. Supplemental Risk Assessment Guidance for the Superfund Program. Part 2—Guidance for Ecological Risk Assessments. Draft Final. EPA 901/5-89-001.

U.S. EPA. 1989c. Ecological Assessment of Hazardous Waste Sites: A Field and Laboratory Reference. EPA 600/3-89/013.

U.S. EPA. 1989d. Risk Assessment Guidance for Superfund, Volume II, Environmental Evaluation Manual. Interim Final. EPA 540/1-89/001.

U.S. EPA. 1989f. Terrestrial Ecosystem Exposure Assessment Model (TEEAM). EPA/600/3-88/038.

U.S. EPA. 1989g. Supplemental Risk Assessment Guidance for the Superfund Program. Part 1—Guidance for Public Health Risk Assessments. Draft Final. EPA 901/5-89/001.

U.S. EPA. 1989h. Exposure Factors Handbook. EPA/600/8-89/043.

U.S. Federal Highway Administration (U.S. FHA), 1977. Administrative Action Environmental Statement for the Burlington M 5000(1) Interstate 89 Connector Project.

U.S. Public Health Service (U.S. PHS). 1987a. Toxicological Profile for Benzene. Draft. 182 pp.

U.S. Public Health Service (U.S. PHS). 1987b. Toxicological Profile for Benzo(a)pyrene. Draft. 134 pp.

U.S. Public Health Service (U.S. PHS). 1987c. Toxicological Profile for Benzo(b)fluoranthene. Draft. 76 pp.

U.S. Public Health Service (U.S. PHS). 1987d. Toxicological Profile for Dibenz(a,h)anthracene. Draft. 86 pp.

U.S. Public Health Service (U.S. PHS). 1987e. Toxicological Profile for Chrysene. Draft. 80 pp.

U.S. Public Health Service (U.S. PHS). 1987f. Toxicological Profile for Benzo(a)anthracene. Draft. 88 pp.

U.S. Public Health Service (U.S. PHS). 1989. Toxicological Profile for Toluene. Draft. 115 pp.

Vermont Department of Fish and Wildlife. 1978. Lake Champlain Fisheries Study, Fish Population Inventory Final Report. Project F-12-R.

Vermont Department of Fish and Wildlife. 1984. Job Performance Report—Inventory, Evaluation, and Management of the Fisheries and Habitat of the Public Waters of Vermont: July 1, 1983—June 30, 1984. Project F-12-R-17.

Williamson, R. D. 1983. Identification of urban habitat components which affect eastern gray squirrel abundance. *Urban Ecology* 7:345–356.

Wren, C. 1987. Toxic substances in furbearers. In *Wild Furbearer Management and Conservation in North America*, M. Novak, J. A. Baker, M. E. Obbard, and B. Malloch, (eds.), pp. 930–936. Ontario: Ministry of Natural Resources.

10

Middle Marsh Ecological Assessment: A Case Study

Peter M. Boucher

Metcalf & Eddy, Inc.

INTRODUCTION

The Sullivan's Ledge landfill Superfund site is a quarry that was used between the 1930s and the 1970s as a disposal area for a variety of industrial wastes, including capacitors and transformers containing high levels of PCBs. Previous investigations (Ebasco Services Inc. 1989) focused primarily on the Sullivan's Ledge disposal area, and revealed that PCBs were present at high concentrations, of up to approximately 1000 mg/kg, in the soils of the disposal area. Since there is a stream immediately adjacent to the site, and much of the surface of the disposal area was unvegetated and had highly erodible soils, it is not surprising that the stream and other ecologically sensitive areas downgradient of the site, namely Middle Marsh and an adjacent wetland, were also contaminated with PCBs. Figure 10-1 illustrates the project site, including the former disposal area, the adjacent wetland, Middle Marsh, and an unnamed stream, which bisects Middle Marsh flowing north.

The site was studied in Phase I and II remedial investigations completed in 1987 and 1989 (Ebasco Services Inc. 1987 and 1989), and the conclusion reached that significant remediation was required to protect human health. The decision for the disposal area, outlined in a June 29, 1989 Record of Decision, was to excavate and solidify the surface soils and then return the stabilized soil to its original place, which would be capped and monitored. The sediments in the adjacent stream were found to be contaminated with PCBs for a distance of 1500 feet downstream of the disposal site. The remediation selected for these sediments was similar to that for the soils at the disposal area; excavation, encapsulation, and burial under an impermeable cap.

The soils adjacent to the stream in areas downgradient of the disposal area were also found to be contaminated. However, the remediation decision for these soils was not as simple or as straightforward as for the disposal area and sediments in the stream channel. The contamination outside the stream channel was not as severe, generally less than 20 mg/kg PCBs. Also, the nature and extent of contamination in the area adjacent to the stream seemed to be much more variable in time and space, which brought into question the seriousness and persistence of threats due to contamination in these areas. Even using the higher concentrations found outside of the stream channel, previous studies showed that the threat to human health was only marginal.

The final and, from an ecological standpoint, most important factor complicating the remediation decision downstream of the site was the existing conditions in the area. The stream supports extensive bordering vegetated wetland areas, including forested wetlands, locally known as Middle Marsh and the adjacent wetland. This wetland is a remnant of a much larger forested wetland system filled in as part of the depression-era construction of the surrounding New Bedford municipal golf course by the Civilian Conservation Corps. Middle Marsh is potentially an important ecological resource for all of the classical values associated with wetlands, including flood storage, wildlife habitat, retention of sediments, and nutrient regulation. The area also falls under the jurisdiction of a long list of site-specific ARARs such as executive orders, the Clean Water Act, state wetland regulations, and state special-status species regulations. Initial observations and investigations indicated that the wetland was achieving all expected ecological functions that are protected by the ARARs. Thus, the remediation decision was a classic dilemma of: should the wetland be destroyed to save it?

The 1989 Record of Decision initially included a plan of no action for Middle Marsh. However, the EPA withdrew the no action decision because of concerns raised by federal and state agencies over potential long-term impacts to National Resource Trustee species and other resources. The U.S. Department of Interior (DOI) and the Massachusetts Department of Environmental Protection raised concerns that if the PCB-contaminated sediments in Middle Marsh were not excavated, they may continue to pose a long-term threat to a variety of aquatic and terrestrial organisms that inhabit the Middle Marsh area. Given the ecological value of the contaminated forested wetland and streambank habitat, and the potential long-term threat to aquatic and terrestrial organisms, the EPA reconsidered the no action alternative, and determined that the remediation decision for Middle Marsh could not be made with the information collected as part of the disposal site RI/FS (Metcalf & Eddy 1991a and 1991b). Middle Marsh and the adjacent wetland were, thereafter, studied as a separate operable unit. There was just not a clear enough picture of the nature, extent, and severity of the contamination in Middle Marsh to formulate a decision acceptable to all parties involved. Further, previous investigations did not address the adjacent wetland. Also, the fate and transport of the PCBs in Middle Marsh were not sufficiently documented to evaluate future conditions under a range of remediation alternatives. Finally, the ecological implications of the contaminated soils outside of the stream bed, and the ramifications of large-scale excavation as part of remediation on the wetland functions, were probably the least understood variables and could not be evaluated based on the ecological information available.

A program was needed to investigate in detail each of these variables and attributes, so that the remediation decision could be made with a clear appreciation of the trade-offs between physical destruction and contamination. This case study describes this program, which consisted of an ecological assessment, and the use of the assessment in formulating the Record of Decision for Middle Marsh remediation. Specifically, this chapter describes the ecological work conducted during the Middle Marsh remedial investigation, as part of the feasibility study, and in support of the Record of Decision. The work was completed between October of 1989 and September of 1991, when a

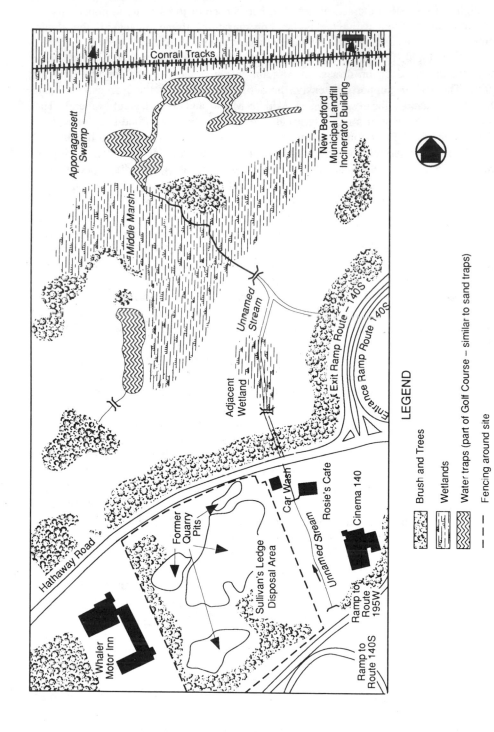

Conrail Tracks

Apponagansett Swamp

Middle Marsh

New Bedford Municipal Landfill Incinerator Building

Unnamed Stream

Adjacent Wetland

Exit Ramp Route – 140S

Entrance Ramp Route 140S

Hathaway Road

Whaler Motor Inn

Former Quarry Pits

Sullivan's Ledge Disposal Area

Car Wash

Rosie's Cafe

Cinema 140

Unnamed Stream

Ramp to Route 195W

Ramp to Route 140S

LEGEND

Brush and Trees

Wetlands

Water traps (part of Golf Course – similar to sand traps)

Fencing around site

296

Record of Decision was signed for this site. This case study is intended to provide an example of an ecological assessment; a source of guidance for ecological risk assessors; a method to assess wildlife exposure to PCBs, which could be adapted to other xenobiotic compounds; and an awareness of the uncertainties which can be inherent in ecological risk assessments.

OVERVIEW OF THE ECOLOGICAL ASSESSMENT

The objective of the ecological assessment was to evaluate the potential for long-term ecological threat to aquatic and terrestrial organisms inhabiting these wetland areas. The specific tasks were to 1) define the ecological resources present and the condition of the wetlands, 2) assess the potential for ecological exposure to site contaminants, 3) identify appropriate remediation goals in terms of wildlife protection and habitat preservation, 4) determine how the remediation alternatives would affect the wetland ecosystem, and 5) provide information for potential mitigation, specifically, wetland areas and functions that must be preserved or replaced. The remedial investigation and ecological assessment included a wetlands assessment; wildlife observations; sediment, soil, surface water, and pore water sampling; hydraulic and hydrologic modeling; and biological sampling. These data were used in the ecological risk assessment, which included a hazard assessment of site contaminants, identification of exposure pathways through the development of an ecological food chain pathway model, an exposure assessment to determine the potential dietary exposure of upper-level consumers, and the development of remediation criteria. Using biological tissue data collected by the EPA (Charters 1991), toxicological data on the effects of PCBs on wildlife, site-specific contamination data, and a site-specific ecological food chain pathway model, the food chain (dietary) exposure of selected wildlife species was estimated. Appropriate remediation goals were identified based on the actual or potential threat posed to biota from contamination. The variety of habitat types present, and the possible exposure pathways in the different areas, were considered, potentially dictating different remediation objectives and criteria for each area. This was also viewed as an opportunity to single out ecological resources currently providing significant benefit, yet not vulnerable to a serious risk, and thus not requiring remediation. Effects levels as input to cleanup levels were calculated for sensitive birds and mammals that feed on wetland, aquatic, and terrestrial food sources, taking into account such factors as measured food source PCB concentrations, composition of diet, and feeding range. Critical functions provided by the wetland were evaluated so that those areas could be preserved or replicated, should a disruptive remedial alternative be selected.

Specific ecological determinations and steps that were accomplished to achieve the overall objectives included:

1. The identification, mapping, and evaluation of ecological resources, such as vegetated wetlands, open water habitat, and upland habitats, through wetland and habitat delineation and functional assessment.
2. The identification of potential wildlife receptors, such as birds, mammals, reptiles,

---

FIGURE 10-1.   Middle Marsh study area.

and amphibians, through wildlife observations, biological evaluations conducted by the EPA (Charters 1991), and a habitat evaluation.

3. The construction of a food chain pathway model of Middle Marsh, to facilitate the understanding of the structure and function of ecological resources, and the identification of exposure pathways for organisms at various levels of the food chain.

4. The evaluation of the transfer of contaminants through the food chain; the identification of indicator species for detailed evaluation; and the development of effects levels for input to remediation criteria.

The feasibility study was conducted with a concern for wildlife protection, habitat preservation, and the human environment. Environmentally sensitive excavation and treatment techniques that would minimize impacts to adjacent wetlands and aquatic habitat were evaluated. For example, such removal methods as vacuuming or air conveyance, and such treatment methods as bioremediation, which was initially viewed as a remediation method that may not require excavation were evaluated. In the development of the alternatives and the proposed plan, a detailed analysis was conducted of potential wetland, wildlife, water quality, and other environmental impacts and extensive mitigating measures were incorporated, including the preparation of a wetland and habitat restoration plan.

ECOLOGICAL ASSESSMENT APPROACH

The overall ecological endpoint was to preserve and maximize the ecological value of the site. The future use of the site was designated as continued open space within a public recreation area, and thus the protection of the ecological value of this remnant of an important regional wetland was deemed an appropriate endpoint. Through preliminary site visits, the site's potential to provide food, breeding habitat, and protective cover for a variety of wildlife species was recognized. The potential danger for long-term impacts from future remediation by excavation, and for adverse effects on downstream areas was also recognized. It was critical to minimize any remedial action by targeting only those areas where exposure exceeded site-specific criteria, and to give special consideration to any special habitats of limited distribution in the wetland (i.e., a densely vegetated stand of tussock sedge or cattails adjacent to the unnamed stream). To accomplish this, both contaminant concentration and habitat type were considered. This allowed the establishment of the appropriate degree of protection for each habitat, considering the ecological value of the area, the operative exposure pathways for the habitat type, and the selected endpoints. In this manner, remediation criteria based on the protection of a sensitive species or pathway would not necessarily be applied across the board, but rather could be applied to individual areas, considering their value, the degree to which the effects level was exceeded, and the applicability of the species and pathways to the individual area.

Under this approach, distinct areas within Middle Marsh that would potentially warrant individual consideration based on vegetation were first identified. Initially, three to four distinct habitat types were envisioned, based on different vegetation, functional values, exposure pathways, and levels of contamination. The permanently inundated aquatic habitat within the wetland was considered distinct from the rest of the wetland. It was also thought that the vegetated wetland areas could be subdivided into two or three habitat types for separate consideration in the ecological assessment. The division would be based on vegetation type, such as forested wetland, emergent wetland, or disturbed areas, as indicated by the presence of the opportunistic species Common reed (*Phragmites australis*), as these areas were expected to illustrate different ecological functions and values, as well as potentially different elevations, hydrology, and soil types.

As the investigation proceeded, it was concluded that all of the vegetated wetland areas supported the same contaminant transfer mechanisms, exposure pathways, and ecological endpoints, and that these functions were indistinguishable for the different wetland types. Consequently, in the final analysis, only two distinct habitats were quantitatively evaluated: aquatic and wetland/terrestrial. During the course of the investigation, a qualitative differentiation of the various wetland/terrestrial areas, for use in evaluating remediation, was conducted, but individual cleanup levels were not established for distinct nonaquatic areas.

Throughout the ecological assessment, the aquatic and wetland/terrestrial areas were evaluated separately. For some elements of the assessment, such as the nature and extent of the contamination and the hazard assessment, the evaluations for the two areas are the same. However, for other elements, such as endpoints, exposure assessments, and remediation criteria, the assessments are habitat-specific and are presented accordingly. This approach required a detailed understanding of the ecology of the areas, the distribution of the contaminants, and the transport mechanisms. These phenomena were investigated as part of the ecological assessment, and are described in a later section of this chapter.

ECOLOGICAL ENDPOINTS

The development of ecological endpoints for the Middle Marsh assessment was an iterative process. At project outset, little information about wetland functions, habitat types, biological communities, or levels of contaminants was available. Previous investigations had included only a limited assessment of PCB concentrations in Middle Marsh and the adjacent wetland. Since PCBs were thought to be the primary contaminant, and it was suspected that dietary exposure would be the critical exposure pathway, an overall site endpoint was established, which was to be protective of the food chain through maintaining concentrations in environmental media (i.e., soil, sediment), and in the tissues of species in the food chain, at levels below those harmful to upper-level consumers such as red-tailed hawk, snapping turtles, and carnivorous mammals.

As we advanced through the ecological assessment, however, this endpoint was modified. Following the hazard identification, toxicity assessment, and assessment of exposure pathways, this endpoint was split into separate endpoints for aquatic and wetland/terrestrial areas. These endpoints were both ultimately designed to protect a sensitive species, mink (*Mustela vison*), that was shown to be particularly sensitive to PCBs, and experienced severe general and reproductive health effects at relatively low dietary concentrations. For aquatic areas, the refined assessment endpoint was the maintenance of a benthic invertebrate community in which tissue concentrations in the food species of secondary consumers would not exceed harmful levels. For wetland/terrestrial areas, the assessment endpoint was the maintenance of all life functions of carnivores that feed in these areas. For both exposure pathways, it was determined that endpoints would be achieved by maintaining tissue concentrations in the food of mink and other mammals and birds at a point below that which caused reproductive failure. Specific measurement endpoints for these species were developed during the risk assessment. In addition, after considering the potential impacts of remediation of the wetland through excavation, an additional assessment endpoint was developed, that all wetland functions must be maintained in the event of a disruptive remediation. Measurement endpoints were developed during the evaluation of potential impacts of different remediation schemes.

DESCRIPTION OF EXISTING CONDITIONS

As part of the RI/FS process, Middle Marsh was to be evaluated for potential remediation, possibly involving extensive disturbance of valuable wetland resources. So that these potential disruptions could be evaluated, quantified, and mitigated, a detailed physical, chemical, and biological characterization of Middle Marsh was prepared. The objective was to become so familiar with Middle Marsh that the potential improvements or adverse effects of future remedial actions or no action in specific areas of the wetland, as defined by vegetation, contamination levels, and exposure pathways could be thoroughly understood. To this end, wetland and habitat delineation, wildlife observations, hydrologic and hydraulic modeling, and environmental sampling were conducted. These studies are described in the following subsections.

Ecological Resources

Habitat Delineation

A comprehensive wetland and habitat investigation was conducted for Middle Marsh in order to establish the types of ecological resources on the site, to determine what wildlife species would be present and their potential for ecological exposure, and to identify appropriate goals for habitat preservation. It included a literature review of available maps and aerial photography, and a field delineation conducted in accordance with the Federal Manual for Identifying and Delineating Jurisdictional Wetlands (Federal Interagency Committee for Wetland Delineation 1989) based on vegetation, soils, and hydrology. Wetland characterization and delineation of habitats within Middle Marsh was conducted in accordance with the U.S. Fish and Wildlife Service's Classification of Wetlands and Deepwater Habitats of the United States (Cowardin et al. 1979). The wetland and habitat delineation, illustrated in Figure 10-2, revealed that Middle Marsh is predominantly palustrine broad-leaved deciduous forested wetland, with internal areas of persistent emergent wetland and scrub–shrub emergent wetland.

The aquatic habitat was limited to the bisecting unnamed stream and a large tributary along the northwest edge of Middle Marsh. To precisely define the aquatic habitat within the wetland, and to determine the presence of areas dominated by obligate aquatic invertebrates, qualitative aquatic community sampling was conducted. These investigations showed that permanently inundated aquatic habitat was limited to the area of a large tributary of the unnamed stream. This one-acre area, delineated in Figure 10-3, was characterized by obligate aquatic organisms, including amphipods, freshwater clams (Sphaeriidae), isopods, Alderfly larvae (*Sialus sp.*), Cranefly larvae (*Tipula sp.*), midge larvae (Chironomidae), tadpoles, and leeches (Hirudinea). It was determined that these areas maintain a self-sustaining aquatic community; serve, at least seasonally, as feeding areas for higher aquatic species (i.e., small fish) and small mammals; and contribute plant and animal material to the stream on a continuing basis. Thus, this area could potentially support small fish, tadpoles, mollusks and crayfish: the food species of several indigenous predatory mammals and possibly birds.

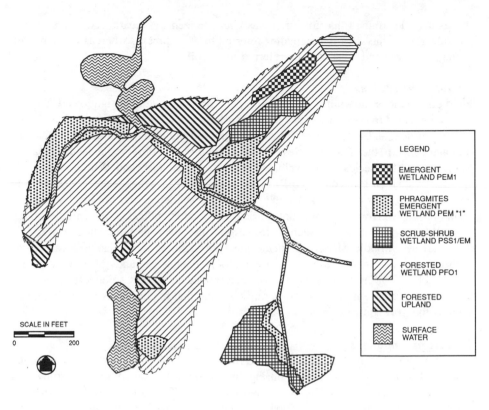

FIGURE 10-2.   Habitat delineation in Middle Marsh and adjacent wetland.

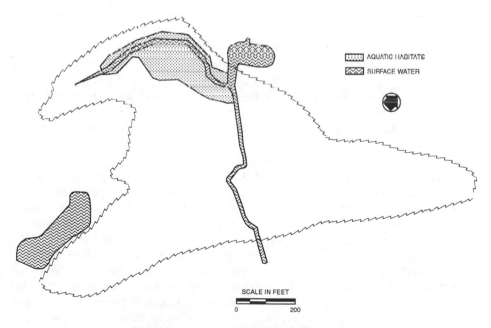

FIGURE 10-3.   Aquatic habitat in Middle Marsh.

It was also determined that this area was characterized by aquatic sediments. Thus, the substrate in this area is hereinafter referred to as sediment, whereas the substrate in vegetated wetland areas will be referred to as soil.

Wildlife Observations

Field observations for wildlife or their sign, and previous site investigations (Charters, 1991), indicated frequent incidence of songbirds and small mammals, as well as frogs, snakes, turtles, and invertebrates in streambank and inundated habitats. A species of regional concern, the spotted turtle *(Clemmys guttata),* was observed in Middle Marsh. Several wildlife species were evaluated using the U.S. Fish and Wildlife Service's Habitat Evaluation Procedure (HEP) (U.S. FWS 1980). Species were selected based on geographic appropriateness, habitat cover types available, and life history require-ments (eg. insectivorous, semiaquatic). Selected models indicated that Middle Marsh is well suited to amphibians such as frogs, and insectivorous birds such as warblers and sparrows. Middle Marsh was determined to be suited to semiaquatic carnivores such as mink *(Mustela vision),* which commonly inhabit wetlands of all kinds (Lin-scombe et al. 1982). Middle Marsh was determined to be less suitable to mammals that require permanent water, such as muskrat, and that require mass producing trees, such as the gray squirrel.

Tissue Concentrations

As part of the additional studies of Middle Marsh, the EPA (Charters 1991) conducted biological and chemical sampling during June and September 1989. The study was conducted, in part, due to the concerns of the U.S. Fish and Wildlife Service for waterfowl, passerine birds, and food for anadromous fish utilizing the habitat. The study consisted of the collection of sediment, soil, surface water, and biota samples. Biota sampling consisted of benthic invertebrates, small mammals, amphibians, earth-worms, and plants. Figure 10-4 illustrates biota sampling stations and the types of samples collected at each station. All samples in this study were analyzed for pesticides and PCBs. Aroclor 1254 was the only contaminant found in the tissue data. Table 10-1 summarizes the tissue data collected at ten stations in Middle Marsh.

Aroclor 1254, the principal contaminant of Sullivan's Ledge, was found in samples of sediment, soil, unfiltered water, small mammals, benthic invertebrates, earthworms, and frogs. Particularly noteworthy results included PCB concentrations in the green frog *(Rana clamitans melanota),* which ranged from 0.19 to 0.73 milligrams per kilogram (mg/kg). Two short-tail shrews *(Blarina brevicauda)* caught at the relatively uncontaminated background station ERT 4 had PCB concentrations of 0.38 and 0.98 mg/kg. Concentrations in meadow voles *(Microtus pennsylvanicus)* caught at the east bank station had PCB concentrations of 0.36, 0.88, and 1.6 mg/kg. Concentrations in deer mice *(Peromyscus maniculatus)* at the east and west bank stations ranged from undetected to 1.0 mg/kg PCB. Concentrations in white-footed mice *(P. leucopus),* which were found only at the west bank station, were 0.68, 0.68 and 0.84 mg/kg PCB. Concentrations of PCBs in earthworms, a common food species of some birds, ranged from undetected at ERT 4 to 2.3 and 1.8 mg/kg at the east bank and west bank, respectively. Aroclor 1254 levels were found to be below the method detection limit (MDL) of 100.0 μg/kg in all plant tissues sampled, and in benthic invertebrates from five of the seven sites sampled. PCBs were detected, however, in benthos at ERT 2 and ERT 3 at concentrations of 0.35 and 0.4 mg/kg, respectively. These data

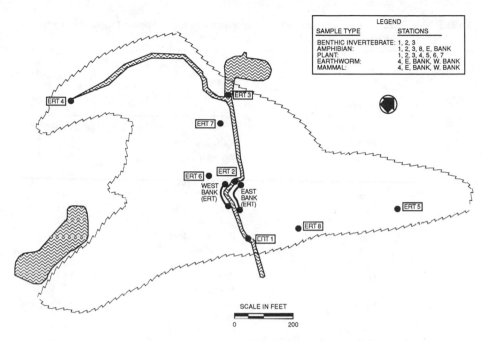

FIGURE 10-4.   Biota sampling stations (Charters 1991).

from tissues of common food species of birds and mammals indicated potential endangerment to upper-level consumers feeding on these species.

Contaminant Transport Hypotheses

The first step in conducting the additional ecological assessment studies was to develop a testable hypothesis concerning the transport of PCBs to and within the wetland. Once the hypothesis was confirmed, the distribution of PCBs could be understood, and the areas of maximum contamination, and thus the greatest potential ecological risk, could be delineated with an efficient sampling program. An understanding of the transport and behavior of PCBs within the different media in the wetland (i.e., soil, sediment, pore water, and standing water) would also be useful in the ecological risk assessment, in employing our approach of subdividing the wetland and minimizing the remediation area, and in assessing various remediation strategies. Also, an understanding of the transport mechanisms would allow a relatively accurate prediction of the area and habitat type that would be affected under various cleanup scenarios, thus allowing a comparison of benefit versus destruction, before the decision was made.

The unnamed stream adjacent to the disposal area appeared to be the most likely vehicle for the transport of PCBs to Middle Marsh, through erosion of contaminated soil particles, particularly smaller size particles, from the disposal area to the stream during rain events. Immediately adjacent to the disposal area, the stream gradient (and thus the velocity of flow) was high, keeping contaminated soil particles in suspension. However, the gradient of the unnamed stream in Middle Marsh flattens out, allowing the stream to overflow into Middle Marsh and allowing particles kept in suspension

TABLE 10-1.   PCB Concentrations in Biota Samples Collected in Middle Marsh (Charters, 1991).

Station	Species Sampled		PCB (Aroclor 1254) in Tissue (mg/kg)	
ERT 1	Benthos		0.1	U
	Green frog	*Rana clamitans melanota*	0.25	
	Rose hips	*Rosa multiflora*	0.1	U
	Grass seed heads	*Phaloris arundinacea*	0.1	U
ERT 2	Benthos		0.35	
	Green frog	*Rana clamitans melanota*	0.27	
	Rose hips	*Rosa multiflora*	0.1	U
	Grass seed heads	*Phaloris arundinacea*	0.1	U
ERT 3	Benthos		0.4	
	Green frog	*Rana clamitans melanota*	0.68	
	Green frog	*Rana clamitans melanota*	0.24	
	Rose hips	*Rosa multiflora*	0.1	U
	Grass seed heads	*Phaloris arundinacea*	0.1	U
ERT 4	Benthos		0.1	U
	Earthworm		0.1	U
	Meadow vole	*Microtus pennsylvanicus*	0.1	U
	Short-tailed shrew	*Blarina brevicauda*	0.38	
	Short-tailed shrew	*Blarina brevicauda*	0.98	
	Rose hips	*Rosa multiflora*	0.1	U
ERT 5	Benthos		0.1	U
	Grass seed heads	*Phaloris arundinacea*	0.1	U
ERT 6	Green frog	*Rana clamitans melanota*	0.19	
	Rose hips	*Rosa multiflora*	0.1	U
	Grass seed heads	*Phaloris arundinacea*	0.1	U
ERT 7	Benthos		0.1	U
	Green frog	*Rana clamitans melanota*	0.73	
East Bank	Earthworm		2.3	
	Green frog	*Rana clamitans melanota*	0.39	
	Meadow vole	*Microtus pennsylvanicus*	0.36	
	Meadow vole	*Microtus pennsylvanicus*	0.38	
	Meadow vole	*Microtus pennsylvanicus*	1.6	
	Deer mouse	*Peromyscus maniculatus*	0.64	
	Deer mouse	*Peromyscus maniculatus*	0.1	U
	Deer mouse	*Peromyscus maniculatus*	0.44	
West Bank	Earthworm		1.8	
	Deer mouse	*Peromyscus maniculatus*	0.27	
	Deer mouse	*Peromyscus maniculatus*	1	
	Deer mouse	*Peromyscus maniculatus*	0.28	
	White-footed mouse	*Peromyscus leucopus*	0.84	
	White-footed mouse	*Peromyscus leucopus*	0.68	
	White-footed mouse	*Peromyscus leucopus*	0.68	
ERT 8	Green frog	*Rana clamitans melanota*	1.02	

U = undetected at detection limit indicated

by the water velocity in the stream to settle out. A similar logic was used to develop a hypothesis for the mechanics of PCB distribution within the wetland. During most hydraulic conditions, the channel of the unnamed stream within Middle Marsh would contain most of the flow. During storms, however, the channel would overflow, inundating much of the wetland area. Indeed, our field observations confirmed that when the channel overflowed, suspended sediments were deposited in wetland areas. It appeared that there were two likely areas of deposition: one was topographical depressions adjacent to the unnamed stream within Middle Marsh, where suspended sediments would be deposited in small frequent storms; the other was in upgradient areas of Middle Marsh near where the stream enters the wetland, where during larger storms that flood Middle Marsh, downstream flow would cease due to a backwater effect, and the remaining suspended sediments would be deposited. The first area would receive deposition relatively frequently and thus could have high concentrations of PCBs. The second area would receive less frequent, but larger amounts of deposition during larger storms.

We designed a program consisting of three critical elements to test these PCB transport and distribution hypotheses. The first two steps were to define and reproduce the hydrologic and hydraulic conditions in the Middle Marsh watershed through numerical modeling. The hydrologic modeling considered rainfall and drainage basin characteristics to predict the flow and volume of water to the study area over time. The hydraulic modeling described the elevation of the water as it passed through the wetlands. The objective of the hydraulic modeling was to determine the frequency and extent to which different areas of Middle Marsh were flooded. Frequently flooded areas would receive regular sediment deposition, and thus potential PCB contamination. This information was needed as the basis for selecting sample locations within the range of elevation intervals in the wetland. The third step was to measure PCB concentrations in the areas of predicted elevated PCB levels, and to identify correlations between contaminant levels and elevation or flooding frequency. These data would be used to support or refute the deposition hypotheses described above, and to support the ecological risk assessment. Our approach for each of these steps is described in the following.

Hydrologic and Hydraulic Modeling

During planning of the field surveys, we found no detailed topographical mapping of the study area. Detailed mapping and cross sections were required for hydraulic modeling, as well as for the mapping of wetland resources, location of sampling stations, and to support remedial planning. Thus, a detailed land survey was conducted, focusing on Middle Marsh, the unnamed stream, adjacent golf course areas and the previously unsampled adjacent wetland area. The principal technical components of this survey included 1) development of a map with 1-foot contours of the areas described above; 2) surveying of 16 stream cross sections located at all hydraulic control structures and in adjacent golf course areas and forested areas, to support hydraulic modeling of the unnamed stream; and 3) establishment of a staked 100-foot grid system to facilitate location of wetland borders and sampling locations. Data collected during the survey were used to produce detailed topographic mapping of the study area. Subsequently, digitized contour data were downloaded into a geographic information system (Arc/Info), and a detailed basemap was generated.

In order to support a detailed hydrologic and hydraulic analysis of Middle Marsh and the adjacent wetland area, hydrologic data were collected during wet weather events. The most definitive data were collected during an 18-hour storm on April 3–4, 1990, when 3.17 inches of rainfall was monitored from start to finish, including measurements of peak flow. The April storm was approximately equal to the storm with a 1-year return frequency, and resulted in the flooding of most of Middle Marsh and the golf course areas north of Middle Marsh.

During the hydrologic surveys, flows were measured at nine cross sections in the unnamed stream. All measurements were taken within the effective cross section (i.e., shallow areas of bank overflow were not considered). Monitoring at each station included measuring the effective cross section of flow. Elevation or stream stage measurements were collected by measuring the distance from the top of a stake of known elevation to the water surface. A rain gage was used to collect information on the amount of rainfall. Flow and gage readings were taken approximately every two hours at each station during the sampling event, in order to develop a hydrograph for the storm. The objective of the surveys was to measure peak wet weather flow data with which to calibrate the hydrologic model.

For this study, a Soil Conservation Service hydrologic model (TR-20) was used to predict flows in the unnamed stream for 24-hour storms with return periods ranging from less than 1 year up to 100 years. We selected this model because it is capable of providing a detailed analysis of separate subcatchments or subwatersheds, and discrete reaches of the stream. Stream flow data collected during the study were used, along with drainage catchment and culvert data, to develop a hydrologic model of wet weather surface runoff from subcatchments tributary to the study area. Model inputs for hydrograph and watershed characteristics were developed using guidance set forth in the Soil Conservation Service National Engineering Handbook (USDA 1981). Areas of subcatchments, hydraulic lengths, and slopes were developed from the most recent USGS map of the area and city drainage plans. The model was calibrated using stream flow and rainfall data collected during the April 3–4, 1990 wet weather (storm) event. This storm was used because it was the largest storm monitored, and because of the volume of data collected. In addition, data collected for this storm included distinct peak flow measurements. The calibrated model was then used to develop hydrographs for more severe, less frequent storm events with return periods between 1 and 100 years. Peak flows calculated for the 1, 2, 5, 25, 50, and 100-year 24-hour storms were 114, 181, 253, 420, 471, and 555 cubic feet per second (cfs), respectively.

Next, a detailed hydraulic analysis of these storms was conducted for the study area including Middle Marsh, the adjacent wetland, and surrounding areas. Cross sections of the study area were developed from surveyed stream cross sections, to-pographic mapping of overbank wetland areas, and field measurements for all culverts, weirs, and bridges. Hydrologic modeling predictions for 24-hour 1, 2, 5, 25, 50, and 100-year storms were used as flow input to the hydraulic model. However, following an analysis of these storms, it was revealed that the majority of Middle Marsh and much of the surrounding area would flood during all storms analyzed, and that the modeling of these storms did not provide adequate resolution of flooding frequency for areas within Middle Marsh. Thus, it was necessary to evaluate smaller storms with return periods between 0 and 1 year. Since there are no established literature values for storms below a 1-year return period, flow values for smaller storms were selected by extrapolating from a plot of rainfall versus return frequency on log-probability paper. Peak flows estimated using this method for the 3-month, 6-month, and 9-month storms were used to evaluate the frequency with which the unnamed stream overtops its banks and floods topographical depressions in wetland areas near the banks of the unnamed stream.

Flood elevations for the 3, 6, and 9-month, and 1, 25, and 100-year storms were superimposed on the basemap of the Middle Marsh area, as shown in Figure 10-5. The analysis revealed that, due to the low, flat topography of Middle Marsh, the banks of the unnamed stream overflow very frequently, flooding extensive areas of Middle Marsh. The storm with a 3-month average return frequency floods approximately 6 acres of the wetland, extending into the extremities of Middle Marsh on both sides of the unnamed stream. The 6 and 9-month 24-hour storms flood only small additional increments of the wetland, due to the rising topography at the edges of the wetland. The 9-month storm was predicted to inundate Middle Marsh and approximately 3 acres of golf course fairways adjacent to the hazards or ponds between Middle Marsh and the railroad embankment. These storms also flooded large areas of the adjacent wetland. The 25 and 100-year storms were predicted to inundate the entirety of Middle Marsh and significantly larger areas of the golf course.

Sediment and Water Sampling

Hydrologic and hydraulic modeling and field observations indicated that contaminant concentrations would be most elevated in areas of frequent flooding and sediment deposition (Boucher et al. 1990). Accordingly, we designed a "smart" sampling plan to test our hypotheses, by selecting stations over the range of elevations in the wetlands and at varying distances from the unnamed stream. Sampling locations for PCBs focused on areas near the stream, where flooding was more frequent and where deposition would be expected, rather than expending financial resources on relatively uncontaminated areas. In addition, we collected samples in the vicinity of previous hot spots and within the different habitat types present, to see if there was a correlation between PCB concentrations and types of vegetation, and to see if minor variances in topography too subtle to be detected by survey but evident in alterations in vegetation told us anything about PCB distribution. We also sampled the adjacent wetland area, which had not been sampled during previous investigations, and was shown by hydraulic modeling to be a frequently flooded area.

Specifically, the sampling program was intended to:

- Provide additional chemical data on water and substrate media, to better define the horizontal and vertical extent of contamination in Middle Marsh.
- Determine the location of "hot spots."
- Examine the relationship of contaminant concentrations with the frequency of flooding and resultant sediment deposition.
- Define the partitioning of contaminants between the sediment and their interstitial or pore water.
- Determine contaminant concentrations in the previously unsampled adjacent wetland area.

In sampling Middle Marsh, we faced more than just the traditional encumbrances of environmental sampling. Middle Marsh is remotely located within a golf course. Thus there was no access road, no trailer, no running water, no restrooms, and no phone. With nowhere to store equipment, we rented golf carts, which we used at the beginning and end of each day during the two week sampling period to mobilize and demobilize most of our equipment from the panel vans we used as home base. Most of our

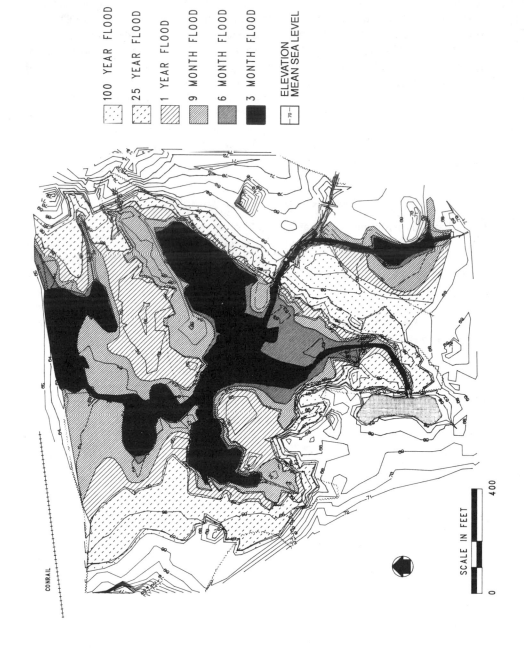

100 YEAR FLOOD
25 YEAR FLOOD
1 YEAR FLOOD
9 MONTH FLOOD
6 MONTH FLOOD
3 MONTH FLOOD

ELEVATION
MEAN SEA LEVEL

70

CONRAIL

SCALE IN FEET

0   400

paperwork and sample packaging was conducted under makeshift tents. It rained on 5 of the 10 days, and was windy on most others, which presented a serious challenge while handling bottle labels, bottle tags, bubble wrap, and chain of custody forms. The rain caused temporary flooding of many of our sampling stations. Other natural hazards included ticks and snapping turtles.

During 1990, ninety-eight (98) substrate samples, seventeen (17) pore water samples, and fourteen (14) surface water samples were collected. Figure 10-6 illustrates the stations sampled during 1990. Stations were located by triangulation using a 200-foot measuring tape and the grid system established during the topographical survey. At each station, the sample was collected at the location representing the most common elevation, soil type, and vegetation, and the location was sketched and photographed. Each sample was documented by recording: 1) the description of the sample (physical characteristics such as odor, color, presence of dead vegetation, surface sheens, etc.); 2) the approximate depth of the sample collection; 3) information on the sample type (grab or composite); and 4) the date and time of sample collection. Each sample was immediately labeled and tagged (as required), preserved if necessary, stored on ice at 4° C, and shipped to the assigned laboratory for analysis.

Surface substrate samples were collected using a hand auger. Vegetative cover was removed before the sample was taken. Organic matter was included in the samples, but any recently deposited organic matter, such as loose, individually distinguishable leaves or twigs, was removed. The top six (6) inches of substrate were collected as composites from multiple core holes within a sampling area of ten-foot radius, to ensure sufficient sample volume. Subsurface cores were collected with a hand auger from the top two (2) feet of substrate. The cores were collected in four (4) individual six-inch segments of increasing depth for each two-foot core. The borings were removed from the hand auger with a stainless steel trowel and collected into a stainless steel bowl, where they were composited prior to being placed into sample jars.

Pore water samples were collected from the water that seeped into the two-foot core holes from which core samples were taken. The integrity of the core hole was maintained by inserting a perforated PVC well point. Samples were extracted from the well point using a PVC bailer. Surface water samples were collected directly into sample containers. Samples for metals analysis were filtered using a laboratory cleaned and dedicated 0.45 micron filter. Filtration for PCB analysis was performed by the laboratory. Both total (unfiltered) and dissolved (filtered) metals and PCB concentrations were of interest for the ecological risk assessment.

In conducting the sampling described above, we followed a Quality Assurance Project Plan and a Sampling and Analysis Plan that we developed prior to sampling and which were reviewed and approved by the EPA. The sampling operation was the subject of an internal audit for compliance with the above procedures, as well as health and safety. Samples collected in Middle Marsh were analyzed following procedures approved by the EPA Region I QA/QC. The substrate was analyzed for TCL (Target Compound List) PCBs, pesticides, semivolatile and volatile organics, and TAL (Target Analyte List) metals. The substrate was also analyzed for total organic carbon to facilitate comparison of PCB levels with the EPA's interim sediment quality criteria. Water was analyzed for metals (filtered and unfiltered), volatiles, and semivolatiles.

FIGURE 10-5.   Extent of flooding by selected storms in Middle Marsh and adjacent wetland.

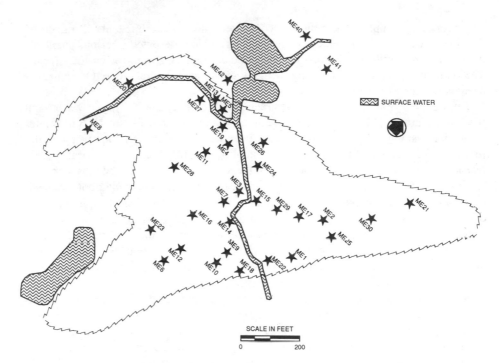

FIGURE 10-6.   Middle Marsh sampling stations.

Filtered and unfiltered pore water and standing water samples were analyzed for PCBs, using a modified PCB procedure in order to achieve the very low detection limits required for comparison with the PCB ambient water quality criterion of 0.014 μg/l.

Nature and Extent of Contamination

This section describes contaminant concentrations and patterns in Middle Marsh and the adjacent wetland area. Contamination of surface substrate, pore water, and surface water by PCBs, pesticides, volatiles, semivolatiles, and metals is addressed. PCBs are emphasized, since they were found in the highest concentrations, with respect to background levels. However, the distribution of heavy metals, such as lead and zinc, is also emphasized, as these contaminants were also elevated in Middle Marsh.

Surface Soils

PCB Aroclor 1254 was the only Aroclor detected in the study area. This is consistent with the results of previous studies (Ebasco 1987 and 1989; Charters 1991). PCBs were not detected at background stations, indicating the unnamed stream as the source of contamination. To facilitate the identification of contamination patterns, a geographical information system (Arc/Info) was used to generate maps of surface concentrations of PCBs. Figure 10-7 shows individual and contoured PCB concentrations above 5 mg/kg in Middle Marsh and the adjacent wetland, assuming that the stream influences the distribution of sediment equally on both sides of the stream. Consistent with the hypothesis developed prior to sampling, the most elevated PCB concentrations in Middle Marsh were found near the unnamed stream and in the most upgradient areas. As predicted, PCB concentrations in both areas decreased with increasing elevation, and thus increased with frequency of flooding. Pesticides and volatile organic compounds were generally undetected in both sediment and water samples. Semivolatile organics were found in surface samples at most of the stations sampled in Middle Marsh and the adjacent wetland. Semivolatile organics detected included: polyaromatic hydrocarbons (PAH), phenols, furans, phthalates, 1,4'-dichlorobenzene, and benzoic acid. However, substrate concentrations were within the range of literature values reported near highways (Butler et al. 1984), and water concentrations were near or below detection limits.

Metals analysis was performed on thirty-four samples from both wetland areas. Sampling indicated that aluminum, calcium, sodium, potassium, and barium and were generally consistent with site-specific background levels (Metcalf & Eddy 1991a; Ebasco 1987) and regional background (Shacklette and Boerngen 1984). Somewhat elevated levels of calcium, sodium, and potassium at some stations may have been due to differences in substrate characteristics or the influence of ash that may have been deposited in the disposal area. Manganese (22.3–1870 mg/kg) and iron (2360–167,000 mg/kg) were widely distributed in Middle Marsh and the adjacent wetland and exceeded site-specific background levels. Iron appears to be related to the disposal site, as evidenced by comparison to site-specific background levels (2490 to 12,100 mg/kg), and by the dark orange color of the sediments in the unnamed stream downstream of the disposal area.

A number of heavy metals were detected at levels above background. Chromium, copper, lead, vanadium, and zinc were elevated above site-specific background levels in Middle Marsh. In addition, literature values (Shacklette and Boerngen 1984) indicate

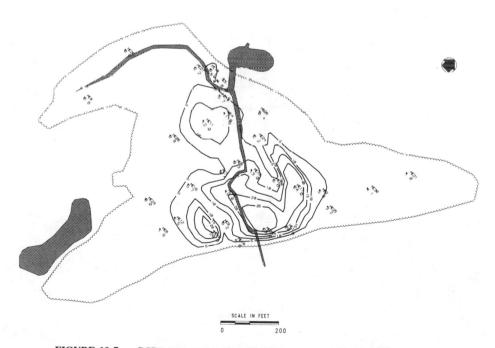

SCALE IN FEET

0   200

FIGURE 10-7.   PCB concentrations in Middle Marsh and adjacent wetland (mg/kg).

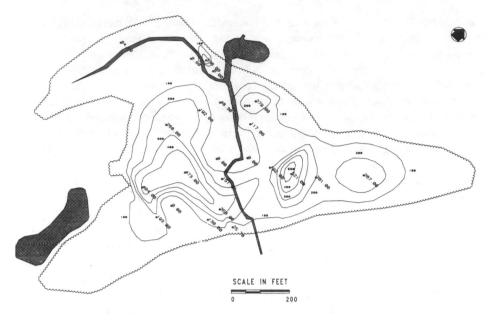

FIGURE 10-8.   Lead distributions in Middle Marsh (mg/kg).

that lead and zinc concentrations exceed regional background levels. There was no pattern in the distribution of chromium and copper. In contrast, lead, vanadium, and zinc were present in a pattern similar to that of PCB. Figures 10-8 and 10-9 depict the distribution of lead and zinc in Middle Marsh. Lead and zinc were also found in the adjacent wetland, but concentrations were much lower than those in Middle Marsh and were generally within site-specific background levels.

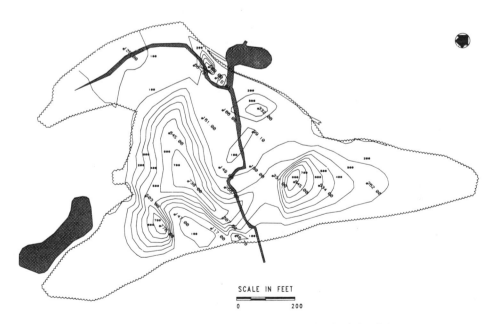

FIGURE 10-9.   Zinc concentrations in Middle Marsh (mg/kg).

Pore Water and Surface Water

Pore water and surface water samples were collected when present at the core sampling stations for comparison with ambient water quality criteria. Table 10-2 presents PCB concentrations (Aroclor 1254) found in the pore water and surface water in both filtered and unfiltered samples. In filtered samples, dissolved PCB concentrations ranged from undetected, at a detection limit of 0.05 µg/l, to 4.4 µg/l. These values are low compared to the solubility of PCB in water, which ranges from 3 to 60 µg/l, but exceed the ambient water quality criterion for PCB of 0.014 µg/l (U.S. EPA 1980). In unfiltered samples, dissolved and particulate-associated PCB concentrations ranged from 1.8 µg/l to 29 µg/l.

In summary, PCB contamination was limited to the two wetland areas and is clearly a function of the deposition of contaminated sediments. In areas of frequent flooding and deposition, PCB concentrations were generally in the range of 10 to 30 mg/kg, and approached zero (<5 mg/kg) outside the frequently inundated areas, confirming the sediment transport and deposition hypotheses described previously. Sampling revealed that hot spots are difficult to reproduce and may be redistributed by storms, and that PCB concentrations may vary widely within a small area. Testing in various habitat cover types showed that PCB concentration was more a function of elevation and flooding frequency than vegetation type. Several heavy metal levels, including lead and zinc, were also elevated, and the pattern of distribution was similar to that of the PCBs. Semivolatile organic compound levels were higher in Middle Marsh relative to the adjacent wetland area, but were generally within the range of background concentrations in both areas. Volatiles, semivolatiles, and metals were found in the pore and surface water samples at levels near or below detection limits. Heavy metals

TABLE 10-2.   Results of PCB Water Analysis (Aroclor 1254).

Station	Surface Water (µg/l)		Pore Water (µg/l)	
	(unfilt.)	(filt.)	(unfilt.)	(filt.)
ME01	0.08J*	0.08J*	1.5J*	0.92J*
ME02	0.039J*	0.022J*	0.78J*	0.56J*
ME03	0.1J*	0.05U	0.87J*	0.64J*
ME04	0.05U	0.05U	1.7J*	1.1J*
ME11	0.05U	0.05U	1.6J*	0.05U
ME14			0.27J*	0.088J*
ME15	0.05U	0.19J*	0.05U	4.4J*
ME17	0.05U	0.05U	1.1J*	1.1J*
ME23	0.05U	0.05U	0.17J*	0.04J*
ME24	0.061J*	0.05U	1.7J*	0.05U
ME29	0.083J*	0.05U	0.68J*	0.45J*
ME36			0.12J*	0.069J*
SL01	0.98J*	0.01U	3.5J*	0.02U
SL04	2.00J*	0.01U	1.8J*	0.7J*
SL14			3.6J*	0.84J*
SL15	1.5J*	0.01U	7.6J*	1.4J*
SL17	1.7J*	0.077J*	29J*	10J*

J = Estimated value

U = Undetected at detection limit

* = Value exceeds EPA Ambient Water Quality Criterion of 0.014 µg/l

were below ambient water quality criteria. However, PCBs (Aroclor 1254) were detected in filtered pore and surface water samples at levels above the ambient water quality criterion for PCBs of 0.014 µg/l.

ECOLOGICAL EXPOSURE ASSESSMENT

Through our field investigations, we found PCBs in Middle Marsh and the adjacent wetland at well above background levels. PCBs also occurred in the pore water at levels above the ambient water quality criterion, and in the tissues of aquatic insects, frogs, earthworms, and rodents. The next step in the ecological assessment was to determine what, if any, damage or threat the PCBs and other contaminants posed to these resources. Thus, we proceeded to conduct an ecological exposure assessment for wildlife species that inhabit Middle Marsh. The assessment consisted of 1) a hazard assessment, in which site contaminants that could pose a threat to ecological resources were identified; 2) a toxicity assessment, in which the contaminants of concern were researched to review their environmental chemistry and to identify potential toxic endpoints; 3) identification of exposure pathways; 4) quantification of exposure through dietary exposure calculations; and 5) development of clean-up levels or remediation criteria for sediments and soils. The approach described here includes the basic steps used in the conduct of human health risk assessments. However, as discussed in Chapter 4, there is no standardized ecological risk assessment approach, few dose-response studies on aquatic species and wildlife for many compounds, and no set cleanup levels or acceptable ranges for exposure or risk. The approach was site-specific and was based on available data, established environmental quality criteria (e.g., interim sediment quality criteria), a site-specific food chain pathway model, and site-specific bioaccumulation factors calculated from tissue data and sediment and soil PCB levels. Literature data from laboratory studies on the effects of PCBs on wildlife were used to develop dietary exposure criteria for selected upper-level consumers. This approach was used for different habitats, so that the disruption of wetlands and other sensitive ecological resources could be minimized based on the threat in the specific area, rather than to the wetland as a whole.

Hazard Assessment

In the hazard assessment, chemicals that could pose a hazard to aquatic and wetland/terrestrial wildlife were identified through comparison to background levels, ambient water quality criteria, sediment quality criteria, and ecological effects levels. However, there was a paucity of established effects level data for the effects of contaminated sediment, soil, and water on wildlife, and few analogous studies where exposure criteria or contaminant levels of concern were established. Thus, in the screening of contaminants, we relied heavily on background levels, and water and ecological benchmarks for sediment.

Volatile organic compounds and pesticides were detected infrequently and at levels below detection limits, and were not considered a threat to wildlife in the study area. Semivolatile organic compounds (SVOCs), especially polycyclic aromatic hydrocarbons (PAHs), may cause a variety of health effects in wildlife; however, semivolatiles in both wetland areas appeared to be within the range of background concentrations (from the literature) that are typically found in soils near highways (Butler et al. 1984). Semivolatiles were found at levels near or below detection limits in water samples, indicating that exposures of wildlife to SVOCs in pore water and surface water do not represent pathways of concern. Further, sediment and soil levels were below interim sediment quality criteria (U.S. EPA 1988a) established for fluoranthene, pyrene, benzo(a)pyrene, and benzo(a)anthracene.

Several heavy metals detected in sediment and soil were above background levels, including copper, chromium, iron, lead, vanadium, manganese, and zinc. Concentrations of these metals were compared to sediment toxicity data summarized by Long and Morgan (1990), who established Effects Range-Low values (ERLs), defined as the low range of sediment contamination that has been observed to cause toxic effects on some aquatic species. After comparison with these values, lead and zinc were identified as the only heavy metals for which the concentrations found in Middle Marsh and the adjacent wetland could cause toxicity to some aquatic species. Long and Morgan (1990) found that sediment lead concentrations of 35–110 mg/kg, and sediment zinc concentrations of 50–125 mg/kg resulted in sublethal effects in aquatic biota. These values are substantially below the maximum lead and zinc concentrations in Middle Marsh of 845 and 521 mg/kg, respectively. However, comparison of pore water metals data to ambient water quality criteria (U.S. EPA 1986) revealed that dissolved (filtered) metals concentrations were near or below ambient water quality criteria for lead, zinc, and other metals, perhaps due to the binding of metals to sediments as sulfides, resulting in low bioavailability for uptake by plants and animals. Thus heavy metals were not evaluated as a hazard to site biota.

PCB sediment and soil concentrations in Middle Marsh and the adjacent wetland were substantially above background concentrations; filtered pore water PCB concentrations exceeded the ambient water quality criterion (U.S. EPA 1980), and (as described later) throughout the exposure assessment for the aquatic area, sediment PCB concentrations exceeded an organic carbon-normalized environmental quality criterion developed by EPA for hydrophobic contaminants in sediments, known as the interim sediment quality criterion (U.S. EPA 1988). Accordingly, the remainder of the risk assessment focused on the ecological risk associated with PCBs in the sediments, soils, water, and biota in Middle Marsh.

Toxicity Assessment

In the hazard assessment, PCBs in Middle Marsh were identified as a potential threat to wildlife, using background levels and established sediment and water criteria as indicators of potential endangerment to higher aquatic species, as well as to wetland/terrestrial wildlife that feed in aquatic areas. Next, an understanding of the environmental chemistry of PCBs, and a relation of the concentrations of PCBs found in the wetland areas to concentrations known to be harmful to ecological receptors was needed. The following subsubsections summarize that was learned about the origin, characteristics, behavior, and environmental toxicity of PCBs. This information would be used to evaluate the threat of the PCBs, and to select toxic endpoints and sensitive species.

Chemistry of PCBs

The manufacture of polychlorinated biphenyls (PCB) was banned by the EPA in 1978, due to their identification as potential human carcinogens. They are produced commercially through the direct chlorination of biphenyl, which is composed of two connected six-carbon rings, using ferric chloride as a catalyst. PCBs with the same number of chlorines per molecule of biphenyl (e.g., tetrachlorobiphenyls) are referred to as homologs. PCB molecules within a homolog with unique patterns of chlorine substitution are referred to as congeners. Although the ten available sites for chlorine substitution could theoretically allow up to 209 PCB congeners, commercially manufactured PCBs are generally composed of 60 to 80 individual congeners. PCBs were manufactured in the United States by Monsanto under the trade name "Aroclor" and were sold as complex mixtures of congeners varying in the level of chlorination. The lower-chlorinated PCBs are light oily fluids while the heavier ones are honey-like. They have low vapor pressures, low water solubility, high flame resistance, and high stability to oxidation. The various Aroclors were numbered depending on their average degree of chlorination, with the third and fourth digits indicating the percent atomic weight of chlorine. For example, Aroclor 1254, the primary contaminant found at the Sullivan's Ledge site and in Middle Marsh, is 54 percent chlorine by weight. In the environment, PCBs are hydrophobic, with a high affinity for organic materials such as soils and sediments. Because of the high affinity of PCBs for particulate matter, and other fate properties, the sediments in surface water bodies are considered a sink for PCBs (Doskey and Andren 1981; Griffin and Chian 1980; Steel et al. 1978; Nisbet and Sarofim 1982). Their widespread use in electronics, hydraulics, and even for dust control in many locations has resulted in widespread environmental contamination.

Environmental Risk of PCBs

There is currently some debate over the chronic human health risk associated with PCBs and, further, the human carcinogenicity of certain Aroclors (Safe, 1985). However, there is considerable evidence of the ecological effects of PCBs through food chain bioaccumulation of the higher-chlorinated congeners by higher organisms such as fish-eating birds and carnivorous mammals. PCBs are lipophilic, and are thus readily passed up the food chain. Numerous species of biota have been shown to be susceptible to the chronic and lethal effects of PCB exposure. In general, toxicity and persistence in the environment increases with the degree of chlorination. In addition, the higher molecular weight congeners are more resistant to metabolization by higher organisms, and tend to bioaccumulate in fatty tissues such as the brain and liver. In recognition of the ecological effects of PCBs, the EPA (1980) developed an ambient water quality criterion of 0.014 μg/l and an interim mean sediment quality criterion (U.S. EPA, 1988a) of 19.5 μg PCB/gram carbon. These criteria are based on Aroclor 1254, a mixture of approximately 70 different congeners.

For the ecological exposure assessment, an understanding of the potential effects of PCBs on biota in Middle Marsh was needed. This would be the basis for selection of appropriate species for the risk assessment and the development of measurement endpoints. A literature search was conducted and toxicological data compiled, such as dose-response relationships, No Observed Effect Levels (NOEL), Lowest Observed Effects Levels (LOEL), Apparent Effects Thresholds (AET), Maximum Allowable Tissue Concentrations (MATC) and LD_{50}s, the dose found to cause mortality in fifty percent of a sample population. The data found is summarized in the following, and is used in the exposure assessment to identify potential exposure pathways, sensitive species, and exposure criteria.

The toxicological literature was reviewed to obtain information on dietary exposure levels that have been shown to result in acute or chronic effects in wildlife species. Given the paucity of information on chronic toxicity for common wildlife species, toxicology data were extracted for laboratory animals and other wildlife species similar to those that occur in Middle Marsh. This would help to identify general trends in the sensitivity of groups of similar species. Laboratory animals exposed to PCBs in their diet showed increased evidence of cancer; reproductive impairment; pathological changes such as lesions on the liver, stomach, and skin; and immunological impairment (Eisler 1986; U.S. Public Health Service 1987; Platonow and Karstad 1973). Relatively low levels of PCBs in the diet of a variety of wildlife species have been shown to cause reproductive impairment, behavioral changes, and mortality in sensitive species (Eisler 1986; Montz et al. 1982; Heinz et al. 1984; Platonow and Karstad 1973).

Toxicity data for a wide variety of aquatic and wildlife species and laboratory animals were summarized by Eisler (1986). Tables 10-3 and 10-4 list LD_{50} values and lethal and sublethal effects data for wildlife species taken from Eisler (1986) and a number of other studies. Lethal levels in birds, as determined by experimental data (Heath et al. 1972; Stickel et al. 1984), appear to be consistent among different species. Reproductive effects were observed in chickens fed with 5 mg/kg PCB in the diet (Heinz et al. 1984). Table 10-4 indicates that reproductive failure in other bird species occurs at dietary levels of PCB between 5 and 10 mg/kg (Heinz et al. 1984; Peakall et al. 1972; Tori and Peterle 1983).

TABLE 10-3. LD_{50} Toxicity Values for PCB (Aroclor 1254) (Eisler, 1986).

Species	Dietary Exposure Period	LD_{50}	Reference
Birds			
Northern bobwhite	5 days on/ 3 days off	604 mg/kg	Heath et al. 1972
Mallard	5 days on/ 3 days off	2,699 mg/kg	Heath et al. 1972
Ring-necked pheasant	5 days on/ 3 days off	1,091 mg/kg	Heath et al. 1972
European starling	4 days	1,500 mg/kg	Stickel et al. 1984
Red-winged blackbird	6 days	1,500 mg/kg	Stickel et al. 1984
Brown-headed cowbird	7 days	1,500 mg/kg	Stickel et al. 1984
Mammals			
Mink	9 months	6.7 mg/kg	Ringer 1983
	Single dose	4.0 g/kg body wt.	Aulerich and Ringer 1977; Ringer 1983
White-footed mouse	3 weeks	>100 mg/kg	Sanders and Kirkpatrick 1977
Rat	6 days	>75 mg/kg	Hudson et al. 1984
	Single dose	0.5 g/kg body wt.	Hudson et al. 1984
Raccoon	8 days	>50 mg/kg	Montz et al. 1982
Cottontail rabbit	12 weeks	>10 mg/kg	Zepp and Kirkpatrick, 1976

TABLE 10-4.   Lethal and Sublethal Effects of PCB (Aroclor 1254) on Wildlife.

Species	Exposure Period	Dietary Exposure (mg/kg)	Effect	Reference	Cited In:
Ringed turtle-doves	3 months	10	Delayed reproductive impairment	Heinz et al. 1984	Eisler 1986
			Hatchability of second clutch severely impaired	Peakall et al. 1972	Eisler 1986
Mourning doves	6 weeks	10	Delayed reproductive behavior	Tori and Peterle 1983	Eisler 1986
Chickens		5	Reproductive impairment	Heinz et al. 1984	Eisler 1986
Mink	4 months	1.0	Reduced reproduction	Ringer et al. 1972	U.S. EPA 1980
	8 months	2	High death rate of kits	Aulerich and Ringer 1977	
	4 months	5	Depressed reproduction	Ringer et al. 1972	U.S. EPA 1980
	160 days	0.64	Severe reproductive effects	Platonow and Karstad 1973	
	105 days	3.57	Reproductive failure, extreme weakness, and death	Platonow and Karstad 1973	
		1.0	Death	Fleming et al. 1983	Eisler 1986
White-footed mice	3 weeks	25–100	Reduced aestivation periods	Sanders and Kirkpatrick 1977	Eisler 1986
Raccoons	3 weeks	25–100	Reduced aestivation periods	Montz et al. 1982	
Rats	43 days	1,000	Resulted in 75 percent death rate	Tucker and Crabtree 1970	U.S. EPA 1980

Mink (*Mustela vison*) are highly susceptible to relatively low dietary levels of PCB Aroclor 1254. Platonow and Karstad (1973) reported that dietary concentrations of 3.57 mg/kg caused death for all mink in 105 days and that 0.64 mg/kg of PCBs over 160 days cause death, extreme weakness, and reproductive failure. Studies summarized by Eisler (1986) indicate that raccoons are less sensitive than mink to Aroclor 1254 but that mammals, in general, appear to be more sensitive than birds (Eisler 1986). Montz et al. (1982) found that daily dietary levels of 50 mg/kg of PCB over an eight-day period had an observable effect on raccoons, reducing their blood cholesterol and sleeptime and increasing microsomal enzyme production.

Site-specific studies by the EPA (Charters 1991) found PCBs in the body tissues of aquatic invertebrates, earthworms, amphibians, and small mammals in Middle Marsh. Since these species serve as food for upper-level consumers such as the American robin and mink, it was recognized that there was potential endangerment to wildlife from bioaccumulation of PCBs in Middle Marsh. Based on the toxicity of PCBs to wildlife, its potential for bioaccumulation in the food chain, and the site-specific studies showing accumulation of PCBs in site biota, we proceeded with the ecological exposure assessment.

Selection of an Indicator Species

We engaged in much discussion regarding the appropriate endpoint for the terrestrial ecosystem. The mink (*Mustela vison*) was selected because of its sensitivity to PCBs, and because Middle Marsh provided a suitable mink habitat. The rationale for its inclusion is described below.

Mink were included because Middle Marsh provides their basic habitat and food requirements. There is extensive bank habitat where mink feed and nearby wooded areas with potential denning sites. Mink are expected to use the site because they have historically occurred in the region (DeGraaf and Rudis 1983), and have recently been sighted in nearby areas, including the Apponagansett Swamp and as road kills in neighboring Dartmouth, Massachusetts (O'Reilly 1991). While it is true that the Middle Marsh system is not considered to be "optimum" mink habitat, as defined by Allen (1986) and as indicated by the habitat assessment conducted for Middle Marsh, it is nevertheless suitable for mink inhabitation as defined by the presence of all life requisites. Allen (1986) stated that "the species is tolerant of human activities and will inhabit suboptimum habitats as long as an adequate food source is available." Mink food preferences are varied, and can be classified as: 1) aquatic (e.g., fish, frogs, and crayfish); 2) semiaquatic (e.g., waterbirds and water-associated mammals); and 3) terrestrial (e.g., rabbits and rodents) (Allen 1986). The importance of each group depends upon availability and season (Linscombe et al. 1982). Our observations and the site-specific studies conducted by the EPA (Charters 1991) indicate that Middle Marsh and the adjacent wetlands have relatively high populations of these prey types, in particular high numbers of frogs and small rodents.

Our research identified the specific attributes of preferred mink habitat, confirming that mink feed in streambank areas and den in the roots of trees near streams and in nearby upland areas (Linscombe et al. 1982). During the site visit in August 1991, a certified wildlife biologist positively identified and photographed the tracks of mink on a muddy bank of the unnamed stream in Middle Marsh. In addition, we found and photographed a number of potentially suitable mink den sites near the unnamed stream and in upland areas conclusively demonstrating the presence of mink on site.

A concern was that Middle Marsh would be used by a small number, perhaps as few as one mink at a time, based on home range requirements, and thus cleanup of Middle Marsh may not be warranted to protect one mink. However, it was realized that, in the larger picture, the rationale for the cleanup would not be to protect one mink, but to restore the area as viable mink habitat, where mink and other species sensitive to PCBs may exist and breed. The objective would be to restore Middle Marsh such that it would support all life functions for a balanced indigenous population, including top-level predators such as the mink, other mustelids, and other sensitive species for which there are no toxicological data.

The mink is a top consumer and carnivore, and thus a prime candidate for consumption of PCB-contaminated food, and it was known that the mink's food sources in Middle Marsh were contaminated (Charters 1991) and that mink are particularly sensitive to PCBs. Further, the use of mink, a species known to be sensitive to PCB, is consistent with EPA guidance. As stated in the EPA document "Risk Assessment Guidance for Superfund—Environmental Evaluation Manual (U.S. EPA 1989):

"Ecologists will often use professional judgment to select a particular organism as an 'indicator species', that is, a species thought to be representative of the well-being and reproductive success of other species in a particular habitat. The indicator species may also be chosen because it is known to be particularly sensitive to pollutants or other environmental changes."

Evaluation of Wetland/Terrestrial Areas

The following subsubsections evaluate exposure in areas of Middle Marsh and the adjacent wetland that were characterized by a wetland/terrestrial and semiaquatic community. These areas support wetland vegetation, but lack permanent standing water and have limited potential for an aquatic pathway for the transfer of contaminants. Wetland and terrestrial species such as terrestrial insects, small mammals, and birds are not in intimate contact with surface water or pore water. For terrestrial species, direct soil contact and food chain exposure are predominant. In soil-dwelling organisms such as earthworms and mice, dermal contact may play a significant role. However, in upper-level consumers, PCB uptake is due primarily to food chain (trophic) bioaccumulation. Typically, fish-eating birds and other piscivorous (fish-eating) and carnivorous (meat-eating) species tend to have the highest exposure and body burdens of PCBs.

Identification of Exposure Pathways
Pathways were identified through the knowledge gained from our wildlife observations, tissue data collected by the EPA, researching the life histories and dietary habitats of observed species, and detailed knowledge of the behavior of PCBs in the environment and the food chain. It was recognized that energy is passed in a series of steps that comprise the food chain. In the process, lipophilic substances, such PCBs and pesticides, can be passed, and if they are not metabolized or excreted, they can accumulate in the food chain, with the highest concentrations potentially in top-level carnivores.

To evaluate exposure of upper-level consumers feeding in these areas, it is important to understand the type of exposure that occurs in the lower levels of the food chain, which are generally represented by invertebrates. In every ecosystem, there are two pathways for energy transfer—the grazing food chain, which originates in the autotrophs or photosynthetic plants, and the detrital food chain, which originates in the decomposition of detritus (Krebs 1978). The grazing food chain in Middle Marsh is generally limited to seed-eating birds, ducks, mice, and turtles, and does not present a significant threat for the transfer of PCBs, which do not accumulate in plants or seeds. However, the detrital food chain could represent a significant pathway. In wetland/terrestrial and shallow-water ecosystems such as Middle Marsh, much of the energy and thus the contaminant transfer to organisms at higher trophic levels flows from species that consume detritus. The detritivores in Middle Marsh, such as insects and other invertebrates, spend most, if not all of their life cycle in these contaminated soils. The passage of PCBs from terrestrial insects to primary consumers, such as small mammals and frogs, to top predators, such as mink and raccoon, was recognized as an important potential pathway.

To define exposure pathways and select the most significant pathways for more detailed evaluation, a conceptual food chain model was developed to represent potential trophic relationships between the species expected to be present in Middle Marsh. Figure 10-10 illustrates the food chain model developed to identify potential pathways and evaluate the ecological threat of PCBs through bioaccumulation. The model is composed of both biotic and abiotic components, to show the role of soil and detritus.

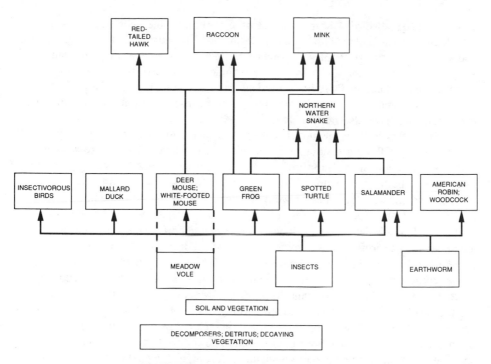

FIGURE 10-10.   Food chain pathway model.

The model, together with site-specific soil and tissue data, and literature information on the habitat and feeding requirements of selected species, assisted in: 1) identifying critical food chain pathways; 2) selecting target species for protection; 3) determining ecological assessment endpoints for remediation; 4) evaluating the impacts of remediation on the wetland area; and 5) identifying appropriate mitigating measures.

Individual species and groups of species (i.e., insectivorous birds, insects) to be included in the model were selected based on observed abundance at the site, presence of suitable habitat for the species, and likelihood of exposure. Abundance was judged based on our sightings and on trapping conducted by the EPA (Charters 1991). Habitat suitability was based on the evaluation conducted as part of the wildlife observations for selected species using U.S. Fish and Wildlife Service Models. Species with frequent or constant exposure to soil and water, such as earthworms, insects, and selected representative small mammals, were included in the model. Conversely, species were excluded from the model if they were assumed to have little or no exposure to site contaminants or if they were shown to have very high tolerances to the contaminants (e.g., snapping turtle, *Chelydra serpentina*) (Olafsson et al. 1983).

The American robin (*Turdus migratorius*) and American woodcock (*Philohela minor*) are included because they are carnivorous and their principal food source is earthworms (Terres 1980), which have been shown by Charters (1991) to carry body burdens of PCBs up to 2.3 mg/kg in Middle Marsh. The raccoon is included because its tracks were observed and its food species include indigenous species such as small mammals, frogs, worms, and reptiles (Whitaker 1980). Similarly, mink may utilize aquatic food sources such as fish, crayfish, tadpoles, and mollusks when an aquatic feeding area is available, as well as small mammals and other terrestrial animals, such as mice and small birds, during a substantial portion of the year (Linscombe et al. 1982). Small mammals such as mice were included because they burrow in the soil and are frequent prey of reptiles and other small mammals, such as raccoons and mink. Amphibians such as the green frog were included because of their abundance, site-specific data indicating PCB body burdens, and because they are frequent prey of reptiles and mammals. Birds that feed on terrestrial insects such as beetles, pill bugs, and centipedes have also been included.

Based on this food chain model, the species or groups with the highest potential for uptake of PCBs and resultant adverse chronic or lethal effects were selected for detailed evaluation in the exposure assessment. These included insectivorous birds, carnivorous (earthworm-eating) birds, and high trophic level mammals, including raccoon and mink. Other species were excluded due to the unavailability of effects data or based on a judgement that exposure of the species would be minimal. For example, although the red-tailed hawk (*Buteo jamaicensis*) was observed on-site on a number of occasions, and is known to consume small mammals, it was not included as a target species because its home range is 0.5–2.2 square miles. Thus, Middle Marsh comprises a maximum of 4 percent of the hawk's range, greatly reducing the percent of its diet that would come from Middle Marsh. Although the snapping turtle is a top-level carnivore and was frequently observed in Middle Marsh, it is not a target species due to its high level of body fat and associated resistance to PCBs and other lipophilic contaminants (Olafssen et al. 1983). The spotted turtle is largely herbivorous and, based on site-specific plant tissue data indicating undetected PCB concentrations, has not been included.

Development of Exposure Criteria

Given the threat of significant adverse effects to mammals and birds from bioaccumulation, dietary exposure criteria were developed for the selected species, based on research conducted during the toxicity assessment. Reproductive effects were seen in birds at dietary levels between 5 and 10 mg/kg PCB. The more conservative value of 5 mg/kg, the LOEL, was selected as an exposure criterion for birds. Mammals were shown in the toxicity assessment to be more sensitive than birds, thus the exposure criterion was expected to be lower. Mink were shown to be the most sensitive species, and the LOEL of 0.64 mg/kg in the diet was retained as an exposure criterion for this species. Protective dietary levels for raccoon were not found in the literature, but probably lie between the selected exposure criteria for birds (5 mg/kg) (Heinz et al. 1984) and mink (0.64 mg/kg) (Platonow and Karstad 1973). ICF–Clement (1988) used a dietary level of 1 mg/kg PCB as an exposure criterion for mammals, and this level was adopted for raccoons and other omnivorous mammals.

Development of Bioaccumulation Factors

Based on information from the toxicity assessment, it was determined that bioaccumulation from direct and dietary exposure were the primary exposure pathways. Therefore, a literature search for bioaccumulation factors (BAFs) and analogous studies of other contaminated sites in which animal tissues were collected and from which bioaccumulation factors could be calculated was conducted and used to evaluate exposure in Middle Marsh. The literature revealed that although the movement of PCBs in aquatic food chains is well documented, and bioaccumulation factors were readily available, there is a paucity of literature and studies on the bioaccumulation of PCBs in wetland and terrestrial food chains. Thus, only the site-specific tissue data collected by Charters (1991) were used to develop bioaccumulation factors for food species such as small mammals, earthworms, and frogs. The bioaccumulation factors developed for these species were calculated as the ratio of PCB in the species tissue to the level in the soil, as follows:

$$\text{Soil} * \text{BAF} = \text{Animal tissue PCB level}$$

$$\text{which yields: BAF} = \frac{\text{Animal tissue PCB level}}{\text{Soil PCB level}}$$

This method accounts for all types of exposure, including direct contact, inhalation, soil ingestion, and trophic magnification or food exposure, and assumes that the exposure level is directly proportional to the level in the soil. This information was used to back-calculate remediation effects levels for the soil that are protective of wildlife, especially targeted upper-level consumers, by maintaining their food supply at or below LOELs.

Substrate-organism bioaccumulation factors (BAFs) for earthworms, frogs, small mammals, and aquatic insects were based on averaged soil and tissue concentrations. For each food species (e.g., earthworms), the tissue data and the soil PCB concentrations from the stations at which the species were sampled were used to calculate a BAF for each station, using the equation described previously. The BAFs from each station were averaged to derive a site-specific bioaccumulation factor. Table 10-5 and the following paragraphs describe how the BAFs were derived or selected for each food species.

BAFs of 0.16 to 0.29 were calculated for earthworms for the east and west banks, respectively. In the exposure assessment, the more conservative value of 0.29 was used to evaluate the dietary levels of earthworm-eating birds, such as the American robin and woodcock. Originally the intention was to average the data; however, for earthworms, there were only two data points. It was decided to select the higher value, because of the low confidence in averaging only two values. For small mammals, the BAF was based on an average of eleven tissue levels from mice and voles captured at the east and west bank stations.

Since upper-level consumers likely feed on different species of small mammals, an average BAF of 0.07 was calculated for mice and voles for stations near the unnamed stream. The average site-specific BAF for soil frog tissue was 0.22. Using sediment and benthic invertebrate data from the unnamed stream, a maximum BAF of 0.19 was derived to represent the uptake of PCBs by aquatic insects.

TABLE 10-5.   Bioaccumulation Factors for
              Terrestrial Food Chain Pathways
              in Middle Marsh (Charters,
              1991).

Pathway	BAF
Soil—Earthworm	0.16–0.29
Soil—Frog	0.22
Soil—Meadow vole	0.02
Soil—White footed mouse	0.07
Soil—Deer mouse	0.08–0.09
Sediment—Aquatic insects	0.19

Wetland/Terrestrial Remediation Criteria

The bioaccumulation factors, exposure pathways, and exposure criteria developed previously are used to evaluate wildlife exposure for the targeted species. Thus, the levels that resulted in no effects on reproduction or other biological characteristics necessary to maintain self-sustaining populations in the marsh were established. Normally, these effects levels would be used in combination with human health, economic, and engineering considerations to develop clean-up levels in Middle Marsh. However, since this RI/FS was a follow-up to an earlier effort, which had already considered these aspects, and the maintenance of populations in the marsh had already been established and accepted as a site objective, the ecological effects levels became the clean-up levels.

A relationship between the effects levels and the environmental media was used, similar to that used to develop the ambient water quality criterion for PCB:

$$C_{media} = \frac{LOEL}{BAF}$$

where: C_{media} = Concentration of PCB in environmental media (e.g., sediment, soil, water) (mg/kg)

LOEL = Dietary lowest observed effect level (mg/kg)

BAF = Bioaccumulation factor from the media to the food species consumed (unitless)

The following paragraphs outline the development of potential clean-up levels for carnivorous (earthworm-eating) and insectivorous birds, raccoons, and the selected indicator species, mink, taking into account feeding range and percent diet. Protection of these species satisfies the endpoint of protection of carnivorous mammals and birds from the reproductive and lethal effects of PCBs.

American robin were sighted frequently at Middle Marsh and consume numerous earthworms, which are also abundant at the site. The soil clean-up level for the American robin and other carnivorous birds (such as the woodcock) was based on a protective dietary level of 5 mg/kg PCB and a BAF of 0.29 for earthworms derived from site-specific worm tissue data and soil concentrations (Charters 1991). Assuming that earthworms comprise 75 percent of these species diets, and that Middle Marsh is 90 percent of their feeding range, a soil clean-up level of 25.5 mg/kg is indicated by the following equation:

$$\text{Soil cleanup level} = \frac{LOEL}{BCF} = \frac{5 \text{ mg/kg PCB}}{(0.29)\,(0.75)\,(0.9)} = 25.5 \text{ mg/kg PCB}$$

Insectivorous birds, such as sparrows, warblers, and swallows that prefer other terrestrial insects, such as beetles, pill bugs, or emerged aquatic insects were also considered. In the absence of site-specific data on terrestrial insects, a soil clean-up level was calculated for insectivorous birds using a BAF of 0.19 developed for aquatic invertebrates (Charters 1991). Assuming that terrestrial insects comprise 100 percent of the bird's diet, and that Middle Marsh is 90 percent of the feeding range, a soil cleanup level of 29.2 mg/kg is indicated by the following equation.

$$\frac{\text{Soil}}{\text{cleanup level}} = \frac{5 \text{ mg/kg PCB}}{(0.19)\ (1.0)\ (0.9)} = 29.2 \text{ mg/kg PCB}$$

Protection of mink and raccoon was selected as a measurement endpoint for the achievement of the assessment endpoint for carnivorous mammals. Upper trophic level carnivorous and omnivorous mammals in Middle Marsh and the adjacent wetland include raccoon and mink. Given that mammals are considerably more sensitive than birds, and could be critical in the development of remediation criteria, an extensive literature review on the life histories of these species, in terms of habitat, food requirements, feeding behavior, and home range requirements was conducted.

The field observations and literature research indicated that mink using the site may either live, breed, and feed on-site, or live off-site and feed on-site. Densely vegetated wetlands such as Middle Marsh are the preferred habitat of mink (Allen 1986). There is an abundance of preferred mink prey available, in the form of small mammals, frogs, and small birds (Linscombe et al. 1982; Allen 1986). Mink prefer aquatic food sources to terrestrial food sources, when both options are equally available (Linscombe et al. 1982). In Middle Marsh, aquatic food sources for the mink include small fish, crustaceans, newts, mollusks, and tadpoles. Mink will also consume a significant number of frogs, when available. However, during winter, when the stream may be partially frozen and when frogs are hibernating, mink will feed largely on small mammals (Linscombe et al. 1982). Thus, we proceeded to develop criteria to ensure that mink would not be exposed to small mammals with tissue levels of over 0.64 mg/kg.

As discussed in the toxicity assessment, serious adverse health effects can appear in mink in 160 days or less. Thus, it was decided not to calculate the mink's dietary exposure as an annual average but to address seasonal changes in the mink's diet that could influence its exposure. Accordingly, the mink's winter diet was evaluated as a potential acute exposure period. In addition, it was recognized that the short time period for manifestation of health effects could be a significant threat to mink young, who are confined to a limited area from late April/mid-May until fall (Linscombe et al. 1982).

Although the 13-acre Middle Marsh is at the lower end of the mink's home range size, together with surrounding areas such as the golf course ponds, it is of sufficient size to support mink because of its dense vegetation and abundant prey. Gerell (1970) and Allen (1986) reported that most minimum home ranges documented in the literature can be attributed to situations of dense cover and/or high prey abundance. Mink often concentrate their feeding in core areas within their home range. These core areas are usually characterized by high prey densities, and are in relatively close proximity to streams (Allen 1986). Given the existence of the stream, which could represent a core feeding area for mink, and the apparent susceptibility of the female mink to the lethal and chronic reproductive effects of PCB exposure, it was determined that the use of the female mink's home range of 20 acres was appropriate.

The calculation of a clean-up level for mink is outlined below, and is based on the site-specific BAF for small mammals of 0.07 (Charters, 1991) and the dietary exposure criterion of 0.64 mg/kg. Since Middle Marsh comprises 65 percent of the mink's home range of 20 acres, the cleanup level was adjusted accordingly by including 0.65 in the denominator. The equation presented below indicates a cleanup level of 14 mg/kg for soil, based on protection of the indicator species, mink.

$$\frac{\text{Soil}}{\text{cleanup level}} = \frac{0.64 \text{ mg/kg PCB}}{(0.07)(0.65)} = 14 \text{ mg/kg PCB}$$

Raccoon, in comparison, are largely omnivorous, feeding on a mixture of berries, insects, crayfish, eggs, frogs, and rodents (Whitaker 1980). They feed opportunistically and may consume a substantial amount of frogs and mice when readily available (Kaufman 1982), as is the case in Middle Marsh. A BAF of 0.22 for frogs and 0.07 for mice is used, based on Charters (1991), with an exposure criterion of 1 mg/kg (ICF–Clement 1988). The raccoon has a home range of 18–36 acres (Kaufman 1982), so we assumed that Middle Marsh comprises 50 percent of the raccoon's feeding range, and that 30 percent of their diet is composed of equal portions of frogs and mice. A soil clean-up criterion of 45.9 mg/kg was calculated for the protection of raccoon as indicated in the following:

$$\frac{\text{Soil}}{\text{cleanup level}} = \frac{1}{[(0.22)(0.5) + (0.07)(0.5)][0.5][0.3]}$$

$$= 45.9 \text{ mg/kg PCB}$$

Uncertainties

The EPA guidance for both human health and ecological risk assessments requires an evaluation of uncertainties encountered in completing the assessment. For wetland areas of Middle Marsh, these include uncertainties in endpoint selection; the hazard assessment (species sensitivity, taxonomic extrapolations, laboratory-to-field extrapolations); the exposure assessment (bioaccumulation potential for PCBs); and in the development of remediation criteria (selection of food species, home ranges). The following paragraphs elaborate on these uncertainties for the wetland/terrestrial exposure assessment. Additional issues are discussed later, following the aquatic exposure assessment. There were a number of uncertainties in the exposure calculations presented earlier, including the use of bioaccumulation factors, and in the selection of home ranges and exposure criteria. Although the bioaccumulation factors were based on site-specific data, in several instances these data were limited. However, the BAF used to evaluate mink exposure, which yielded the lowest cleanup criterion, was based on an average of data from eleven animals, providing a higher degree of confidence in this calculation.

In the exposure assessment for mink, it was considered that a mink may also spend time feeding in relatively uncontaminated areas, such as the golf course and nearby wooded areas. However, as discussed in the risk assessment, mink have a core area within their home range in which they do most of their feeding. According to Whitaker (1980), mink inhabit areas *along* rivers, creeks, lakes, ponds, and marshes. Thus, in Middle Marsh, their exposure would be weighted toward streambank areas, which are not equally distributed in Middle Marsh and the surrounding area. For Middle Marsh, two intensive sampling programs have demonstrated that the areas of highest contamination are close to the unnamed stream in both Middle Marsh and the adjacent wetland. Thus, adjusting the clean-up level based on the size of Middle Marsh compared to the mink's home range (13/20 = 0.65) was reasonable, and definitely not overly conservative.

Further, it was recognized that the use of the lowest observed effect level (LOEL) of 0.64 mg/kg as an exposure criterion may not ensure achieving the measurement endpoint of mink reproduction. As described above, this level in diet was shown to cause death and reproductive failure in mink, and we were justifiably concerned that a dietary level below 0.64 mg/kg could still cause serious sublethal and even lethal effects in mink and other sensitive species. However, in order to account for some conservatism in other aspects of the exposure assessment, the EPA did not incorporate a safety factor. It is, however, standard EPA practice to adjust an LOEL to a no observed effect level (NOEL) by dividing the LOEL by a safety factor of 10. Instead, as a safety factor, the clean-up levels were applied on a point-by-point (never to be exceeded) basis, rather than reducing the average site contaminant concentration to the clean-up level. This method ensures that the mink's dietary level will not exceed 0.64 mg/kg, which was found to cause reproductive failure and even death, and which is the basis for the ambient water quality criterion and sediment quality criterion for PCB. This method was especially appropriate for Middle Marsh and for mink and other species with feeding habits similar to mink, which concentrate their feeding in a core area. Thus, the mink will be protected regardless of where it spends its time or obtains its food.

Evaluation of Aquatic Areas

The following subsubsection discusses exposure in aquatic areas of Middle Marsh, characterized by sediments, overlying water, and a benthic invertebrate community. In these areas, it was determined that exposure of upper-level consumers, such as mink, was a function of the bioavailability of PCBs in the sediments.

Aquatic Exposure Pathways

In the aquatic environment, sediment-dwelling or benthic organisms occupy a critical position in the food chain, and are in intimate contact with the interstitial (pore) water within the sediments. The benthic invertebrates consist primarily of immature insects that develop in close association with the sediments, and then emerge as terrestrial adults. Mollusks, crustaceans, annelids, and adult insects are also important in the benthic assemblage of Middle Marsh. The benthic segment of the aquatic environment generally feed on organic detritus in the sediments, and thus are susceptible to accumulating contaminants, such as PCBs associated with the organic fraction of the sediments. The benthic animals are important prey organisms of fish, some birds, semiaquatic mammals, and other aquatic invertebrates. Consequently, once PCBs are incorporated into the tissue of benthic organisms, an important exposure pathway through the food chain is created. Also, as many of the benthic organisms are terrestrial as adults, they represent a potential exposure pathway to the terrestrial system, since they provide a food source for insectivorous frogs and birds.

The exposure pathways described here could only occur in aquatic areas that were characterized by permanent inundation and the presence of obligate aquatic species. Previous field observations indicated that there was a one-acre portion of Middle Marsh that was permanently flooded and could support an aquatic invertebrate community, and therefore an aquatic exposure pathway. The pathway would be from sediment to pore water to benthic invertebrates and, subsequently, to higher levels of the food chain.

Interim sediment quality criteria were developed by the EPA (1988a) to protect benthic invertebrates and other aquatic life (U.S. EPA 1988b) by maintaining contaminant concentrations in the pore water or interstitial water of sediments at levels below the ambient water quality criteria. These limits were established to protect benthic and epibenthic invertebrate populations, many of which are critical in the aquatic food chain. Thus, sediment quality criteria apply only to permanently inundated aquatic environments that support obligate aquatic species, such as larval insects, mollusks, and crustaccans. These populations are in intimate contact with pore water, as many of them respire cutaneously or through gills. Many benthic organisms are filter feeders, and pump large volumes of water during respiration and while feeding on water-borne organic material. The PCB sediment quality criterion was derived from the corresponding ambient water quality criterion for PCB, which was developed to safeguard against bioaccumulation in the aquatic food chain and to ultimately protect against chronic reproductive effects in upper-level consumers sensitive to PCBs. The calculations for PCB were based on the protection of mink (*Mustela vison*), a species found to be particularly sensitive to PCBs.

The equilibrium partitioning method is currently the most widely used method of calculating sediment quality criteria (see Chapter 7 for a discussion of the method). The calculations are based on two interrelated assumptions: 1) that the pore water concentration is controlled by partitioning between the liquid phase and the solid phase, and 2) that the toxicity and exposure of benthic organisms is a function of the pore water concentration, rather than the total concentration in the sediment. Hydrophobic chemicals, like PCB, tend to bind to colloidal organic particles. Thus, there is an inverse relationship between the organic content of sediments and the bioavailability of hydrophobic contaminants such as PCB. Partitioning is dictated by a partitioning coefficient that theoretically allows calculation or prediction of pore water concentrations from known sediment concentrations, provided that the organic content of the sediment is known. The partitioning coefficient is based on a contaminant characteristic, the octanol-water coefficient, and a sediment characteristic, the organic carbon fraction.

Use of the partitioning coefficient allows back-calculation of sediment levels that, within certain probability, will maintain the pore water at below the chronic ambient water quality criterion. For PCBs, the chronic ambient water quality criterion for freshwater is 0.014 µg/l (U.S. EPA 1980), and is based on the protection of wildlife. Specifically, the criterion is based on the protection of mink. Platonow and Karstad (1973) found that mink fed on a diet containing 0.64 mg/kg Aroclor 1254 for 160 days either died, were extremely weak, or experienced reproductive failure. This value, together with a bioaccumulation factor of 45,000 for salmonid fish accumulating PCBs directly from water (U.S. EPA 1980), yields the calculation 0.64/45,000 = 0.014 µg/l PCB. The EPA used the equilibrium partitioning method to back-calculate the sediment quality criterion of 19.5 µg PCB/gram carbon.

In 1988, the U.S. EPA (1988a) published interim mean and 95 percent confidence limit sediment quality criteria for 13 chemicals. There is a wide range of experimentally derived equilibrium partitioning coefficients. To account for this range, the U.S. EPA (1988a) calculated the pore water concentration equivalent to the Water Quality Criterion for PCBs for each available coefficient. This set of values yielded a mean concentration of 19.5 µg PCB/g carbon, and a 95% confidence interval of 3.87 to 99.9 µg PCB/g carbon. Given the site-specific data indicating that bioaccumulation is occurring in Middle Marsh (Charters 1991), and due to the presence of aquatic habitats in portions of Middle Marsh with elevated PCB concentrations, it was decided

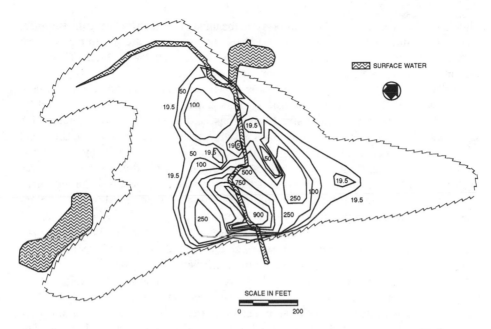

FIGURE 10-11.   Organic carbon-normalized PCB concentrations in Middle Marsh (μg PCB/g carbon).

that the equilibrium partitioning method was appropriate for this site and the decision was made to evaluate the interim PCB mean sediment quality criterion as an indicator of wildlife endangerment in aquatic areas, and to evaluate it as a remediation criterion. In order to enable consideration of equilibrium partitioning-derived sediment quality criteria in the exposure assessment as a predictor of pore water concentrations, and as an indicator of potential ecological exposure, the sediments and soils in Middle Marsh were also analyzed for total organic carbon (TOC) and TOC-normalized PCB concentrations were plotted in micrograms of PCB per gram of organic carbon. Figure 10-11 illustrates PCB levels that exceed the interim mean sediment quality criterion for PCB of 19.5 μg PCB/g carbon.

Aquatic Remediation Criteria

The objective in establishing remediation criteria was to ensure the existence of a healthy ecosystem, as indicated by conditions suitable for an unaffected, reproducing mink population. To achieve this objective, all potential food sources for mink must be free from PCB contamination that could inhibit reproduction. Data presented by Linscombe et al. (1982) demonstrates, for example, the variability in mink diet between seasons, and from location to location, based on food availability. Since the dietary level of 0.64 mg/kg can cause reproductive impairment in less than a year (160 days) (Platonow and Karstad 1973) in mink, the LOEL of 0.64 mg/kg was treated not as an average annual exposure criterion, but as a short-term exposure criterion; in other words, the PCB level in the mink's diet should never exceed this level. This would protect the mink during extended periods of the year when its diet would be primarily aquatic based. Thus, the aquatic pathway must be protective of the endpoint when it may be the only food source. It was determined, based on the field observations and the mix of species collected by the EPA (Charters 1991), that the summer diet of mink is likely dominated by aquatic species. Frogs, tadpoles, and crayfish are abundant in

Middle Marsh, and fish have been observed in the unnamed stream. During the summer, many of the mink's food species, such as frogs and crayfish, are in intimate contact with the sediment, and feed on aquatic insects contaminated with PCB. Therefore, the aim was to ensure safe concentrations in these aquatic food sources.

The operative aquatic exposure pathway in Middle Marsh extends from sediment to pore water to benthic invertebrates. These aquatic insects are then consumed by intermediate consumers, such as small fish, frogs, and other larger invertebrates such as crayfish. These intermediate consumers are ultimately the aquatic prey of mink and other secondary consumers. To achieve the selected endpoint, PCB levels in these food species of mink must not exceed 0.64 mg/kg. Since the sediments are the source of PCBs in the aquatic area, methods of determining acceptable levels in the sediments were evaluated, including the equilibrium partitioning approach used in the development of the interim sediment quality criteria, and the use of BAFs to quantify uptake of PCBs from sediments by aquatic organisms, as was shown earlier for species collected by the EPA (Charters 1991). Based on the methods used to derive the ambient water quality criterion for PCBs, the equilibrium partitioning method was used to calculate the sediment concentration, which would result in an acceptable PCB level in the mink diet. Accordingly, the interim sediment quality criterion of 19.5 μg PCB/gram carbon was evaluated as an indicator of areas where bioaccumulation could occur in the aquatic food chain. The total organic carbon data was used to calculate TOC-normalized PCB concentrations and to identify areas that exceeded the criterion of 19.5 μg PCB/g carbon. Through graphic information system-based contouring of these TOC-normalized values, a 0.35 acre portion of the aquatic area (Figure 10-3) was identified as an area of potential need for cleanup.

We used the bioaccumulation characteristics of PCBs in sediments to establish remediation criteria and not just the aquatic toxicity aspects of the compound. Thus, the endpoint used in the PCB SQC was taken from the rationale used in the development of the ambient water quality criterion for PCBs: to protect against bioaccumulation in the food chain. For Middle Marsh, since the endpoint was the protection of mink and their food, the SQC was appropriately used as an indicator of areas in which the pore water could result in an aquatic food web with PCB concentrations over 0.64 mg/kg, and not to protect the structure of the benthic community. This approach was evaluated considering site-specific data, and was found to be substantiated. PCB concentrations in the pore water of Middle Marsh exceeded the ambient water quality criterion of 0.014 μg/l. Further, the interim mean sediment quality criterion of 19.5 μg PCB/gram carbon, which was designed to keep the pore water at this level, was exceeded by a factor of up to five in the aquatic area shown in Figure 10-2. Site-specific data was shown that the benthos in Middle Marsh bioaccumulate PCB from the sediments at levels of 0.35 to 0.4 mg/kg, as seen at ERT 2 and ERT 3 (Figure 10-4), two stations that also exceeded the upper SQC. Potential mink food sources, such as fish, frogs, and crayfish, feed on these benthic animals. It was assumed that levels of 0.4 mg/kg in the benthos would, with high probability, be accumulated to levels higher than 0.64 mg/kg after passing through several additional food chain steps and concentrating to higher levels in such species as frogs, crayfish, and small fish. Sediment in the unnamed stream in excess of the upper PCB interim sediment quality criterion (U.S. EPA 1988a) resulted in benthic tissue concentrations of approximately 0.4 mg/kg (Charters 1991). The upper SQC is exceeded in much of the aquatic area. These benthic tissue concentrations are close to the levels in mink diet that have been shown to produce reproduction inhibition (0.64 mg/kg) (Platonow and Karstad 1973). A diet of benthos (or the adult insects resulting from the benthic larvae) at the measured levels of PCB by fish, crayfish, or frogs could result in tissue concentrations above the levels shown to be harmful to mink.

There are a variety of uncertainties inherent in the sediment quality criteria calculations. For example, the partitioning coefficients were developed empirically, and have not yet been field validated for universal applicability. Although pore water samples were collected, the rationale was not to validate the SQC model, but rather to obtain a range of values for use in the ecological risk assessment. Further, there is considerable debate on whether the equilibrium partitioning method accounts for the exposure of organisms to particulates. Dissolved PCB levels in Middle Marsh pore water exceeded the ambient water quality criterion; however, unfiltered or total PCB levels exceed these by approximately an order of magnitude. Indeed, many benthic and epibenthic organisms are filter-feeders, and may ingest significant quantities of particles of sediment.

Perhaps one of the least protective factors used in the development of the SQC for PCBs and in the ecological exposure assessment for Middle Marsh is the use of the bioaccumulation factor of 45,000 for salmonid fish in the development of the AWQC and the SQC. This value was used to represent the level of uptake of PCBs by aquatic species. However, bioaccumulation factors developed in the laboratory for Aroclor 1254 summarized in the ambient water quality criterion document for PCBs (U.S. EPA 1980) range up to 238,000 for the fathead minnow. Whereas the mink in Middle Marsh does not normally feed on salmonids such as trout or salmon, the fathead minnow, and other species with higher BAFs than salmonids, could be a part of the mink diet in Middle Marsh. This indicates that a considerably higher BAF value could have been selected, which would have resulted in lower environmental quality criteria for PCBs (i.e., the AWQC and SQC) and concomitantly lower clean-up levels for the aquatic portion of Middle Marsh. Further, in the aforementioned criterion document, the EPA states that "available information strongly indicates that field bioaccumulation factors for PCBs are probably a factor of 10 higher than the available laboratory BAF values." The higher field values would result from dietary exposure, as would occur for aquatic species in the Middle Marsh. The laboratory values used in criterion development are based on direct and respiratory exposure only, and may not be protective of the selected endpoints.

It is important to note that, for Middle Marsh, the SQC was used as an indicator of potential wildlife exposure and then the results were field-verified. The use of the equilibrium partitioning method, together with the PCB ambient water quality criterion as part of an overall ecological risk assessment, is consistent with EPA guidance. The EPA publication *Guidance for Conducting Remedial Investigations and Feasibility Studies Under CERCLA* (U.S. EPA 1988c) includes the following statement concerning the determination of risk:

"The objective of the RI/FS process is not the unobtainable goal of removing all uncertainty, but rather to gather information sufficient to support an informed risk management decision regarding which remedy appears to be most appropriate for a given site. . . . These choices [as to the appropriate course], like the remedy selection itself, involve the balancing of a wide variety of factors and the exercise of best professional judgement."

The EPA document "Risk Assessment Guidance for Superfund—Environmental Evaluation Manual" (U.S. EPA 1989) contains the following statements on Pages 1–3 and 6–12:

> "Decisions such as those made on Superfund sites are necessarily made with varying degrees of uncertainty. The ecological assessment is intended to reduce the uncertainty associated with understanding the environmental effects of a site and its remediation, and to give specific boundaries to that uncertainty. However, it is important to recognize that ecological assessments of Superfund sites are not research projects: they are not intended to produce absolute proof of damage, nor are they designed to answer long-term research needs."
>
> "Beyond criteria exceedances, however, risk characterization is most likely to be a weight-of-evidence judgement."

In the case of Middle Marsh, the pore water PCB concentrations that exceeded the ambient water quality criterion of 0.014 μg/l, the sediment levels that exceeded the sediment quality criterion, and the elevated PCB concentrations in site biota, including benthic organisms, were part of the "weight-of-evidence" that there was potential endangerment to wildlife in Middle Marsh. In particular, biological tissue data verified that exposure to PCB sediment concentrations exceeding the upper sediment quality criterion resulted in accumulation of PCBs in benthic organisms, which are critical in the aquatic food chain. It was concluded that this could result in food chain bioaccumulation and, ultimately, to exposure of mink and other sensitive species to detrimental dietary concentrations of PCBs.

SUMMARY OF THE ECOLOGICAL EXPOSURE ASSESSMENT

Middle Marsh has been reported as the first Superfund site in which the EPA has determined the necessity of cleanup based on ecological concerns alone (HMCRI 1991). However, this decision was not made lightly without significant investigation or an approach focused on critical areas of the wetland where exposure was significant. The evaluation was also based on specific pathways and receptors, rather than a generic approach of applying a standard "clean-up level" to the entire area, disregarding pathways, ecological function, and the implications of remediation. By using the equilibrium partitioning method for sediments in the aquatic areas, and application of site-specific bioaccumulation factors to the food chain pathway model for wetland/terrestrial areas, clean-up levels of 2.6 mg/kg (19.5 μg PCB/gram carbon) for sediments and 14 mg/kg for soils in wetland/terrestrial areas were developed for the protection of biota at the site using mink (*Mustela vison*) as an indicator species. Specifically, these clean-up levels were designed to protect mink and other potentially sensitive species from chronic health effects from PCB exposure, and to restore the area as a viable mink habitat, where mink and other species sensitive to PCBs may exist and breed. All information on the sensitivity of mink indicates that its use as a measurement endpoint will ensure protection of other sensitive species for which there are no toxicological data. The application of these criteria to Middle Marsh and the adjacent wetland revealed that the clean-up was limited, and focused on areas near the unnamed stream. Given the endpoint of mink protection, these areas made sense intuitively, based on knowledge of the feeding habitats of mink. Figure 10-12 illustrates

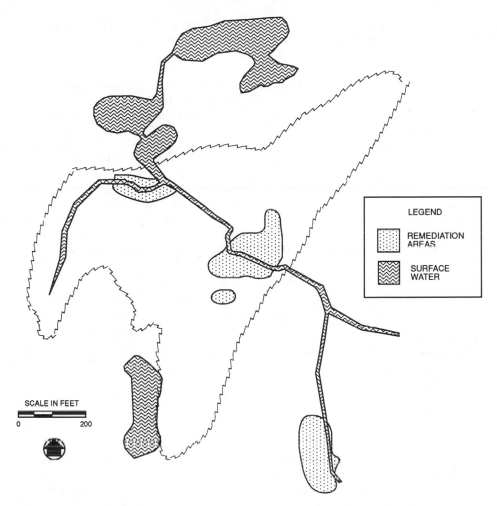

FIGURE 10-12. Middle Marsh aquatic and wetland/terrestrial cleanup areas.

the aquatic and wetland/terrestrial cleanup areas. A graphical approach was used to quantify the extent that aquatic areas of Middle Marsh exceeded the sediment quality criterion, by superimposing Figure 10-3 on Figure 10-11 and delineating aquatic areas in which normalized sediment concentrations exceeded the mean sediment quality criterion. This indicated that approximately one-third of an acre west of the unnamed stream was aquatic and exceeded the mean sediment quality criterion of 19.5 μg PCB/g carbon. Wetland/terrestrial cleanup areas were delineated based on the contoured PCB concentrations illustrated in Figure 10-7. This indicated the need for clean-up in two areas of Middle Marsh and a portion of the adjacent wetland.

Several comments from the public and the potentially responsible parties questioned the clean-up levels as being too conservative. However, under CERCLA, the EPA must ensure that its action provides overall protection of the environment, including the prevention of habitat loss. This is especially critical given the increasing human population and the rate of habitat loss, especially in wetland areas. Through the ecological assessment, it was determined that excavation of these areas was necessary to ensure that mink and other sensitive species could exist and breed. Protection of this habitat is consistent with the recommendations of the EPA's Science Advisory Board (U.S. EPA 1990) who stated that:

> "Ecological systems like the atmosphere, oceans and wetlands have a limited capacity for absorbing the environmental degradation caused by human activities. After that capacity is exceeded, it is only a matter of time before those ecosystems begin to deteriorate and human health and welfare begin to suffer.
>
> In short, beyond their importance for protecting plant and animal life and preserving biodiversity, healthy ecosystems are a prerequisite to healthy humans and prosperous economies. Although ecological damage may not become apparent for many years, society should not be blind to the fact that damage is occurring and the losses will be felt sooner or later by humans. Moreover, when species and habitat are depleted, ecological health may recover only with great difficulty, if recovery is possible at all."

ECOLOGICAL ELEMENTS OF THE FEASIBILITY STUDY

During the remedial investigation and ecological assessment, the sensitivity and value of these wetlands and their vulnerability to potential future remedial action was recognized. To mitigate these potential impacts, and to influence the evaluation of alternatives, concern for the wetlands and wildlife was carried through to the feasibility study, and information developed during the ecological assessment was used directly in the evaluation of potential impacts. The ecological elements of the feasibility study included a detailed review of pertinent wetland and other environmental regulations, identification and evaluation of technologies that minimize damage to wetlands, and the formulation and assessment of comprehensive remediation alternatives sensitive to the environment. Assessment of each alternative included consideration of environmental impacts, such as loss of indigenous vegetation, and the development of mitigating measures. These studies were intended to ensure that no avoidable or irreversible impacts would occur, and that any impacts would be mitigated. The following subsections describe ecological considerations in the alternatives analysis and selection, in the assessment of impacts and development of mitigating measures, and in response to Section 404(b)(1) of the Clean Water Act, which requires selection of the "least environmentally damaging practicable alternative" for projects involving disturbance of wetlands.

Alternatives Analysis

During the Feasibility Study (FS), thirteen potential alternatives were identified. These alternatives can be classified under the general categories of no action, in situ containment, in situ treatment (bioremediation), and excavation followed by treatment. These categories, and a brief summary of the alternatives screening process, are described in the following subsubsections.

The alternatives were evaluated with consideration of the requirements of a variety of wetland-related ARARs, including the Clean Water Act and executive orders, as summarized in the following material. The Section 404(b)(1) guidelines of the Clean Water Act contain mandatory environmental criteria for projects involving the disturbance of wetlands. These guidelines must be considered in the alternatives analysis, and an overriding justification must be provided if unacceptable adverse individual or cumulative impact will occur. Requirements include compliance with applicable state water quality standards, prevention of significant degradation of waters of the U.S., and that all appropriate and practicable steps be taken to minimize impacts to the aquatic ecosystem. Such requirements are defined in the Clean Water Act regulations, such as 40 CFR 230, which specifies that a project involving fill material should be designed and maintained to emulate a natural ecosystem. The restoration should be based on characteristics of a natural ecosystem in the vicinity of the proposed activity, to ensure that the restored area will be maintained physically, chemically, and biologically by natural processes. Further, the Executive Orders (E.O.) 11988, Floodplain Management, and E.O. 11990, Protection of Wetlands, require that actions in floodplains or wetlands restore and preserve the natural and beneficial values of the wetland and floodplain areas. E.O. 11990 requires that actions in wetlands "consider the maintenance of natural systems including conservation and long-term productivity of existing flora and fauna, species and habitat diversity and stability, [and] hydrologic utility."

No Action
No action would leave aquatic and wetland habitats intact, but would leave contaminants in place. No action was considered unacceptable because of site-specific tissue data from organisms, such as small mammals, frogs, and earthworms, demonstrating that PCBs are taken up by these common food species of larger mammals such as mink, raccoon, and birds. Through the ecological exposure assessment, it was determined that dietary exposure to these organisms could cause chronic reproductive effects in sensitive wildlife species, and could reduce the diversity of species at the site and affect the structural and functional integrity of the wetland community. Therefore, the no action alternative was screened out because the PCB-contaminated sediment, if not remediated, would pose an imminent and substantial endangerment to biota at the site.

In situ Containment
In situ containment, which involves the capping of contaminants within the wetland, was screened out due to technical feasibility, and because it would result in permanent loss of wetland habitat and considerable loss of flood storage capacity. Given the availability of other alternatives, in situ containment was determined to involve avoidable wetland and floodplain impacts, and to be inconsistent with wetland and floodplain-related ARARs, including Executive Orders 11990 and 11988, as well as the Clean Water Act and state wetland regulations.

In situ Bioremediation
This alternative was screened out for a variety of reasons. For PCBs, this technology has not been proven in the field, and there is considerable uncertainty as to its effectiveness on soil-bound and highly chlorinated PCBs, such as those present in Middle Marsh. Since PCBs have a high affinity for organic material such as soil and sediment, this uncertainty is compounded by the presence of highly dense organic silt in Middle Marsh, which sequesters the PCBs, making them unavailable to bacteria. Further, in such environments, PCB-degrading bacteria have no competitive advantage for growth substrate and cannot outcompete other indigenous bacteria. There are few, if any, contractors that offer this technology.

This alternative was initially considered because of its potential to offer a means of reducing PCBs in Middle Marsh without excavation. However, given the uncertainties outlined above, and the likely disturbance of wetlands necessary to enhance the aerobic phase of degradation, including the installation of flash boards and pipe systems for the distribution of nutrients and oxygen, and mixing of the soil to enhance aeration, this alternative was found to offer no significant advantages over other alternatives, and was eliminated.

Excavation

Ten of the alternatives considered involved excavation combined with various combinations of treatment and disposal technologies. Although excavation would require temporary disturbance of approximately two acres of wetland, it was determined that excavation was necessary and that, through a variety of measures, these impacts could be minimized, and to a large extent mitigated. The elements of the selected remediation included: 1) site preparation; 2) excavation of four areas of Middle Marsh and an adjacent wetland; 3) dewatering the excavated material, and disposal under the cap to be constructed at the Sullivan's Ledge Disposal Area (Operable Unit 1); 4) restoration of the wetlands; 5) long-term environmental monitoring; and 6) institutional controls.

Under 40 CFR 230.10(a), the least environmentally damaging alternative must be selected for projects involving alteration of wetlands. To ensure that this requirement is met, project proponents must address two questions:

1. Do the alternatives accomplish the objectives?
2. Are alternatives available?

The following paragraphs describe how these questions were responded to in the 404(b)(1) analysis.

Following the evaluation of alternatives, it was concluded that excavation was necessary to accomplish the objectives of the project by attaining the ecological endpoints. Through the RI process, it was demonstrated that the actual or threatened releases of hazardous substances from Middle Marsh and the adjacent wetland, if not addressed by implementing the response action of excavation and disposal, may present an imminent and substantial endangerment to biota through aquatic and wetland/terrestrial exposure pathways. Following a detailed evaluation of remedial technologies and remedial alternatives, an environmentally sensitive plan was developed to excavate approximately two acres of wetland, with disposal at the Sullivan's Ledge disposal area. Although the remediation would result in some direct short-term impacts to Middle Marsh, it was concluded that disturbance of wetlands and floodplains was the only practicable alternative for this site that would address the PCB contamination in the Middle Marsh study area, while minimizing the adverse impact on the terrestrial and aquatic ecosystem. If not excavated, the contaminants in the sediments and soils of the wetland would continue to pose unacceptable environmental risks, and could be the cause of any absence of a viable and diverse ecosystem in the wetlands.

The second question seeks to ensure that feasible alternatives that avoid wetland impacts have not been overlooked. Although this question is normally addressed in the context of construction projects, it was again concluded for the Middle Marsh project that, given the site-specific PCB concentrations, and the potential endangerment to the aquatic and wetland/terrestrial organisms that inhabit the Middle Marsh area, excavation of wetlands was the only practicable alternative that would address PCB contamination in the Middle Marsh study area, while minimizing adverse impact on the terrestrial and aquatic environments. Based on the extensive technology and alternatives evaluations conducted during the feasibility study, it was determined that there was no other practicable alternative with lesser impacts on the aquatic environment. Potential alternatives to excavation included no action, in situ containment, and in situ bioremediation. However, it was determined that neither the no action remedy nor in situ remedies utilizing bioremediation or containment would be able to achieve the overall purpose of the project, which is to reduce risk to environmental receptors at the site. No action was screened out due to the continued long-term threat to biota. In situ containment and in situ bioremediation were screened out due to technical infeasibility and environmental impact.

Impacts and Mitigating Measures

The Clean Water Act also requires, for projects involving alteration of wetlands, a comprehensive analysis of potential impacts to wetlands, wildlife, water quality, endangered species, and downstream habitats. Thus, as part of the feasibility study, concerns about the impact of remediation on wildlife and habitats in Middle Marsh were addressed, and were very deliberately examined throughout the RI/FS process to assess the natural resources present at the site, to evaluate potential short- and long-term impacts, and to mitigate those impacts. These studies, which included detailed sampling, were intended to ensure that only appropriate areas of primary concern were targeted for clean-up, so that large areas did not have to be disturbed and that any impacts would be mitigated.

Wetland Impacts

The remedial action plan involved excavation of approximately two acres of wetland, a relatively small amount compared to the total of 14.5 acres of wetland on the site. Although the remediation would result in some direct short-term impacts to Middle Marsh, it was determined through the alternatives analysis process that disturbance of wetlands and floodplains was the only practicable alternative that would address PCB contamination and protect the wetland and aquatic ecosystem. It was also determined that these limited impacts could be mitigated, and an extensive conceptual mitigation plan was developed as part of the FS (Metcalf & Eddy 1991b). Following site cleanup activities, impacted wetlands would be backfilled with clean soil and organic material such as peat moss, organic silt, and shredded trees and vegetation. The areas would be graded, stabilized, and then planted with vegetation appropriate to the type of wetland affected. During implementation of the remedy, steps would be taken to minimize the destruction, loss, and degradation of wetlands, including the use of sedimentation basins or silt curtains to prevent the downstream transport of contaminated sediments. In the feasibility study, a plan was devised to place most of the required access roads within wetland areas to be remediated, minimizing damage to adjacent areas. It was also specified that excavation of certain areas be conducted using hand-held shovels and wheelbarrows to transport excavated material, thus minimizing the need for access roads. A commitment was made that the clean-up would be conducted properly through the development of detailed specifications for performance of the work, proper equipment, experience of the contractor, mitigation, and employment of an appropriate specialist for wetland restoration. In considering the impacts of excavation, it was recognized that it would take several years to reestablish dense vegetation in the remediation areas, which compose approximately 14 percent of Middle Marsh and the adjacent wetland, and that excavation would involve the removal of trees from several areas of forested wetland habitat. However, we were confident that the ecological forces and conditions that created forested wetland in this area would persist following remediation and that the trees planted as part of the mitigation scheme and natural succession would reestablish forested wetland in these limited areas and that, without PCBs, Middle Marsh would offer suitable habitat for a wider diversity of species.

All wetland and upland areas would be restored, to the maximum extent feasible, to similar hydrologic and botanical conditions existing prior to excavation. As described previously all access roads, both within and outside of wetlands, and the staging area, will be accompanied with mitigating measures such as sandbags, haybales, swales, and culverts to maintain existing runoff patterns and to prevent erosion and sedimentation in any wetland area. In addition, all access areas would be removed from the site and disturbed areas restored to their original condition following remediation. The details of the restoration plan would be developed during design, at which time the least disruptive and environmentally correct restoration program would be developed. At that time, the factors important to successful restoration of wetland areas, such as replacement of hydric soils, hydraulic control, and vegetation reestablishment would be identified. These mitigation commitments ensured that the cleanup would meet or attain all applicable or relevant and appropriate federal and state requirements that apply to the site, including Section 404 of the Clean Water Act; Floodplain Management and Protection of Wetlands, Executive Orders 11988 and 11990, respectively; and Massachusetts Department of Environmental Protection Wetlands Protection Regulations.

Wildlife Impacts

Concerns were raised during the public comment period, that the proposed remediation might destroy the wildlife it was to protect. However, the project would directly affect approximately two acres of wetland, a relatively small amount compared to the total 14.5 acres between Middle Marsh and the adjacent wetland. Furthermore, following remediation, the wetland and aquatic organisms that inhabit the surface soils and sediments would quickly repopulate the disturbed areas, and wildlife, including mink, would return as well.

Section 404(b)(1) also requires an analysis of potential impacts to endangered species. Although no federally listed species were observed on-site, it was evaluated, under the Massachusetts wildlife and rare species regulations, whether the proposed plan would pose a substantial danger of destroying the habitat of the spotted turtle or the Mystic Valley amphipod, two species of special concern in Massachusetts. After researching the preferred habitat and breeding cycles of these species, it was concluded that the proposed plan would not jeopardize the continued existence of critical habitat of any endangered or threatened species. The spotted turtle courts in the period between March and May and nests in dry areas in June. Their young, or hatchlings, emerge in late August–September or overwinter in the nest until spring. Mitigating measures introduced to reduce impacts to the spotted turtle population included a detailed survey of the remediation areas, to catch and translocate any adults to uncontaminated areas of the wetland, restriction of heavy equipment to defined work areas, and control of turbidity and erosion. Short-term impacts could include displacement, noise disturbance, and short-term habitat loss. However, although the spotted turtle was seen in Middle Marsh during the RI, it was seen in wet, swampy areas far to the north of the unnamed stream. With the exception of one small aquatic area, the planned excavation would only be conducted in relatively dry, grassy, vegetated wetland areas, which are not the preferred habitat of the spotted turtle. The areas to be excavated would be isolated using hay bales, sandbags, and silt fences, to minimize damage to adjacent areas. Any turtles found in the excavation areas would be located by biologists and moved to safe areas. A representative of the Massachusetts Natural Heritage and Endangered Species Program (Copeland 1991) concurred that "the spotted turtle can adapt to short-term changes in its habitat, with proper planning, execution, and design of the proposed work."

In addition, the Mystic Valley amphipod, another Massachusetts species of special concern, is known to exist in southeastern Massachusetts. It was not observed at the site, and it was determined that, if it was, it would not be adversely affected by the excavation. It must be noted that the proposed action would only directly affect approximately 14 percent of the total wetland in the study area. Thus, the proposed action would not jeopardize the continued existence of the spotted turtle or the Mystic Valley amphipod and would not jeopardize critical habitat.

Water Quality Impacts

Section 404(b)(1) also requires an analysis of whether the proposed action will result in significant degradation of the waters of the United States. For this project, it was determined that the dredging, dewatering, and replacement of dredged material would not result in such degradation, and that the action would not violate any applicable water quality standard or toxic effluent standard under CWA §307. Compliance would be achieved through implementation of extensive mitigating measures, including activated carbon treatment of sediment dewatering effluent. In addition, mitigating measures would be implemented during and after dredging, to ensure that the replacement areas would be stable, would not erode, and would continue to perform the wetland functions of nutrient, sediment, and toxicant removal and stabilization. The area would be restored as close as is practical to re-excavation conditions such that there would be no long-term adverse impacts to wildlife, recreation, aesthetics, or economic values. The proposed action would preserve and enhance ecosystem integrity.

Summary of the 404(b)(1) Analysis

In the 404(b)(1) analysis, it was stressed that the Middle Marsh restoration would not disrupt wetlands for the purposes of development, and would not permanently fill wetlands, relying on restoration of areas altered by remediation to mitigate wetland damages. In addition, it was stressed that the excavation would occur only in areas in which the environment was significantly degraded, with the objective of reversing the degradation. In the larger picture, this project would not contribute to water quality degradation in the area, but rather would contribute to ongoing efforts to restore this wetland area as a valuable habitat and to control the degradation of downstream areas, including the Apponagansett Swamp, which receives surface flow and sedimentation from Middle Marsh. The fill for access roads would be temporary and would be placed, to the extent practical, within areas to be excavated. The only permanent fill would be beneficial fill that would be placed for the purposes of replacing PCB-contaminated substrate in the wetland. In fact, it was anticipated that excavation and wetland restoration in several areas that were characterized by mounded material and Common reed (*Phragmites*) would result in an improvement in habitat quality through removal of mounds of contaminated material and replacement with vegetation of higher habitat value.

In the final analysis, the objective for this project was the long-term restoration of the area as viable habitat where mink and other species sensitive to PCBs could exist and breed. The excavation and restoration of PCB-contaminated areas in Middle Marsh would restore it such that it would support all life functions for a balanced indigenous population of wildlife. Through the RI/FS and ecological assessment process, the direct impacts of taking this action were carefully balanced against the long-term ecological damage of no action, and it was determined that the restoration plan described earlier would not cause other significant adverse environmental impacts or significant degradation. It was determined that the planned excavation would not result in other significant environmental impacts, would not cause or contribute to significant degradation of water quality or habitat, and would enhance ecosystem integrity. Ultimately, the least damaging, most protective, practical alternative for restoration of Middle Marsh was selected.

ACKNOWLEDGMENTS

The RI/FS upon which this case study is based was funded by the U.S. EPA under contract number 68-W9-0036. Ms. Jane Downing was the Remedial Project Manager from EPA Region I. The views and opinions of the authors expressed herein do not necessarily state or reflect those of the U.S. EPA.

REFERENCES

Allen, A. W., 1986. Habitat suitability index models: Mink. U.S. Wildlife Service Biological Report 82 (10.127).

Aurlerich, R. J. and R. K. Ringer, 1977. Current status of PCB toxicity to mink, and effect on their reproduction. *Arch. Environm. Contam. Toxicol.* 6:279–292.

Boucher, P. M., J. T. Maughan, and J. Downing, 1990. Ecological assessment and modeling of a contaminated wetland. *Proceedings of the 11th National Conference,* Hazardous Materials Control Research Institute.

Butler, J. D., V. Butterworth, S. C. Kellow, and H. G. Robinson, 1984. Some observations on polycyclic aromatic hydrocarbon (PAH) content of surface soils in urban areas. *Sci. Total Environ.* 33:75–85.

Charters, D. W., 1991. Environmental Assessment, Middle Marsh Sullivan's Ledge Site, New Bedford, Massachusetts, Final Report. U.S. EPA, Environmental Response Branch.

Cowardin, L. M., V. Carter, F. C. Golet, and E. T. LaRoe, 1979. Classification of Wetlands and Deepwater Habitats of the United States. U.S. Department of the Interior, Fish and Wildlife Service, FWS/OBS-79/31.

DeGraaf, R. M., and D. D. Rudis, 1983. New England Wildlife: Habitat, Natural History, and Distribution. U.S. Department of Agriculture. Northeastern Forest Experiment Station. General Technical Report NE-108.

Doskey, P. V. and A. W. Andren, 1981. Modeling the flux of atmospheric polychlorinated biphenyls across the air/water interface. *Environ. Sci. Tech.* 15(6):705–710.

Ebasco Services Inc., 1987. Phase I Remedial Investigation Report, Sullivan's Ledge Site, New Bedford, Massachusetts. EPA Contract No. 68-01-7250.

Ebasco Services Inc., 1989. Volume I Draft Final, Remedial Investigation, Sullivan's Ledge Site, New Bedford, Massachusetts. EPA Contract No. 68-01-7250.

Eisler, R. (ed.) 1986. Polychlorinated Biphenyl Hazards to Fish, Wildlife, and Invertebrates: A Synoptic Review. U.S. Fish and Wildlife Service Biological Report 85 (1.7).

Federal Interagency Committee for Wetlands Delineation, 1989. Federal Manual for Identifying and Delineating Jurisdictional Wetlands. U.S. Army Corps of Engineers, U.S. Environmental

Protection Agency, U.S. Fish and Wildlife Service, and U.S.D.A. Soil Conservation Service, Washington, D.C. Cooperative Technical Publication.

Fleming, W. J., D. R. Clark, Jr., and C. J. Henry, 1983. Organochlorine pesticides and PCB's: A continuing problem for the 1980s. *Trans. North Am. Wildl. Natur. Resour. Conf.* 48:186–199.

Gerell, R., 1970. Home ranges and movements of the mink in southern Sweden. *Oikos* 21:160–173.

Griffin, R. A. and E. S. K. Chian, 1980. Attenuation of water-soluble polychlorinated biphenyls by earth materials. U.S. EPA, Municipal Environmental Research Laboratory, Office of Research and Development, Cincinnati, OH, EPA-600/2-80-027.

Heath, R. G., J. W. Spann, E. F. Hill, and J. F. Kreitzer, 1972. Comparative dietary toxicities of pesticides to birds. *In* Polychlorinated Biphenyl Hazards to Fish, Wildlife and Invertebrates: A Synoptic Review, R. Eisler. U.S. Fish and Wildl. Serv. Biol. Rep. 85(1.7). 72 pp.

Heinz, G. H., D. M. Swineford, and D. E. Katsman, 1984. High PCB residues in birds from the Sheboygan River, Wisconsin. *In* Polychlorinated Biphenyl Hazards to Fish, Wildlife and Invertebrates: A Synoptic Review, R. Eisler. U.S. Fish and Wildl. Serv. Biol. Rep. 85(1.7). 72 pp.

HMCRI, 1991. EPA decides on cleanup decision based solely on ecosystem, not health dangers. Hazardous Material Control Research Institute, *FOCUS* 7:12, Dec. 1991.

Hudson, R. H., R. K. Tucker, and M. A. Haegele, 1984. Handbook of toxicity of pesticides to wildlife. *In* Polychlorinated Biphenyl Hazards to Fish, Wildlife and Invertebrates: A Synoptic Review, R. Eisler (ed.). U.S. Fish and Wildl. Serv. Biol. Rep. 85(1.7). 72 pp.

ICF–Clement Associates, Inc., 1988. Endangerment Assessment for the F. O'Connor site in Augusta, Maine. EPA Contract No. 68-01-6939, Document Control No. 319-ES1-RT-FKJL-1.

Kaufman, J. H., 1982. Raccoon and Allies (*Procyon lotor* and Allies). In *The Wild Mammals of North America: Biology, Management, Economics*. J. A. Chapman and G. A. Feldhamer (eds.), pages 567–585. Baltimore, MD: Johns Hopkins Univ. Press.

Krebs, C. J., 1978. Ecology: The Experimental Analysis of Distribution and Abundance. New York. Harper & Row.

Linscombe, G., N. Kinler, and R. J. Aulerich, 1982. Mink (*Mustela vison*). In *Wild Mammals of North America: Biology, Management, Economics*. J. A. Chapman and G. A. Feldhamer (eds.), pages 629–643. Baltimore, MD: Johns Hopkins Univ. Press.

Long, E. R. and L. G. Morgan, 1990. The potential for biological effects of sediment-sorbed contaminants tested in the national status and trends program. National Oceanic and Atmospheric Administration, Technical Memorandum NOS OMA 52.

Metcalf & Eddy, 1991a. Final Remedial Investigation: Additional Studies of Middle Marsh, Volume I. EPA Contract No. 68-W9-0036.

Metcalf & Eddy, 1991b. Final Feasibility Study Report of Middle Marsh, New Bedford, Massachusetts. U.S. EPA Contract No. 68-W9-0036.

Montz, W. E., W. C. Card, and R. L. Kirkpatrick, 1982. Effects of polychlorinated biphenyls and nutritional restriction on barbituate-induced sleeping times and selected blood characteristics in raccoons (*Procyon lotor*). *Bull. Environ. Contam. and Toxic.* 28:578–583.

Nisbet, I. C. T. and A. F. Sarofim, 1972. Rates and routes of transport of PCBs in the environment. *Environmental Health Perspectives* 1:1.

Olafsson, P. G., A. M. Bryan, B. Bush, and W. Stone, 1983, Snapping turtles—A biological screen for PCBs. *Chemosphere* 12 (11/12):1525–1532.

O'Reilly, M., 1991. Letter from M. O'Reilly, Dartmouth, MA to J. Downing, EPA on May 30, 1991.

Peakall, D. B., J. L. Lincer, and S. E. Bloom, 1972. Embryonic mortality and chromosomal alterations caused by Aroclor 1254 in ringdoves. *In* Polychlorinated Biphenyl Hazards to Fish, Wildlife and Invertebrates: A Synoptic Review, R. Eisler. U.S. Fish Wildl. Serv. Biol. Rep. 85(1.7). 72 pp.

Platonow, N. S. and L. H. Karstad, 1973. Dietary effects of polychlorinated biphenyls on mink. *Can. J. Comp. Med.* 30:391–400.

Ringer, R. K., 1983. Toxicology of PCBs in mink and ferrets. In *PCBs: Human and Environmental Hazards,* F. M. D'Itri and M. A. Kamrin (eds.). Woburn, MA: Butterworth Publishers.

Ringer, R. K., R. J. Auleric, and M. Zabik, 1972. Effect of dietary polychlorinated biphenyls on growth and reproduction of mink. U.S. EPA Ambient Water Quality Criteria for Polychlorinated Biphenyls. EPA 440 5-80-68.

Safe, S., S. Bandiera, T. Sawyer, L. Robertson, L. Safe, A. Parkinson, D. E. Thomas, D. Ryan, L. Reik, W. Levin, M. A. Denomone, and T. Fujita, 1985. PCBs: Structure–Function Relationships and Mechanism of Action. *Environmental Health Perspectives* 60:47–56.

Sanders, O. T., and R. L. Kirkpatrick, 1977. Reproductive characteristics and corticoid levels of female white-footed mice fed *ad libitum* and restricted diets containing a polychlorinated biphenyl. *Environ. Res.* 13:358–363.

Shacklette, H. T. and J. G. Boerngen, 1984. Element Concentrations in Soils and Other Surficial Materials of the Conterminous United States. USGS Professional Paper 1270. pp. 105.

Steen, W. C., D. F. Paris, and G. L. Baugham, 1978. Partitioning of selected polychlorinated biphenyls to natural sediments. *Water Research* 12:655–657.

Stickel, W. J., L. F. Stickel, R. A. Dyrland, and D. L. Hughes, 1984. Aroclor 1254 residues in birds: Lethal levels and loss rates. *Arch. Environ. Contam. Toxicol.* 13:7–13.

Terres, J. K., 1980. *The Audubon Society Encyclopedia of North American Birds.* New York: Alfred A. Knopf, 1109 pp.

Tori, G. M. and T. J. Peterle, 1983. Effects of PCBs on mourning dove courtship behavior. *In* Polychlorinated Biphenyl Hazards to Fish, Wildlife and Invertebrates: A Synoptic Review, R. Eisler. U.S. Fish and Wildl. Serv. Biol. Rep. 85(1.7). 72 pp.

Tucker, R. K. and D. G. Crabtree, 1970. Handbook of toxicity of pesticides to wildlife. U.S. EPA Ambient Water Quality Criteria for Polychlorinated Biphenyls. EPA 440 5-80-68.

U.S. Department of Agriculture (USDA), 1981. National Engineering Handbook, Section 4, Hydrology. Soil Conservation Service.

U.S. EPA, 1980. Ambient Water Quality Criteria for Polychlorinated Biphenyl. Office of Water Regulations and Standards. EPA 440-5-80-68.

U.S. EPA, 1986. Quality Criteria for Water. Office of Water Regulations and Standards. EPA 440-5-86-001.

U.S. EPA, 1988a. Interim Sediment Criteria Values for Nonpolar Hydrophobic Organic Compounds. Office of Water, Criteria and Standards Division.

U.S. EPA, 1988b. Draft Briefing Report to the EPA Science Advisory Board on the Equilibrium Partitioning Approach to Generating Sediment Quality Criteria.

U.S. EPA, 1988c. Guidance for Conducting Remedial Investigations and Feasibility Studies Under CERCLA, Interim Final. EPA/540/G-89/004.

U.S. EPA, 1989. Risk Assessment Guidance for Superfund—Environmental Evaluation Manual, Interim Final. Office of Emergency and Remedial Response. EPA/540/1-89/001A. OSWER Directive 9285.7-01.

U.S. EPA, 1990. Reducing Risk: Setting Priorities and Strategies for Environmental Protection: SAB-EC-80-021.

U.S. Fish and Wildlife Service (U.S. FWS), 1980. Habitat Evaluation Procedures (HEP) ESAM 102. Division of Ecological Services.

U.S. Public Health Service, 1987. Toxicological Profile for Selected PCBs. Oak Ridge National Laboratory, Oak Ridge, TN.

Whitaker, J. O., 1980. *The Audubon Society Field Guide to North American Mammals.* New York: Alfred A. Knopf. 775 pp.

Zepp, R. L. Jr., and R. L. Kirkpatrick, 1976. Reproduction in cottontails fed diets containing PCB. *In* Polychlorinated Biphenyl Hazards to Fish, Wildlife and Invertebrates: A Synoptic Review, R. Eisler (ed.). U.S. Fish and Wildl. Serv. Biol. Rep. 85(1.7). 72 pp.

APPENDIX
Scientific Names

Common Name	Scientific Name
PLANTS	
American elm	*Ulmus americana*
Black locust	*Robinia pseudo-acacia*
Black-eyed susan	*Rudbeckia serotina*
Border meadow rue	*Thalictrum venulosum*
Box elder	*Acer negundo*
Bulrush	*Scirpus* spp.
Burreed	*Sparganium* spp.
Canada buffalo-berry	*Shepherdia canadensis*
Cattail	*Typha* spp.
Chicory	*Cichorium intybus*
Clover	*Trifolium* spp.
Common mullein	*Verbascum thapsus*
Common reed	*Phagmites australis*
Cottonwood	*Populus deltoides*
Duck potato	*Sagittaria* spp.
Duckweed	*Lemna* spp.
Goldenrod	*Solidago* spp.
Gray birch	*Betula populifolia*
Green ash	*Fraxinus pennsylvanica*
Honeysuckle	*Lonicera* spp.
Horsetail	*Equisetum* spp.
Quaking aspen	*Populus tremuloides*
Queen Anne's lace	*Daucus carota*
Red maple	*Acer rubrum*
Red-osier dogwood	*Cornus stolonifera*
Reed canary grass	*Phalaris arundinacea*
Sensitive fern	*Onoclea sensibilis*
Silver maple	*Acer saccharinum*
Speckled alder	*Alnus rugosa*

(continued)

APPENDIX   (*continued*)

Common Name	Scientific Name
Staghorn sumac	*Rhus typhina*
Swamp white oak	*Quercus bicolor*
Wildrice	*Zizania aquatica*
Willow spp.	*Salix* spp.
Wood nettle	*Laportea canadensis*
FISH	
Atlantic (landlocked) salmon	*Salmo salar*
Banded killifish	*Fundulus diaphanus*
Black crappie	*Pomoxis annularis*
Blacknose shiner	*Notropis heterolepis*
Bowfin	*Amia calva*
Brown bullhead	*Ameiurus nebulosus*
Chain pickerel	*Esox niger*
Channel catfish	*Ictalurus punctatus*
Common carp	*Cyprinus carpio*
Eastern mudminnow	*Umbra pygmaea*
Emerald shiner	*Notropis atherinoides*
Fathead minnow	*Pimephales promelas*
Golden shiner	*Notemigonus crysoleucas*
Lake sturgeon	*Acipenser fulvescens*
Largemouth bass	*Micropterus salmoides*
Longnose gar	*Lepisosteus osseus*
Northern pike	*Esox lucius*
Pumpkinseed	*Lepomis gibbosus*
Rock bass	*Ambloplites rupestris*
Smallmouth bass	*Micropterus dolomieu*
Sunfish spp.	*Lepomis* spp.
Walleye	*Stizostedion vitreum*
White perch	*Morone americana*
White sucker	*Catostomus commersoni*
Yellow perch	*Perca flavescens*
BIRDS	
American black duck	*Anas rubripes*
American crow	*Corvus brachyrhynchos*
American goldfinch	*Carduelis tristis*
American kestrel	*Falco sparverius*
American robin	*Turdus migratorius*
American tree sparrow	*Spizella arborea*
Bald eagle	*Halioeetus leucocephalus*
Barn swallow	*Hirundo rustica*
Belted kingfisher	*Ceryle alcyon*
Black-capped chickadee	*Parus atricapillus*
Black-crowned night heron	*Nycticorax nycticorax*
Blue jay	*Cyanocitta cristata*
Blue-winged teal	*Anas discors*
Brown-headed cowbird	*Molothrus ater*
Canada goose	*Branta canadensis*
Canvasback	*Aythya valisineria*
Cedar waxwing	*Bombycilla cedrorum*
Chestnut-sided warbler	*Dendroica pensylvanica*
Chimney swift	*Chaetura pelagica*
Common goldeneye	*Bucephala clangula*

APPENDIX   *(continued)*

Common Name	Scientific Name
Common grackle	*Quiscalus quiscula*
Common yellowthroat	*Geothlypis trichas*
Double-crested cormorant	*Phalacrocorax auritus*
Downy woodpecker	*Picoides pubescens*
Eastern kingbird	*Tyrannus tyrannus*
Eastern phoebe	*Sayornis phoebe*
Empidonax flycatcher	*Empidonax* spp.
European starling	*Sturnus vulgaris*
Gray catbird	*Dumetella carolinensis*
Great black-backed gull	*Larus marinus*
Great blue heron	*Ardea herodias*
Great horned owl	*Bubo virginianus*
Green-backed heron	*Butorides striatus*
Green-winged teal	*Anas crecca*
Herring gull	*Larus argentatus*
Hooded merganser	*Lophodytes cucullatus*
House finch	*Carpodacus mexicanus*
House sparrow	*Passer domesticus*
Killdeer	*Charadrius vociferus*
Least flycatcher	*Empidonax minimus*
Mallard	*Anas platyrhynchos*
Merganser spp.	*Mergus* spp.
Mourning dove	*Zenaida macroura*
Northern cardinal	*Cardinalis cardinalis*
Northern flicker	*Colaptes auratus*
Northern mockingbird	*Mimus polyglottos*
Northern rough-winged swallow	*Stelgidopteryx serripennis*
Olive-sided flycatcher	*Contopus borealis*
Pileated woodpecker	*Dryocopus pileatus*
Purple finch	*Carpodacus purpureus*
Purple martin	*Progne subis*
Quail	*Colinus virginianus*
Red-tailed hawk	*Buteo jamaicensis*
Red-winged blackbird	*Agelaius phoeniceus*
Ring-billed gull	*Larus delawarensis*
Ring-necked duck	*Aythya collaris*
Rock dove (Domestic pigeon)	*Columba livia*
Ruffed grouse	*Bonasa umbellus*
Scaup spp.	*Aythya* spp.
Song sparrow	*Melospiza melodia*
Spotted sandpiper	*Actitis macularia*
Starling	*Sturnus vulgaris*
Tree swallow	*Tachycineta bicolor*
White-breasted nuthatch	*Sitta carolinensis*
Wood duck	*Aix sponsa*
Woodcock	*Philohela minor*
Yellow warbler	*Dendroica petechia*

MAMMALS

Badger	*Taxidea taxus*
Beaver	*Castor canadensis*
Big brown bat	*Eptesicus fuscus*
Bobcat	*Felis rufus*
Coyote	*Canis latrans*

(continued)

APPENDIX   (*continued*)

Common Name	Scientific Name
Deer mouse	*Peromyscus maniculatus*
Domestic cat	*Felis domesticus*
Domestic dog	*Canis familiaris*
Eastern chipmunk	*Tamias striatus*
Eastern cottontail	*Sylvilagus floridanus*
Eastern gray squirrel	*Sciurus carolinensis*
Ermine	*Mustela erminea*
Gray fox	*Urocyon cinereoargenteus*
Ground squirrel	*Citellus spp.*
Hairy-tailed mole	*Parascalops breweri*
House mouse	*Mus musculus*
Little brown myotis	*Myotis lucifugus*
Long-tailed weasel	*Mustela frenata*
Masked shrew	*Sorex cinereus*
Meadow vole	*Microtus pennsylvanicus*
Mink	*Mustela vison*
Muskrat	*Ondatra zibethicus*
Northern short-tailed shrew	*Blarina brevicauda*
Norway rat	*Rattus norvegicus*
Porcupine	*Erethizon dorsatum*
Raccoon	*Procyon lotor*
Red fox	*Vulpes vulpes*
River otter	*Lutra canadensis*
Southern red-backed vole	*Clethrionomys gapperi*
Star-nosed mole	*Condylura cristata*
Striped skunk	*Mephitis mephitis*
White-footed mouse	*Peromyscus leucopus*
White tailed deer	*Odocoileus virginianus*
Woodchuck	*Marmota monax*
REPTILES AND AMPHIBIANS	
Bullfrog	*Rana catesbeiana*
Common snapping turtle	*Chelydra serpentina*
Eastern American toad	*Bufo americanus*
Eastern garter snake	*Thamnophis sirtalis*
Eastern milk snake	*Lampropelis triangulum*
Eastern smooth green snake	*Opheodrys vernalis*
Green frog	*Rana clamitans*
Map turtle	*Graptemys geographica*
Northern brown snake	*Storeria dekayi*
Northern dusky salamander	*Desmognathus fuscus*
Northern leopard frog	*Rana pipiens*
Northern ribbon snake	*Thamnophis sauritus septentrionalis*
Northern ringneck snake	*Diadophis punctatus*
Northern spring peeper	*Hyla crucifer*
Northern two-lined salamander	*Eurycea bislineata*
Northern water snake	*Nerodia sipedon*
Painted turtle	*Chrysemys picta*
Pickerel frog	*Rana palustis*
Red-spotted newt	*Notophthalmun viridescens*
Spotted salamander	*Ambystoma maculatum*
Wood frog	*Rana sylvatica*

Index